METHODS IN ALCOHOL - RELATED NEUROSCIENCE RESEARCH

METHODS & NEW FRONTIERS IN NEUROSCIENCE

Series Editors
Sidney A. Simon, Ph.D.
Miguel A.L. Nicolelis, M.D., Ph.D.

Published Titles

Apoptosis in Neurobiology
Yusuf A. Hannun, M.D., Professor/Biomedical Research and Department Chairman/
    Biochemistry and Molecular Biology, Medical University of South Carolina
Rose-Mary Boustany, M.D., tenured Associate Professor/Pediatrics and Neurobiology,
    Duke University Medical Center

Methods for Neural Ensemble Recordings
Miguel A.L. Nicolelis, M.D., Ph.D., Professor/Department of Neurobiology,
    Duke University Medical Center

Methods of Behavioral Analysis in Neuroscience
Jerry J. Buccafusco, Ph.D., Professor/Pharmacology and Toxicology,
    Professor/Psychiatry and Health Behavior, Medical College of Georgia

Neural Prostheses for Restoration of Sensory and Motor Function
John K. Chapin, Ph.D., Department of Physiology, State University of New York
    Health Science Center
Karen A. Moxon, Ph.D., Department of Electrical and Computer Engineering,
    Drexel University

Computational Neuroscience: Realistic Modeling for Experimentalists
Eric DeSchutter, M.D., Ph.D., Department of Medicine, University of Antwerp

Methods in Pain Research
Lawrence Kruger, Ph.D., Professor Emeritus/Neurobiology, UCLA School of Medicine

Motor Neurobiology of the Spinal Cord
Timothy C. Cope, Ph.D., Department of Physiology, Emory University School of Medicine

Nicotinic Receptors in the Nervous System
Edward Levin, Ph.D., Associate Professor/Department of Pharmacology and Molecular
    Cancer Biology and Department of Psychiatry and Behavioral Sciences,
    Duke University School of Medicine

Methods in Genomic Neuroscience
Helmin R. Chin, Ph.D., NIMH, NIH Genetics Research
Steven O. Moldin, Ph.D, NIMH, NIH Genetics Research

Methods in Chemosensory Research
Sidney A. Simon, Ph.D., Professor/Department of Neurobiology,
    Duke University Medical Center
Miguel A.L. Nicolelis, M.D., Ph.D., Professor/Department of Neurobiology,
    Duke University Medical Center

The Somatosensory System: Deciphering the Brain's Own Body Image,
Randall Nelson, Ph.D., Professor of Anatomy and Neurobiology,
University of Tennessee College of Medicine
New Concepts in Cerebral Ischemia
Rick Lin, Ph.D., Professor/Department of Anatomy, University of Mississippi
Medical Center
DNA Arrays: Technologies and Experimental Strategies
Elena Grigorenko, Associate Professor/Department of Physiology and Pharmacology,
Bowman Gray School of Medicine, Wake Forest University

METHODS IN ALCOHOL - RELATED NEUROSCIENCE RESEARCH

Edited by
Yuan Liu
David M. Lovinger

CRC PRESS

Boca Raton London New York Washington, D.C.

Library of Congress Cataloging-in-Publication Data

Methods in alcohol-related neuroscience research / [edited by] Yuan Liu, David M. Lovinger
    p. cm. (Methods & new frontiers in neuroscience)
    Includes bibliographical references and index.
    ISBN 0-8493-0203-X (alk. paper)
    1. Alcohol--Physiological effect--Research--Methodology. 2.
 Brain--Pathophysiology--Research--Methodology. 3. Alcoholism--Molecular
aspects--Research--Methodology. I. Liu, Yuan, Ph.D. II. Lovinger, David M. (David
Michael), 1959- III. Methods & new frontiers in neuroscience series.
    [DNLM: 1. Alcohol-Induced Disorders, Nervous System--physiopathology. 2.
Neurology--methods. 3. Research--methods. WL 140 M592 2002]
    QP801.A3 M45 2002
    616.86′1′0072—dc21

                                           2001052454
                                              CIP

Visit the CRC Press Web site at www.crcpress.com

U.S. Government work
International Standard Book Number 0-8493-0203-X
Library of Congress Card Number 20021052454
Printed in the United States of America 1 2 3 4 5 6 7 8 9 0
Printed on acid-free paper

The opinions expressed in this book are solely those of the authors and do not reflect the policy or official opinions of the NIAAA, NIH or any other agency of the Federal Government.

Dedication

*We dedicate this volume to the memory
of Dr. Thomas V. Dunwiddie (1951-2001),
a leader in alcohol research and a beloved colleague.*

Methods and New Frontiers in Neuroscience

Our goal in creating the Methods and New Frontiers in Neuroscience Series is to present the insights of experts on emerging experimental techniques and theoretical concepts that are, or will be, at the vanguard of neuroscience. Books in the series cover topics ranging from methods to investigate apoptosis to modern techniques for neural ensemble recordings in behaving animals. The series also covers new and exciting multidisciplinary areas of brain research, such as computational neuroscience and neuroengineering, and describes breakthroughs in classical fields such as behavioral neuroscience. We want these to be the books every neuroscientist will use to get acquainted with new methodologies in brain research. These books can be given to graduate students and postdoctoral fellows when they are looking for guidance to start a new line of research.

Each book is edited by an expert and consists of chapters written by the leaders in a particular field. Books are richly illustrated and contain comprehensive bibliographies. Chapters provide substantial background material relevant to the particular subject. Hence, they are not only "methods books." They contain detailed "tricks of the trade" and information as to where these methods can be safely applied. In addition, they include information about where to buy equipment, and Web site addresses that are helpful in solving both practical and theoretical problems.

We hope that, as the volumes become available, the effort put in by us, the publisher, the book editors, and individual authors will contribute to the further development of brain research. The extent to which we achieve this goal will be determined by the utility of these books.

Sidney A. Simon, Ph.D.
Miguel A.L. Nicolelis, M.D., Ph.D.
Series Editors

Preface

Alcohol abuse and alcoholism are widespread problems that plague societies throughout the world. The enormous costs of the effects of alcohol on individuals, families, and society arise from many sources: the toxic effects of acute alcohol consumption, the danger of injury to oneself and others during intoxication, the detrimental effects on health from long-term alcohol abuse, the monetary costs of these health problems, and the damage to families and social groups brought about by alcoholism and associated behavioral and psychiatric disorders. It has become increasingly evident through contemporary research that alcohol abuse and alcoholism arise, at least in part, from causes beyond the control of the individual. For example, it is now widely appreciated that genetic factors contribute strongly to the likelihood of occurrence of alcoholism.

The ultimate goal of alcohol-related research is to develop better prevention, treatment and cures for the undesirable effects of alcohol on the brain and behavior, and for the causes that contribute to alcohol-related disorders. Investigation of the underlying mechanisms of neural effects of alcohol that contribute to craving, intoxication, tolerance, dependence, withdrawal, and alcohol consumption is at the heart of scientific attempts to understand and combat alcohol abuse and alcoholism. Neuroscience research in this area has made remarkable progress in the last two decades, and the advances are due, in great part, to the wide array of powerful biomedical, bioengineering and computational biological techniques that are now being employed. These techniques include a great variety of approaches, ranging from gene mapping and examination of molecular interactions of alcohol at the sub-cellular level to recording of neural activities in freely behaving animals, and imaging alcohol effects on the living human brain. To date, however, no comprehensive text has covered most of these recently developed research methods in the alcohol-related research field.

The primary goal of this volume is to provide an up-to-date technical guide for investigators interested in pursuing alcohol-related neuroscience research at the molecular, cellular, or systems levels.

The research methods described in this volume are organized into several different levels: genetics (Chapter 1 on QTL mapping and Chapter 2 on knockout and knockin techniques), molecular and cellular (Chapter 3 on chimeragenesis and mutagenesis, Chapter 4 on single cell RNA profiling, Chapter 5 on antibody-based techniques, Chapter 6 on bilayer techniques, and Chapter 7 on channel kinetic analysis), and system levels (Chapters 9 and 10 on multi-electrode recordings, Chapters 11 and 12 on *in vivo* microdialysis, and Chapters 13 and 14 on imaging techniques).

Chapters are written by experts in each area, and are focused specifically upon the application of these methods to alcohol research. It must be emphasized that we were not able to include all of the relevant techniques in this single text, given limitations on space and the availability of authors. Many methods not explicitly described in this volume are mentioned briefly herein, and we hope that these short descriptions will lead readers to explore the literature further to gain a better appreciation of the diversity of research in this area. We also realize that space limitations in this volume have prevented us from providing all the experimental details to be described. Nevertheless, each chapter contains a selection of extensive references that will allow the reader to find additional details that are important for exploring and understanding a particular method in greater depth. Many of the techniques described here will find application far beyond alcohol research, and are useful in a wide variety of neuroscience sub-disciplines. Close examination of this volume can provide information of great use to young neuroscientists and other investigators interested in entering this larger research field or endeavor.

The chapters in this volume also contain information about specific effects of alcohol on brain proteins, brain function, and behavior. In addition, the references cited in these chapters will allow the reader to gain substantial information about the status of alcohol-related neuroscience research. Thus, this volume not only provides information of a technical nature, but also can serve as an overview of many exciting facets of investigations of alcohol effects on the brain. We hope that the description of research techniques and findings contained in this volume will inspire investigators to enter the fascinating and rapidly expanding alcohol research field. With increased interest and expertise in this area, we should be able to discover new and improved means to combat alcohol abuse and alcoholism.

Yuan Liu, Ph.D.
David M. Lovinger, Ph.D.

Acknowledgments

We would like to thank Dr. Miguel Nicolelis, the series editor of the CRC Methods and New Frontiers in Neuroscience, for suggesting this volume, and Barbara Norwitz, publisher, and Tiffany Lane, editorial assistant, at CRC, for their patience and support throughout the process of this project. We would also like to thank Dr. Randall Stewart for his scientific and editorial input; Drs. Dennis Glanzman and Paul Nicholls, and Ms. Belinda Morin for their editorial suggestions; and Ms. Carol Salter for her help with the table of contents.

Yuan Liu, Ph.D.

David M. Lovinger, Ph.D.

About the Editors

Yuan Liu, Ph.D. was born in the People's Republic of China. During the decade of "Cultural Revolution" all formal education was banned in China, so her middle- and high-school education was self-taught after work (she was assigned as a factory worker) in a dark cellar room that she shared with her mother for 10 years. In 1978, when the revolution finally ended, she passed the national college entrance examination and earned acceptance into Peking University, where she received both Bachelor's and Master's degrees in neurophysiology. Her two theses were on neural mechanisms of mammalian hibernation. In 1985, she joined Dr. John G. Nicholls' laboratory in the Biozentrum, Universität Basel, Switzerland, where she explored biophysical properties of presynaptic Ca^{++} channels, distribution of postsynaptic transmitter receptors and cellular mechanisms underlying synaptic regeneration, and received her Ph.D. in 1989. She was awarded the Grass Fellowship to spend the summer of 1988 at the Marine Biological Laboratory in Woods Hole to learn optical recording techniques. Dr. Liu then took postdoctoral training at SUNY at Stony Brook studying function and structure relationships of postsynaptic receptors using single-channel patch-clamp recording techniques. In 1991, she joined the intramural research program at the National Institutes of Health (NIH), where her main research focus was on synaptic transmission, plasticity and regeneration.

In 1995, Dr. Liu became the program director for Basic Neuroscience Research at the National Institute on Alcohol Abuse and Alcoholism (NIAAA), where she organized four satellite symposia in conjunction with the Society for Neuroscience Annual Meetings on alcohol-related topics including: Approaches for Studying Neural Circuits — Application to Alcohol Research (1996); The "Drunken" Synapse — Studies of Alcohol-Related Disorders (1997); Applications of Gene Knockout Techniques to Alcohol Research (1998); The Cerebellum and Alcohol: Roles in Cognitive and Motor Functions (1999). She edited the proceedings of the 1996 symposium, which was published in *The Journal of Alcoholism: Clinical and Experimental Research* (1998), and co-edited with Dr. Walter A. Hunt *The "Drunken Synapse" Studies of Alcohol-Related Disorders*, which was published in 1999. She also co-organized a workshop with Dr. Mike Eckardt from the NIAAA on Computational Neurobiology Approaches and Alcohol Research. In recognition of her development of an outstanding neuroscience portfolio in alcohol research, Dr. Liu received the NIH Director's Award in 1999.

Since 1999, Dr. Yuan Liu has served as the program director for Channels, Synapses, and Circuits at the National Institute of Neurological Disorders and Stroke (NINDS). Her program focuses on basic and translational research in channels, synapses, and neural circuits that are responsible for normal and abnormal brain functions. She is the NINDS representative to several trans-NIH and intra-governmental agency committees related to computational neuroscience and neuroinformatics.

David M. Lovinger, Ph.D. grew up in Tucson, Arizona, and received his B.A. degree from the University of Arizona in 1981, where his studies focused on biopsychology. Encouraged by his strong interest in the neurobiological basis of learning and memory, he entered the graduate program in Neurobiology and Behavior in the Psychology Department at Northwestern University. Under the tutelage of Dr. Aryeh Routtenberg, Dr. Lovinger focused on understanding the role of protein phosphorylation in hippocampal synaptic plasticity, with an emphasis on the role of protein kinase C. His dissertation was entitled Regulation of the Maintenance of Hippocampal Long-Term Potentiation by Protein Kinase C and Protein F1. He received his Ph.D. from Northwestern in 1987, and then entered the laboratory of Dr. Forest F. Weight at NIAAA for postdoctoral training. The focus of his research expanded to include investigation of ion channel pharmacology and examination of alcohol effects on ligand-gated ion channels. His postdoctoral work included characterization of alcohol effects on glutamate-gated ion channels, serotonin-gated ion channels and glutamatergic synaptic transmission. Dr. Lovinger moved to Vanderbilt University as an assistant professor of molecular physiology and biophysics in 1991. His work since that time has focused on the molecular basis of alcohol actions on ligand-gated ion channels, as well as on the cellular and molecular mechanisms involved in synaptic transmission and plasticity in the basal ganglia. He has risen through the academic ranks to become full professor of molecular physiology and biophysics, pharmacology, and anesthesiology. Dr. Lovinger has been the recipient of several honors and awards, including The Research Society on Alcoholism Young Investigators Award for 1992, and a MERIT award from NIAAA in the year 2000. He previously co-edited a volume entitled Presynaptic Receptors in the Mammalian Brain along with the late Dr. Thomas V. Dunwiddie. Dr. Lovinger will assume leadership of the Laboratory of Integrative Neuroscience in the NIAAA intramural research program in the autumn of 2001.

Contributors

Elfar Adalsteinsson, Ph.D.
Department of Radiology
Stanford University School of Medicine
Stanford, CA

Michael J. Beckstead
Department of Physiology and
  Pharmacology
Wake Forest University School of
  Medicine
Winston-Salem, NC

John Crowley, Ph.D.
Department of Neurobiology
University of Massachusetts Medical
  School
Worcester, MA

T. Michael Gill, Ph.D.
Ernest Gallo Clinic and Research
  Center
Department of Neurology
University of California, San Francisco
Emeryville, CA

Bennet Givens, Ph.D.
Department of Psychology
Ohio State University
Columbus, OH

Rueben A. Gonzales, Ph.D.
Department of Pharmacology
College of Pharmacy
University of Texas
Austin, TX

Gregg E. Homanics, Ph.D.
Department of Anesthesiology
University of Pittsburgh
Pittsburgh, PA

Patricia H. Janak, Ph.D.
Ernest Gallo Clinic and Research
  Center
Department of Neurology
University of California, San Francisco
Emeryville, CA

Terry L. Jernigan, Ph.D.
Laboratory of Cognitive Imaging
University of California, San Diego
La Jolla, CA

Yuan Liu, Ph.D.
National Institute of Neurological
  Disorders and Stroke
National Institutes of Health
Bethesda, MD

Gregory F. Lopreato, Ph.D.
Waggoner Center for Alcohol and
  Addiction Research
University of Texas at Austin
Austin, TX

David M. Lovinger, Ph.D.
Department of Molecular Physiology
  and Biophysics
Vanderbilt University
Nashville, TN

Shao-Ming Lu, Ph.D.
Center for Aging and Developmental
  Biology
Aab Institute of Biomedical Sciences
University of Rochester Medical Center
Rochester, NY

Nandor Ludvig, M.D., Ph.D.
Department of Physiology and
  Pharmacology
State University of New York
Health Science Center at Brooklyn
Brooklyn, NY

S. John Mihic, Ph.D.
Waggoner Center for Alcohol and
  Addiction Research
Institutes for Neuroscience and Cellular
  and Molecular Biology
University of Texas at Austin
Austin, TX

Robert O'Connell, Ph.D.
Department of Neurobiology
University of Massachusetts Medical
  School
Worcester, MA

Abraham A. Palmer, Ph.D.
Oregon Health and Science University
Department of Behavioral
  Neuroscience
Portland, OR

Adolf Pfefferbaum, M.D.
Neuroscience Program
SRI International
Menlo Park, CA

Tamara J. Phillips, Ph.D.
Department of Behavioral
  Neuroscience
Oregon Health and Science University
VA Medical Center and Portland
  Alcohol Research Center
Portland, OR

Donita L. Robinson, Ph.D.
Department of Chemistry
University of North Carolina
Chapel Hill, NC

Daniel D. Savage, Ph.D.
Department of Neuroscience
University of New Mexico
Health Science Center
Albuquerque, NM

Suzanne Sikes
Department of Molecular Physiology
  and Biophysics
Vanderbilt University
Nashville, TN

Edith V. Sullivan, Ph.D.
Department of Psychiatry and
  Behavioral Sciences
Stanford University School of Medicine
Stanford, CA

Amanda Tang, Ph.D.
Department of Pharmacology
College of Pharmacy
University of Texas
Austin, TX

Stavros Therianos, Ph.D.
Center for Aging and Developmental
  Biology
Aab Institute of Biomedical Sciences
University of Rochester Medical Center
Rochester, NY

Steven N. Treistman, Ph.D.
Department of Neurobiology
University of Massachusetts Medical
  School
Worcester, MA

C. Fernando Valenzuela, M.D., Ph.D.
Department of Neuroscience
University of New Mexico
Health Science Center
Albuquerque, NM

Kent E. Vrana, Ph.D.
Department of Physiology and
  Pharmacology
Wake Forest University School of
  Medicine
Winston-Salem, NC

Jeff L. Weiner, Ph.D.
Department of Physiology and
  Pharmacology
Wake Forest University School of
  Medicine
Winston-Salem, NC

Hermes H. Yeh, Ph.D.
Center for Aging and Developmental
  Biology
Aab Institute of Biomedical Sciences
University of Rochester Medical Center
Rochester, NY

Qing Zhou, Ph.D.
Department of Molecular Physiology
  and Biophysics
Vanderbilt University
Nashville, TN

Table of Contents

1 Quantitative Trait Locus (QTL) Mapping in Mice

Abraham A. Palmer and Tamara J. Phillips

CONTENTS

1.1 INTRODUCTION

Complex traits, such as the human disease alcoholism, are influenced by multiple genetic and environmental factors. Animals can be used to model certain components of this disease. However, even the apparently simplified traits modeled in animals are found to be under the influence of multiple genes, each of relatively small effect. Thus, the distributions of these traits in genetically heterogeneous populations tend to be continuous, rather than displaying simple Mendelian inheritance patterns. The genomic loci that influence the quantitative variability in these traits are termed quantitative trait loci (QTL).* The identification of the genes influencing such quantitative traits can be approached using QTL mapping methods that were originally developed by plant geneticists (see Lynch and Walsh, 1998). Initial efforts may be directed at localizing an influential gene (a QTL) to a chromosome or a chromosomal segment. Additional experiments may then be performed, using stringent statistical criteria, to confirm the presence of the QTL. Finer mapping methods are then used to narrow the chromosomal region in which the QTL resides, with the ultimate goal of identifying the causative gene or regulatory element. Single gene mutants, gene expression analyses and sequence information can be used at each stage to provide evidence that a candidate gene is the QTL (Figure 1.1).

QTL mapping depends on the availability of genetic markers with known positions on particular chromosomes. When many such markers are polymorphic between strains, it is possible to determine which chromosome segments were inherited from which ancestors. Multiple markers are required for each chromosome because of crossing-over, a process that intermixes chromosome segments during meiosis. By examining many markers on each chromosome, the relationship between variability in the phenotype of interest and variability in genetic material can be determined — the essence of QTL mapping. Genetic markers are not selected based on any functional significance, but rather because their differences between strains are easily identifiable. In general, these markers correspond to non-coding regions of the genome, meaning that they are not actual genes. The most commonly used markers are regions whose PCR products are of different lengths for different mouse strains. If these concepts are unfamiliar, more thorough discussions are available (Grisel and Crabbe, 1995; Falconer and Mackay, 1996; Lynch and Walsh, 1998).

QTL mapping approaches require numerous comparisons between phenotype and genotype. When so many comparisons are made, the criteria for statistical significance must be stringent to avoid false positive errors. These stringent criteria diminish the power to detect real effects, increasing the probability of false negative errors. Thus, every QTL study grapples with the need to balance these two diametrically opposed considerations. The result is that many hundreds or even a few thousand animals are required to accurately map the location of one or more QTL for a particular trait. Frequently, multiple experiments must be performed, first to provisionally map QTL, and subsequently to confirm and more accurately map their location.

The goal of the current chapter is to serve as an introductory practical guide to the most relevant methods, technologies and approaches to QTL mapping studies with specific reference to alcohol-related basic research. Whenever possible, citations will be given to original theoretical reports, review articles, World Wide Web links and relevant findings based on these methodologies.

---

* QTL stands for quantitative trait locus (singular) or loci (plural); in both cases we will use the abbreviation QTL.

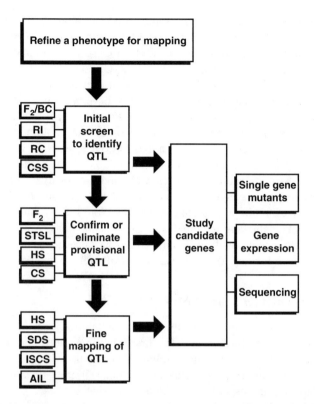

FIGURE 1.1 Flow diagram showing a possible approach to a QTL mapping study. First, a phenotype for mapping is chosen and optimized. An initial screening step may conclusively map QTL and identify provisional QTL. Provisional QTL can then be confirmed or eliminated using additional mapping populations. Fine mapping can be used to more precisely isolate the QTL. At any stage, candidate gene hypotheses can be investigated using methods such as single gene mutants, gene expression assays and sequencing. Abbreviations: F_2/BC = F_2 or backcross (BC) population; RI = recombinant inbred; RC = recombinant congenic; CSS = chromosome substitution strain; STSL = short term selected lines; HS = heterogeneous stock; CS = congenic strain; SDS = small donor segment strain; ISCS = interval specific congenic strains; AIL = advanced intercross line.

1.2 WHY MICE?

The majority of QTL research related to alcoholism has focused on the use of murine models. Justifications for this choice are numerous (Silver, 1995; Crabbe et al., 1999a; McPeek, 2000; Moore and Nagle, 2000; Crawley, 2000; Paigen and Eppig, 2000; Peltonen and McKusick, 2001). First, the genome of the mouse has been heavily studied for several decades, and thus thousands of molecular genetic markers have already been organized into highly refined genomic maps. Second, many genes have already been cloned and mapped in mice, and the relationships between mouse and human chromosome segments (termed "syntenic regions") are known. This trend will continue, with the genomes of several commonly used laboratory strains scheduled to be fully sequenced in the near future (Belknap et al., 2001; Peltonen and McKusick, 2001). Third, a variety of powerful mapping populations, discussed further in section 1.4, have been developed for the mouse, the combinations of which are not available in any other mammalian species (Grisel and Crabbe, 1995; Moore and Nagle, 2000; Williams, et al., 2001). Fourth, behavioral and physiological models for various sub-components of alcoholism have been well established and, in some senses, validated (Crabbe et al., 1999a). Fifth, mice are the most amenable mammalian species to transgenic, knockout and knockin technology; the usefulness of these techniques is discussed with reference to QTL research in section 1.8, and more broadly in Chapter 2 of this book. Sixth, mice are the smallest mammalian species widely used for neuroscience research, and are relatively inexpensive to maintain. Seventh, mice could also be considered to be among the least sentient mammalian species, an important ethical consideration. Finally, mice reach reproductive age quickly, allowing for rapid successive generations, which is necessary for many aspects of genetic analysis. For these reasons, mice are the preferred species for genetic research relating to alcoholism and many other human diseases.

1.3   TRAIT SELECTION AND MEASUREMENT

QTL mapping efforts are tremendously large undertakings, and, as such, require scrupulous planning and forethought. The first stage in a study, and perhaps most crucial, is the choice of trait(s) to examine (Aylsworth, 1998). Most often, the ultimate goal is to identify genes relevant to human disease; thus, the trait must adequately model some aspect of alcoholism or whatever disease is of interest. Obviously, if the trait is ill-defined, unduly complex, or not reliably measurable, a QTL study will not succeed. Advice on measurement of a wide variety of behavioral traits is available in an excellent book by Crawley (2000). Traits specifically relevant to alcoholism have been recently reviewed (Hitzemann, 2000; Li, 2000).

Beyond its relevance to alcoholism, the suitability of a trait for *genetic* analysis should be carefully considered. First and foremost, it is crucial that the trait is heritable, i.e., the variability between subjects must be partially under the control of genes (Falconer and Mackay, 1996; Lynch and Walsh, 1998). Heritability (h^2) is a property not only of the trait being measured, but also of the sample population. Estimates of h^2 may differ markedly across populations composed of different mouse strains, and thus, different genetic material. Informative estimates of h^2 can be obtained from parent-offspring regression, a panel of standard inbred strains, recombinant inbred strains (section 1.4.2), selected lines (section1.4.6), and classical Mendelian crosses, among others (Falconer and Mackay, 1996; Lynch and Walsh, 1998). In practice, a formal estimation of h^2 is rarely obtained prior to beginning a QTL mapping study. Instead, projects are often pursued based on the observation of a statistically significant difference in the trait of interest between the two inbred strains to be used to establish the mapping population (Paigen and Eppig, 2000). This approach carries certain risks, because the presence of a difference between the two genotypes does not prove that the trait is heritable (the difference could be associated with environmental influences). Conversely, the lack of a difference does not prove that informative genetic variability is absent in the two strains. Instead, each strain may possess some genes that increase and some that decrease the trait such that, in the aggregate, the strains exhibit no difference, but possess relevant genetic variability. Whereas choosing strains that maximize trait-relevant genetic variability is clearly important, it is also desirable to employ strains that are genetically well defined, and that are incorporated into some of the complex mapping populations discussed in section 1.4. Practical suggestions include (but are not limited to): A/J, C57BL/6J, DBA/2J, BALB/cByJ, C3H/HeJ, CAST/Ei, NON/Lt, LP/J, SPRET/Ei, ARK/J, BTBR/J, 129S3/SvImJ and FVB/NJ. These strains are mentioned for several reasons. First, numerous molecular genetic marker polymorphisms are already defined in these strains. Additionally, many of them have been included in useful mapping populations, such as recombinant inbred, congenic, chromosome substitution strains, and interval specific congenic strains. Finally, many of them are included in the "A list" among the 40 inbred strains in the newly initiated Mouse Phenome Project (http://aretha.jax.org/pub-cgi/phenome/mpdcgi), which will gather normative phenotypic data for a diverse group of inbred strains (Paigen and Eppig, 2000).

The portion of the total trait variance that is heritable can be enhanced by limiting variability due to environmental and technical error. For example, if human observation is an important factor, averaging the readings of multiple human observers might substantially diminish technical error, and thus enhance the trait's h^2 (Williams et al., 1996). Another source of undesirable variability is environmental factors such as: differences in age, sex, weight, litter size, litter number, pre- and post-natal maternal conditions, exposure to pathogens, housing density, fighting with cage mates; differences in temperature and humidity; inconsistencies in food and tap water quality, activity and construction in the vivarium; differences in animal care staff habits; differences in timing or intensity of room lights; and weekday vs. weekend phenotype measurement, etc. Which of these factors might be important to a specific trait for the specific strains under study is rarely known, but is a frequent source of laboratory mythology and superstition. A deliberate study of the reproducibility of behavioral genetic data recently showed that even when careful attention was paid to these factors, environmental variability powerfully influenced the pattern of genetic contributions to certain traits (Crabbe et al., 1999b). It is probably not realistic to completely standardize all environmental factors across all test subjects. However, it is important to minimize environmental variability as much as possible.

Changes in the way a trait is defined, such as differences in dose, length of measurement, type of measurement, etc., may also alter the estimate of h^2, and are therefore well worth consideration (see Lander and Botstein, 1989; Palmer et al., 2000). For example, in a study that is still in progress, we obtained 12 measurements of prepulse inhibition over the course of a single testing session. An average of these 12 measurements was prepared to represent the trait of interest. Examination of the data demonstrated that the estimated heritability could be increased by a "jack-knifing" procedure, in which the highest and the lowest single value out of the 12 were removed, leaving just 10 values to establish the average. In another study, using the average of two scores per individual for the duration of ethanol-induced loss of the righting reflex reduced environmental variance by 39% (Markel et al., 1995), and has been used to confirm QTL for this measure of ethanol sensitivity (Markel et al., 1997a). In our studies of two-bottle choice drinking, ethanol consumption is measured daily over a period of 4 days, and an estimate of ethanol consumption and preference is developed based on these data (Phillips et al., 1998a). In another study, an estimate of test-retest variability was used to eliminate animals with low reliability for ethanol locomotor stimulant response, to diminish what were presumed to be environmental confounds (Hitzemann et al., 1998). Finally, Melo et al. (1996) used a "consistency test" to identify animals whose drinking values were highly variable, so that these animals could be eliminated prior to QTL mapping analysis. These examples illustrate some of the approaches investigators have used to attempt to limit technical and environmental variability.

Other important considerations are the rate at which a trait can be measured and the number of measurements that can be made simultaneously. Because QTL mapping studies might ultimately involve thousands of measurements, any modifications that make measurement of the trait more efficient will provide enormous returns. However, it is likely that some important alcohol-related traits are, by their very nature, costly and inefficient to measure. It may be to the detriment of the field to neglect the genetic analysis of such traits (Lederhendler and Schulkin, 2000; Belknap et al., 2001). Alcoholism is a lifetime disease, which takes years to evolve to its most extreme form, so it would be surprising if all relevant genes could be identified using exclusively rapid-throughput animal models.

Finally, consideration should be given to measuring multiple independent traits in each subject. For example, Demarest et al. (1999) examined locomotor response to ethanol, and then administered a benzodiazepine the following week to the same subjects, providing an unconfounded measure of ethanol response (the principal question), and a possibly confounded, but still potentially interesting, measure of benzodiazepine response. In some instances, more than one trait can be measured simultaneously. For example, the "grid test" measure of ataxia, as well as certain place-preference tests also provide an independent index of ethanol-induced locomotor activity (Cunningham, 1995; Phillips et al., 1996). If the study involves multiple days of testing, both acute and chronic responses to a treatment may be assessed in the same animals (Phillips et al., 1995). These approaches use valuable animal resources more efficiently, and increase the chances of gaining useful information from a single experiment.

1.4 INITIAL MAPPING POPULATIONS

Following the careful selection and refinement of the trait for a QTL mapping study, a genetically heterogeneous population must be selected. Essentially no QTL mapping project can be taken from phenotype to gene using a single population. Instead, a multistage approach is typically required. Though somewhat arbitrary, the stages of a QTL study can be divided into three steps: initial, confirmatory, and fine mapping. Our group and others have had great success using recombinant inbred strains as an initial mapping population, followed by confirmatory work using F_2 populations and short-term selected lines (Grisel and Crabbe, 1995; Phillips et al., 1998a; Hitzemann et al., 1998; Crabbe et al., 1999a; Grisel, 2000). This is certainly not the only way to proceed, and, depending on whether there is ready access to a recombinant inbred (RI) colony, it may not be the most expeditious. When multiple mapping populations are employed, statistical procedures can be used to combine probability scores from multiple independent experiments (for an example, see Phillips et al., 1998a).

1.4.1 INTERCROSS POPULATIONS

Intercross populations are the most traditional mapping populations and are also the most intuitive. The classical example is an F_2 population, such as those that were the focus of Mendel's work. F_2 populations are created by breeding two different inbred strains to create an F_1 generation. With the exception of sex chromosomes, all F_1 animals are genetically identical to one another, with each being heterozygous at all autosomal loci. These F_1 animals can then be bred with each other to produce an F_2 population (Figure 1.2A). Alternatively, a backcross (BC) population can be created in which the F_1 is backcrossed to one of the progenitor strains (Figure 1.2B). BC populations are most useful when a tendency toward genetic dominance is suspected (Lander and Botstein, 1989).

Breeding of F_2 and BC mice is typically performed in the vivarium of the researcher's institution; F_1 mice can also be produced by the researcher, or, in some cases, are available from commercial vendors (for example B6D2 F_1 are sold by The Jackson Laboratory http://www.jax.org). Once the F_1 mice have been obtained, it may be desirable to breed in such a manner as to balance the distribution of sex chromosomes. F_1 male mice will have an X chromosome from their maternal progenitor. Therefore, when breeding the F_2, one may wish to utilize equal numbers of males derived from each type of maternal ancestor. The potential for maternal effects should also be considered, so the most stringent breeding scheme would be to use an equal number of females born to each type of progenitor mother, and breed these with the two possible types of males, creating four different types of breeding combinations in the final creation of the F_2.

To gain information about dominance effects, testing can be performed on the F_1 generation prior to making a decision about whether to breed an F_2 or a BC population. If the F_1 generation shows a phenotype very similar to one of the two progenitor strains, rather than an intermediate phenotype, this suggests one or more dominant alleles are influencing the phenotype. Thus, the trait may be best studied in a population backcrossed onto the recessive inbred strain. If the phenotype is intermediate between the two progenitors, then creation of an F_2 population is preferable. If assessing the phenotype of interest is problematic prior to breeding, a separate sub-population of F_1 mice, not to be used for breeding, may be tested for the trait of interest, along with the two progenitor strains.

In planning and proposing the creation of an intercross population for the purpose of a QTL search, it is also important to consider the number of animals required in advance. A convenient equation (see Belknap et al., 1996) can be used to determine the number of animals needed based on the acceptable levels of type I and type II error (significance and power), and the estimated contribution of an individual QTL to the total variability of the trait. Another important consideration is the need for substantial amounts of genotyping to be performed at the end of the experiment. This is unavoidable, because each F_2 or BC mouse is genetically unique. Luckily,

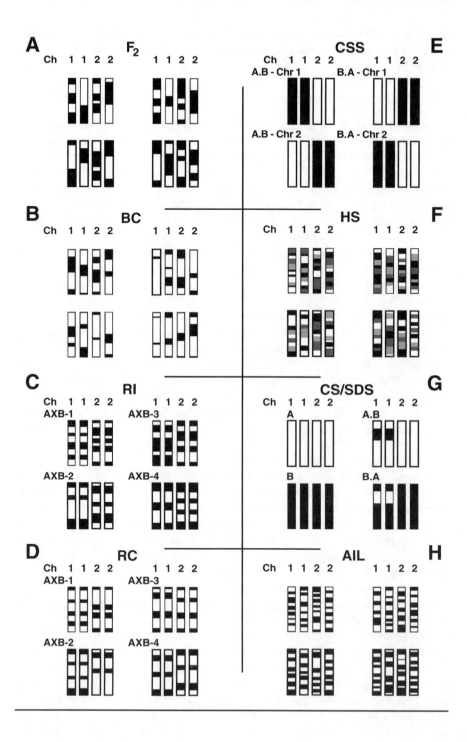

FIGURE 1.2 (Previous page) Schematic representation of two of the possible 20 pairs of chromosomes belonging to the imaginary mouse strains A and B (white and black, respectively). (A) Four different F_2 animals, each of which is unique and can be either heterozygous or homozygous at any given locus. (B) Four different backcross (BC) animals, which are like the F_2 except that strain A comprises approximately 75% of the genetic material, and strain B comprises the remaining 25%. (C) Four different recombinant inbred (RI) strains, which are homozygous at every locus. RI strains are genetically identical from animal to animal and generation to generation. Note additional crossovers (transitions from white to black) accumulated over multiple generations of breeding required to create these animals as compared with F_2 and BC animals. (D) Four different recombinant congenic (RC) strains, like the RI strains except that the genome of strain A comprises approximately 75% of the genome and strain B comprises the remaining 25% of the genome. (E) Four different chromosome substitution strains (CSS). Strains A.B — chr 1 and A.B — chr 2 are composed of a strain A background with the indicated chromosome from strain B introgressed. Strains B.A — chr 1 and B.A — chr 2 are instead comprised of a strain B background with the indicated chromosome from strain A introgressed. (F) Four animals from a heterogeneous stock (HS), each composed of approximately equal contributions from strains A, B, C and D (white, black, light gray and dark gray, respectively). Note the much higher number of crossovers than in the populations shown in panels (A)–(D). The number of crossovers in an HS increases with the number of generations since the inception of the HS, and is typically very large. This large number of crossovers offers an advantage over an F_2 population for fine mapping purposes as discussed in the text. HS, like F_2 and BC, may be heterozygous or homozygous at any given locus .(G) A is pure background strain A; B is pure strain B. A.B is a congenic strain with a donor segment from strain B introgressed onto strain A.B.A is the reciprocal strain. Animals are termed either congenic strains (CS) or small donor segment strains (SDS) depending on the length (in cM) of the introgressed segment, as discussed in the text. Note that these animals are homozygous at every locus. (H) Four individual animals from an advanced intercross line (AIL). Note the large number of crossovers that have accumulated over successive generations. This large number of crossovers offers an advantage over an F_2 population for fine mapping. AIL, like HS, F_2 and BC, may be heterozygous or homozygous at any given locus.

it is not necessary to genotype all subjects. Instead, most of the power for mapping can be obtained by genotyping only those individuals whose scores are in the extremes of the population. For example, in an F_2, by genotyping only those animals in the 5% extremes of both tails of a normal distribution (10% of the total population), one would obtain 44% of the total power available from the population (termed efficiency). By genotyping the extreme 10% of both tails, efficiency rises to 65%. Genotyping 15% of the two extremes yields an efficiency of 78%, and genotyping 25% of both extremes gives an efficiency of 93%, with the remaining half of the population contributing less than 10% of the efficiency. This practice, termed "selective genotyping," is statistically valid as outlined in a paper by Lander and Botstein (1989). Thus, the number of animals to be genotyped should be designed to balance the costs of producing and phenotyping the animals against the costs of genotyping them. However, even if the cost of breeding and phenotyping is high relative to genotyping, it is unlikely that the middle 50% of animals would be worth genotyping, given how little they contribute to the efficiency of the experiment. The amount of genotyping may be further reduced if an initial screening population, such as those discussed in sections 1.4.2, 1.4.3 or 1.4.4, is used to determine which chromosome regions warrant further attention (see Grisel, 2000).

Some phenotypes are influenced by sex, such that high-scoring animals are typically of one sex, and low-scoring animals are of the other. If animals from both extremes of the phenotypic distribution were chosen for selective genotyping without regard to sex, the QTL analysis would be impaired, as trait variation would be partially due to sex, rather than inherited alleles. Therefore, selective genotyping must never sample indiscriminately from both males and females. This problem can be addressed in one of two ways. First, phenotyping or genotyping in an intercross population can be confined to only one sex. To avoid increased costs associated with not using half of the mice produced, male and female animals can be used for two different studies conducted simultaneously (e.g., the studies of Hitzemann et al., 1998 and Demarest et al., 1999). Alternatively, both sexes can be used in the same study. If both sexes are to be used, it is typically best to choose an equal number of mice from each sex for genotyping. Thus, when selecting animals from the phenotypic extremes, 25% of animals will be high-scoring males, 25% low-scoring males, 25% high-scoring females, and 25% low-scoring females. We have used this approach in past investigations (Phillips et al., 1998a); it not only utilizes the available animals more fully, but also allows an initial screen to be conducted for QTL that influences the trait in one sex but not the other (termed sex-specific QTL). Such QTL are theoretically predicted, and their existence has been confirmed experimentally (e.g., Melo et al., 1996). If the initial analyses suggest the presence of sex-specific QTL, it might be necessary to add additional animals to the study to achieve the desired statistical power.

F_2 populations are sometimes used to estimate the correlation between two or more traits measured in the same individuals. For example, one might plot the volume of ethanol consumed in a two-bottle choice paradigm against preference for a tastant solution (e.g., saccharin), in an effort to identify a potential relationship between these two factors. Unfortunately, if the goal is to assess the genetic relationship between these two factors, this approach is potentially flawed (Crabbe et al., 1990). The differences among individuals for the two traits is influenced by both genetic and environmental factors. It is possible that the estimated phenotypic correlation is composed of a genetic correlation with one r-value, and an environmental correlation with another r-value of the opposite sign (complicated by a possible interaction between genetic and environmental factors). In such a case, the larger r-value would be observed empirically, and, if significant, the potentially erroneous conclusion that a genetic relationship exists would be made. Alternatively, if the phenotypic correlation is not significant, one may falsely conclude that no genetic relationship exists, when, in fact, the sources of correlation may have canceled each other out. The number of animals typically tested in a mapping study using an F_2 population is large, making it more likely that significant p values will be obtained. However, these correlations reflect a phenomenon of unknown etiology, and should be reported as phenotypic correlations and not used to prove a genetic relationship.

1.4.2 RECOMBINANT INBRED (RI) STRAINS

Recombinant inbred strains, like intercross populations, are a powerful and commonly used genetic mapping population (Figure 1.2C). Each individual RI strain is created from an F_2 cross of two inbred strains. F_2 progeny from this cross are inbred for many (typically more than 20) generations to create a single isogenic RI strain (for a very helpful figure illustrating this process, see Figure 2 of Grisel and Crabbe, 1995). Many such RI strains are created at once, providing a panel of RI strains, each reflecting a unique recombination of genes from the two parental strains. Each RI strain is somewhat like a clonal line of an individual F_2 animal, but there are important differences between RI and F_2 populations. First, RI mice, by virtue of many generations of inbreeding, are homozygous at all loci, unlike F_2 mice, which are heterozygous at many loci. As a consequence, RI mice cannot elucidate dominance or recessive relationships between alleles. Additionally, F_2 mice all share similar maternal influences, as their mothers are all F_1 mice. In contrast, RI mice of different strains may provide very different maternal care to their offspring. This has the potential for distorting genetic analyses. Finally, due to the additional generations following creation of an F_2, and prior to the completion of inbreeding of the RI strains, additional crossover events occur. More crossovers allow for finer mapping, but require more markers per chromosome. In technical terms, the additional generations lead to reduced linkage disequilibrium, and resultant map expansion (Falconer and Mackay, 1996; Lynch and Walsh, 1998; Williams et al., 2001). This phenomenon is also important for understanding the advantages of advanced intercross lines and heterogeneous stocks, both of which are defined below.

RI strains were not originally created for QTL mapping purposes. However, the development of numerous molecular markers made it possible to create dense genetic maps of these strains. This provides the powerful advantage that once a particular strain is genotyped at a given locus, it is not necessary to genotype that strain at that locus again. This is true because, neglecting spontaneous mutation, the strain's genotype will not change over time. Databases of the genotypes at hundreds of molecular markers across dozens of RI strains are available at http://www.informatics.jax.org/searches/riset_form.shtml. Whereas the number of known markers currently varies with the particular panel of RI strains being studied, some, such as the BXD/Ty, (named for the two progenitor lines C57BL/6J (B6) and DBA/2J (D2); Taylor, 1978), have been genotyped to an average density of about 1 cM resolution (Williams et al., 2001). Just as in an intercross population, the relationship between the trait and the genotype is tested to identify the location of a QTL. Thus, RI strains do not require any genotyping by the investigator — one of their principal advantages as compared with segregating populations.

Another advantage of RI mice is that, because all animals of a given strain are genetically identical, multiple animals from each strain can be tested to more accurately estimate each strain's mean value for the trait of interest. This practice reduces environmental and technical variability (Belknap, 1998; Moore and Nagle, 2000; Williams et al., 2001). This is analogous to using a test-retest strategy in an intercross population. However, in this case, the possible effects of repeated testing are absent. In addition, environmental events unique to a particular animal will not affect the phenotypic estimate as drastically as in an F_2 study. For example, a particular F_2 animal might suffer an infection at a young age, permanently changing the trait of interest. Retesting this animal would not negate this source of environmental variability. However, if multiple mice of the same RI strain were tested, the effects of a similar insult to an individual animal would have only a small effect on that RI strain's mean trait value. Alternatively, evaluation of an RI strain for outliers might detect such an animal, and that individual could subsequently be eliminated. Yet another sometimes unrealized advantage of RI mice is that a trait of interest can be studied under several different conditions to identify environmental influences that might be different across strains, or to study different doses of a drug (Williams et al., 2001).

RI strains also have disadvantages. One is that they are challenging to breed and maintain. Certain RI strains are notoriously poor breeders; others may have inadequate maternal skills. Another disadvantage is the limited power of the existing RI panels to detect QTL. While more powerful than a comparable number of F_2 mice (Belknap, 1998), the largest RI mouse set (BXD) consists of only 36 strains. Power to detect QTL is contingent on the number of strains available (Moore and Nagle, 2000). In contrast, there is no upper limit on the number of intercross animals that can be created, providing theoretically unlimited mapping power. Nevertheless, there is at least one instance where a QTL that contributed a large percentage of total variability was conclusively mapped based on RI mice alone (Janowsky et al., 2001). Due to their limited power, RI mice are more commonly used as an initial screening population. The advantage of this approach is that RI mice require no genotyping but provide good coverage of the entire genome. As a result, subsequent studies can focus only on those regions implicated by the RI mice, dramatically reducing the amount of genotyping needed (Grisel and Crabbe, 1995).

For alcohol research, the BXD RI panel is among the most extensively studied. A number of ethanol-related behavioral traits differ markedly between the two progenitor strains, which suggests that genes relevant to these traits could be mapped. Some of the ethanol-related traits that have been mapped include:

- Acute locomotor response (Phillips, et al., 1995; Hitzemann et al., 1998; Demarest et al., 1999)
- Sensitization to the locomotor response to ethanol (Phillips et al., 1995)
- Severity of withdrawal following acute and chronic ethanol exposure (Crabbe, 1998; Buck et al., 1997)
- Corticosterone response to ethanol (Roberts et al., 1995)
- Voluntary ethanol drinking (Phillips et al., 1994; Rodriguez et al., 1995)
- Ethanol-conditioned taste aversion (Risinger and Cunningham, 1998)
- Conditioned place preference (Cunningham, 1995)
- Hypothermia (Crabbe et al., 1994)
- Grid test ataxia (Phillips et al., 1996)
- Loss of the righting reflex (Rodriguez et al., 1995; Browman and Crabbe, 2000)
- Screen test ataxia (Browman and Crabbe, 2000).

These results form a powerful data set for exploring genetic co-determination of ethanol traits. For example, RI data supported a genetic relationship between severity of the withdrawal syndrome in ethanol-dependent animals and voluntary ethanol consumption; those strains most likely to experience severe withdrawal were least likely to voluntarily consume ethanol (Metten et al., 1998). The ability to examine correlations such as these offers a compelling ancillary benefit to the use of RI strains for QTL mapping.

Another RI panel that has been used in ethanol research is the LS × SS RI set. LS (Long Sleep) and SS (Short Sleep) mice were originally selected from a heterogeneous stock (HS) of mice for long and short duration of loss of the righting reflex (LORR) following ethanol administration (McClearn and Kakihana, 1981; Markel et al., 1996). The HS mice are the product of an eight-way cross of eight standard inbred strains: C57BL, BALB/c, RIII, AKR, DBA/2, IS/Bi, A, and C3H/2 inbred mouse strains (McClearn et al., 1970). Originally, animals from both selected lines were entered into a breeding program to create 40 RI lines, of which 27 survived the inbreeding process* (DeFries et al., 1989). These animals are unlike the BXD RI set in that they contain alleles from eight, rather than two, different strains, complicating genotyping. A marker set for these strains has been published (Markel et al., 1996), and QTL studies have been conducted (e.g., Erwin et al., 1997; Markel et al., 1998). The fact that the LS and SS progenitors were selectively bred for differences in LORR duration from a heterogeneous stock provides maximal genetic variability for this trait. This is advantageous for mapping of this and other traits that may share a common genetic basis (Markel et al., 1998). There is also the potential for greater mapping resolution, though this may require a much different and more complex methodology appropriate for heterogeneous populations (Talbot et al., 1999; Mott et al., 2000), and would require a denser marker map than is presently available. The LS × SS RI panel is being expanded to nearly 100 strains. A 10 cM-density marker map is also being generated for these strains (Johnson and Bennett, personal communication). This new panel will offer substantially greater mapping power due to the larger number of strains available for testing.

It has recently been suggested that one method for extending the power of RI panels is via the combination of several existing RI sets that share one common ancestor. Specifically, the AXB, BXA, BXD, BXH and CXB RI strains all share a C57BL/6 progenitor, and together offer more than 100 inbred strains for study. Genetic analysis of these strains centers on the relationship between a phenotype and the alleles that are either C57BL/6 or "other." Collectively, this RI panel is called BXN (Williams et al., 2001). The primary advantage of this approach is to dramatically enhance the mapping power and map resolution of RI mice by increasing the number available for testing (Williams et al., 2001). However, the presence of multiple alleles derived from multiple inbred strains could complicate mapping. In general, it would be better, though very expensive, to have a single RI panel of 100 strains, constructed from only two inbred strains.

The number of mice tested per strain has an important effect on the mapping power of RI strains. The equations that identify QTL for RI strains do not take into account the number of animals tested, as would be the case in, for example, an ANOVA or a t-test. Instead, these equations simply account for the number of strains measured. Whereas this might give the impression that only a single animal needs to be tested per strain, the reliability with which the true trait value for each strain

---

* Because of the detrimental effects of inbreeding on fitness and fecundity, it is common for a percentage of RI lines to be lost during the inbreeding process.

can be estimated is only as good as the number of samples taken, given the inevitable error associated with measurements in a laboratory setting. Indeed, one of the advantages of using RI strains vs. other mapping populations is that error may be suppressed by making repeated measurements of genetically identical animals. Belknap (1998) has derived a formula to estimate the number of mice that should be tested in an RI study. This formula suggests that, for very low heritability ($h^2 = 0.1$), approximately 10 animals should be used per RI strain, whereas for $h^2 = 0.5$, more than one but fewer than four subjects per strain could be used. These numbers are based on the comparative power of testing RI mice vs. a similar number of F_2 mice. These numbers assume no selective genotyping in the F_2 population (unlikely), and they do not account for the expense of genotyping the F_2 mice, a cost not associated with measuring additional RI mice. In practice, the h^2 is not known with great precision, so an investigator's insight into the variability of a trait must be used to approximate h^2. Perhaps the most parsimonious solution is to simply test more animals than recommended by this method to be confident of the accuracy of the strain mean and to yield data useful for estimating genetic correlations (see above). On the other hand, when working with the extended LS × SS or BXN strains, or other large RI panels, limiting the number of animals tested per strain is imperative.

RI mice can sometimes be obtained from other researchers or core facilities. However, the ultimate source for most mouse RI panels is The Jackson Laboratory (JAX; Bar Harbor, ME). Unless the plan is to test very few animals per strain, a breeding colony must be established at the investigator's facility. To minimize the chance of spontaneous mutation and genetic drift, it is best to replace the RI breeders from an independently maintained colony with mice from JAX after no more than three generations beyond the original stock. This assures that all animals identified as being from a particular strain will be genetically isomorphic with those maintained by JAX. As discussed in section 1.4.1, only one sex may be tested, or both may be tested and analyzed separately. Testing both sexes allows for the detection of sex-specific QTL, but doubles the workload. While it would be possible to order different breeding pairs from an RI line at different times to spread out the burden of breeding, housing and testing, the likelihood of inadvertent temporal differences in the housing and testing procedures would introduce a second (non-genetic) between-groups difference that is likely to substantially confound the findings of any investigation. Therefore, the best scientific plan would be to maintain a colony that contained all of the different RI strains simultaneously, such that between-strain differences would not be due to unwanted temporal effects. This ideal situation is typically not feasible. Multiple passes of different strain combinations are usually the only practical approach to RI studies. We advise testing a few of the progenitor strain mice in as many passes as possible to obtain an index of differences among iterations.

Current information regarding the latest revisions of genetic maps in RI panels is available from JAX on the Web (http://www.informatics.jax.org/searches/riset_form. shtml). A helpful resource maintained by Robert Williams (http://www.ner-venet.org/MMfiles/MMlist.html) contains genotypes of various RI sets, including the BXN set, as well as files prepared for immediate use in Map Manager QTX (see section 1.6).

1.4.3 RECOMBINANT CONGENIC (RC) STRAINS

RC strains are the inbred product of one or more rounds of backcrossing beginning with an F_1 mouse. RC strains can be thought of as a special type of RI strain in which one progenitor strain's genome is disproportionately over-represented and the other is correspondingly under-represented (Figure 1.2D). A set of RC mice created by a cross between BALB/cHeA and STS/A mice has been used to identify QTL relevant to cancer research (Groot et al., 1992; Moen et al., 1991). Similarly, a cross between BALB/cJ and C57BL/6ByJ has been used to map QTL for ethanol preference (Vadasz et al., 2000). The application of RC mice is analogous to the use of RI mice, with the possible advantage that additional backcrossing should provide a convenient avenue for finer mapping via the production of congenic strains (see below).

1.4.4 CHROMOSOME SUBSTITUTION STRAINS (CSS)

Another approach that offers some advantages over RI strains is the construction of a panel of strains termed CSS (previously referred to as consomic strains; Nadeau et al., 2000). To create CSS, an entire individual chromosome from one progenitor strain, rather than a chromosomal segment, is deliberately introgressed onto a second strain by backcrossing and genotyping. The final product is a mouse of one strain that possesses a single pair of chromosomes from another "donor" strain (Figure 1.2E). Two reciprocal panels of CSS are being created using the C57BL/6J and A/J strains for the first, and C57BL/6J and 129/Sv for the second (Nadeau et al., 2000). These panels will include a CSS for each of the mouse chromosomes (including the X and Y chromosomes) transferred onto both possible strain backgrounds. The CSS offer significant advantages over RI strains (Nadeau et al., 2000). The first advantage is that, whereas in RI strains the differences between strains are non-systematic, and therefore redundant from strain to strain, the CSS have no overlap in introgressed material from strain to strain, and therefore maximize contrast. Additionally, by comparing each CSS against its corresponding background, the chromosome(s) possessing a relevant QTL can be immediately determined. Furthermore, a marginally significant QTL can be easily confirmed by testing additional animals from only one CSS vs. the background strain. Dominance of a putative QTL can be assessed by simply crossing a given CSS to its background strain to compare both homozygote states to the heterozygote condition. Additionally, CSS may be useful for identifying interactions between two genetic loci (termed epistasis). Finally, CSS are excellent starting material for the construction of interval-specific congenic strains for the purpose of finer mapping of the QTL to specific chromosomal regions (see section 1.5.1).

1.4.5 HETEROGENEOUS STOCKS (HS)

Another recently advanced strategy for mapping involves the use of HS mice (Figure 1.2F). One HS that has been used for mapping is the product of an eight-way cross of standard inbred strains (McClearn et al., 1970; Talbot et al., 1999; Mott et al., 2000). HS are much more complex than an intercross population, but are analyzed in a conceptually similar manner, with each animal representing a unique recombination of alleles. HS mice have the distinction of possessing a greater assortment of alleles than mapping populations derived from only two inbred strains (Talbot et al., 1999; Mott et al., 2000). Additionally, because they have been maintained for more than 60 generations as an outbred stock, the density of accumulated crossovers is very great, which potentially allows for highly precise mapping, when there is a commensurate high density of markers available. However, a more complex, multi-point analysis is required to effectively analyze these strains (Mott et al., 2000). This is because any individual marker allele may be shared by two or more of the strains contributing to the HS stock, even though ancestral origin of that chromosome segment may not be the same. Therefore, two animals might appear to be genetically identical at a chromosomal region, even though they actually inherited the chromosomal segment from different ancestors. Using their multipoint analysis technique, Mott et al. (2000) have overcome this difficulty and mapped QTL for fearfulness in HS mice. A different HS has been developed that has been used for QTL mapping of ethanol-induced locomotor response (Demarest et al., 2001). The examination of correlations between multiple traits in HS faces the same pitfalls as in intercross populations, and should therefore not be reported as a genetic correlation (see section 1.4.1).

1.4.6 SHORT TERM SELECTED LINES (STSL)

One interesting strategy for confirming QTL nominated by other studies is to use bi-directional selection for the trait under study to determine whether such selection leads to a divergence in the allele frequencies of markers in the region of the suspected QTL. The basis for this method has been demonstrated in practice, and shown to be effective and to agree with QTL results from other studies (Belknap et al., 1997). Animals are selectively bred for a limited number of generations to obtain maximal mapping information and to study genetically correlated responses during early generations, before random fixation of trait-irrelevant alleles becomes prevalent. Because STSL are produced by mass selection, the inbreeding rate may be more rapid than would be seen in longer-term selection lines produced using a rotational breeding scheme (Falconer and Mackay, 1996). Ultimately, about four generations of selection balance the effects of additional generations of selection on marker frequency (which are useful for mapping) and random changes in marker frequency due to genetic drift (which are deleterious to mapping efforts; Belknap et al., 1997).

This approach has been described by Belknap et al. (1997), who used STSL in an attempt to confirm QTL associated with voluntary ethanol-drinking behavior. Generally, the highest- and lowest-scoring mice of each sex are interbred (~6–10 pairs) to create the S1 generation. Similarly, the lowest-scoring mice are interbred to initiate the alternate line. Both the number of breeding pairs and the number of generations of breeding have predictable consequences on cumulative random genetic drift. It is recommended that selective breeding proceed for a small number of generations to reduce fixation of trait-irrelevant genes (see Belknap et al., 1997).

Important genotype information in the putative QTL regions can be obtained from parents and offspring.

The small number of animals used and the limited number of generations of selection are both obvious advantages of this strategy. In addition to decreasing random genetic drift, use of a small number of selection generations may also limit diminished reproductive capacity and avoid the accumulation of too many crossover events, which depresses linkage disequilibrium, and therefore requires additional marker genotyping. However, small numbers of selection generations still allow the most significant QTL to become predominant or even homozygously fixed (Belknap et al., 1997). The relative efficiency of STSL vs. an F_2 population may be similar (Belknap et al., 1997), though this judgment would necessarily depend on the cost of phenotyping vs. genotyping. Additionally, like the RI, RC, and CSS, but unlike F_2, BC and HS populations, STSL can and have been used to determine whether other traits are genetically correlated with the selection trait (Metten et al., 1998).

1.4.7 CONGENIC STRAINS (CS)

RI, RC, F_2 and BC mice represent arbitrary recombinations of the progenitor strain's genetic material. CS, like CSS, are instead purposefully constructed such that they contain primarily one progenitor's genome, but possess a small portion of a second progenitor's genome (Figure 1.2G). CS differ from CSS in that they typically contain less than an entire chromosome from the donor strain. Furthermore, CS are usually produced as a means of testing a specific QTL that has already been provisionally mapped. CS are created by repeated generations of backcrossing of the donor strain onto the background strain. After each generation of backcrossing, the offspring are genotyped to identify those offspring that still possess the desired donor region. These animals are then carried forward to the next backcross generation. After many successive generations, the donor strain's genome is replaced by the background strain's *except* at the region that has deliberately been preserved. After a certain number of backcross generations (about 10), offspring that possess the region of interest are intercrossed to homozygously fix the donor region. The homozygous offspring are identified and can then be maintained without further genotyping. CS are compared to the pure background strain to determine whether the introgressed region influences the trait of interest. Because each animal within a CS is genetically identical, multiple congenic and pure-background animals can be phenotyped and compared, thus providing additional power for detection of the sometimes small effect of the introgressed region. Oftentimes, a reciprocal CS is also produced in which the same region is transferred onto the opposite background. A recent review charts the progress of CS development for finer mapping of alcohol- and drug-abuse-related QTL and gives further methodological and theoretical details (Bennett, 2000).

Methods for the more rapid creation of CS have been developed. Marker-assisted or "speed congenic" animals are created by selecting from each generation those animals that have gained the highest percentage of background markers in each round of backcrossing but still retain the segment of interest from the donor strain (Wakeland et al., 1997; Markel et al., 1997b; Bennett and Johnson, 1998; Bennett, 2000). Selection of these animals will significantly hasten the removal of those regions of the donor's genome that are not of interest. An additional advantage may be gained by starting with an RI or RC strain that possesses the desired chromosomal segment, but is otherwise enriched for the background onto which the segment of interest is to be introgressed. As mentioned, the CSS would also make an advantageous starting point for the creation of a CS (Nadeau et al., 2000).

1.5   FINER MAPPING POPULATIONS

Initial QTL mapping seeks to identify QTL anywhere in the genome. Once one or more QTL are found, it is desirable to define the QTL's map location as precisely as possible. The spectrum of possible approaches to this problem have been reviewed (Darvasi, 1998). In this section, we describe some populations useful for fine mapping. Localization of a QTL to a small chromosome region (preferably of < 1 cM) is necessary before any attempt at positional cloning, or other means of definitive gene identification based on position, is feasible.

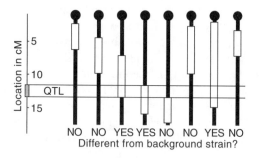

FIGURE 1.3 Eight interval-specific congenic strains (ISCS) are represented by a single chromosome each. Small chromosomal segments from a donor strain (represented as white) have been introgressed onto the background strain (represented as black). Along the y-axis, the map location of these segments is given in cM. Along the x-axis, the results of comparison of each strain to the background strain are given. It is concluded that those strains differing in the target phenotype from the background strain possess the QTL in the introgressed segment. Thus, the map location of the QTL is narrowed to the region defined by the two horizontal lines. ISCS are homozygous for the introgressed region and would be homozygous for the background strain at all other chromosomes.

1.5.1   Interval-Specific Congenic Strains (ISCS)

In section 1.4.7 we discussed the creation of a single CS to test for the presence of a QTL. For finer mapping of suspected QTL, ISCS are useful for more precisely mapping the QTL. In an ISCS panel, the region that has been determined to harbor a QTL is introgressed onto a recipient genetic background via repeated backcrossing (Figure1.3). Each strain of this panel is ultimately destined to possess overlapping segments of this region of potentially differing lengths. Mice are genotyped to ascertain that they have retained the region of interest, and then the requisite animals are backcrossed to the recipient strain. This process is repeated for up to 10 generations to obtain a series of strains of one genetic background, differing only with regard to the segment of donor DNA that has been transferred. Once created, the ISCS are compared to the recipient background strain for the phenotype of interest. As for any CS, differences in the phenotype can be attributed to the introgressed chromosomal region. However, ISCS are particularly useful for fine mapping because the pattern of differences across strains narrows the region harboring the QTL. Figure 1.3 illustrates this point for a panel of 8 ISCS. As with CS, ISCS can be produced more rapidly using marker-assisted strategies (see section 1.4.7).

Once the QTL has been finely mapped using ISCS, a small donor segment (SDS) CS may be generated to further refine the QTL location (Figure 1.2G). Each SDS congenic would possess a very small introgressed region. The ISCS containing the QTL as close as possible to one end of the introgressed segment can be backcrossed twice to the background strain to obtain a segregating BC population. These mice can then be genotyped to identify recombinants due to crossovers close to and on the other side of the QTL, leaving only a very small introgressed segment. Each of these recombinant mice would then serve as a progenitor of a new SDS CS. Again, strains created in this way would be screened for the trait of interest, and identification of a particular strain possessing the altered trait would further narrow the QTL region.

1.5.2 ADVANCED INTERCROSS LINES (AIL)

AIL are an extended intercross of two inbred strains. Like F_2 mice, they are genetically segregating, but due to repeated generations of breeding (to F10, for example) more crossovers have occurred (Figure 1.2H). The result is a smaller and smaller recombination frequency with each successive generation (diminished linkage disequilibrium). In other words, the segments of DNA inherited from each progenitor strain are broken into smaller pieces across generations. This is what provides their advantage for fine mapping. An AIL could be used as an initial mapping population. However, the greater density of recombinations requires that many more markers be used for initial screening. Therefore, the technique is most useful for finer mapping of QTL (Darvasi and Soller, 1995; Le Roy, 1999).

The development of AIL appears to simply require breeding across multiple generations. However, in practice, development is complicated by the tendency of alleles at particular loci to move away from Hardy-Weinberg (H-W) equilibrium over successive generations at a rate faster than that predicted by genetic drift (Falconer and Mackay, 1996). This is a consequence of the unavoidable selection for alleles, and epistatic interactions among alleles, that affect fitness and fecundity. Intervals that shift from H-W equilibrium will appear statistically associated (linked or anti-linked) even if they are on different chromosomes (Williams et al., 2001). Thus, it might be best to genotype periodically during the course of AIL development and to select for under-represented chromosome regions in subsequent generations to retain allelic balance. Alternatively, the problem can be corrected to some extent statistically after an AIL has been generated, but with a regional reduction in power to detect QTL. It should be noted that the problem of maintaining a mapping panel at close to the H-W expectation is common to all multigenerational crosses, including

RI, RC, HS, STSL and AIL. As in all mapping studies, investigators will need to assess the relative gain in precision with the increase in genotyping effort.

1.6 DNA PREPARATION AND GENOTYPING

The collection, isolation, and storage of DNA for genotyping are components of most QTL mapping projects. A DNA sample for genotyping can be obtained from any tissue, but, most commonly, a small segment of tail is used in live animals, and either spleen or tail are used post mortem. An excellent protocol for the isolation of DNA and the amplification of simple sequence length polymorphism (SSLP) markers by PCR is available on the Web at http://www.nervenet.org/papers/DNAIsolation.html, as part of an outstanding article on QTL research, which contains numerous useful links to other Web sites (http://www.nervenet.org/papers/ShortCourse98.html). Research Genetics (http://www.resgen.com), the major commercial source of PCR primers useful for QTL mapping work, maintains an excellent Web site that provides methodological assistance with respect to the selection and application of specific markers.

Markers that are useful for QTL mapping rely on some discernable characteristic of genomic DNA that occurs in only one place in the entire genome, and that differs between strains. Originally, only markers that had visually discernable consequences, such as different coat colors or mutations, were available. Restriction fragment length polymorphisms (RFLP) were one of the earliest types of molecular markers. RFLP used restriction digests of genomic DNA to produce DNA fragments that were of different lengths. The fragments were then resolved on a gel and their positions visualized by hybridization with a labeled probe. More recently, SSLP have become the predominant type of marker in use. SSLP are analyzed using PCR to produce oligodeoxynucleotide products that differ in length from one strain to another. These PCR products can then be visualized with ethidium bromide or by using fluorescent or radioactively tagged primers. Different fluorescent dyes can be used to label different markers, which allows multiple PCR reactions to be conducted in the same tube and visualized on the same gel, substantially increasing the efficiency of genotyping (Iakoubova et al., 2000). Single nucleotide polymorphisms (SNP) are another class of markers that are now becoming more widely used. Under appropriate conditions, PCR primers are extremely sensitive to single base pair mismatches. Therefore, if a single base polymorphism exists between strains, only primers with the appropriate base at the site of the polymorphism will support amplification by PCR. SNP markers take advantage of this sensitivity to determine which polymorphisms are harbored by which strains. In general, all markers are assigned their positions on chromosomes by using linkage analysis, a procedure that is beyond the scope of this chapter, but that has been well articulated elsewhere (Falconer and Mackay, 1996; Lynch and Walsh, 1998).

Several next-generation approaches that offer greater throughput for large scale genotyping operations have been developed. One major approach abandons the goal of identifying the genotype of each animal individually, and, instead, reasons that if high-scoring animals have one genotype and low-scoring animals have another, then the marker must be linked to a QTL. Therefore, it is possible to pool DNA samples from many high-scoring animals (as from an F_2 population), and to compare their PCR products with similarly pooled low-scoring animals, though it may involve some loss of power (Darvasi and Soller, 1994; Lipkin et al., 1998).

A variant of this approach deviates from traditional analysis by using SNP and pooled samples. Pooled DNA from high- and low-scoring animals can be amplified with two different sets of primers, one designed for one of the SNP, and the other designed for the other SNP. The rate of amplification in these pooled samples will reflect the relative concentration of the two possible templates. With a specialized technique termed kinetic PCR, the relative rates of amplification can be monitored, and the frequency of each SNP in the pooled sample from the high- and low-scoring animals can then be estimated. This methodology is complex and costly to set up, but it offers tremendous advantages of speed at only moderate loss of precision (Germer et al., 2000). Oligonucleotide array technology has also been recently applied to the task of identifying SNP in intercross individuals, which again requires substantial setup costs, but yields large gains in efficiency (Lindblad-Toh et al., 2000). In the future, these approaches may become less costly and more widely used (Iakoubova et al., 2000).

An alternative to the expensive and sometimes daunting task of establishing a PCR genotyping facility is the use of a public or private genotyping service. The Marshfield Clinic has an NIH funded facility for high throughput genotyping (http://research.marshfieldclinic.org/genetics). The facility accepts applications for genotyping that are reviewed for their scientific merit. Accepted applicants are not charged for the service. The Center for Inherited Disease Research (http://www.cidr.jhmi.edu), and The Kleberg Genotyping Center (http://imgen.bcm.tmc.edu/molgen/KGC/Home.htm), also accept applications that are reviewed for scientific merit. Applicants are charged a fee based on their funding source and institutional affiliation. The Australian Genome Research Facility also performs genotyping services for a fee (http://www.agrf.org.au). They appear to give preference to Australian based users, but will also consider international applicants. The Genomics Core Facility of the University of Utah provides PCR-based geno-typing, but appears primarily focused on human work at the present time (http://www.hci.utah.edu/groups/genomics). SeqWright (http://www.seqwright.com) and Lark Industries (http://lark.com) both provide fee based genotyping services that can be adapted to the user's needs. Finally, Charles River Laboratories (http://www.theriondna.com/lab-fact.htm) and The Jackson Laboratories (http://www.jax.org), both well known rodent breeding facilities, also provide geno-typing services for a fee.

1.7 STATISTICAL MAPPING METHODS

QTL analyses involve fundamentally simple statistical tools that have been elabo-rated over the course of the last century. Both the simple and the more complex mathematical and statistical models should be understood in general terms by all users, but, like all biostatistical methods, they do not need to be appreciated in their most minute details. Although not required reading, if one wishes to get a general appreciation for the mathematical basis of QTL analyses, an indispensable starting point is the classic text by Falconer (subsequently, Falconer and Mackay), *Introduc-*

tion to Quantitative Genetics, now in its fourth edition (1996). Another newer and impressive text is *Genetics and Analysis of Quantitative Traits* by Lynch and Walsh (1998). Both books address the historical evolution of the important statistics, and also discuss applications of the relevant methodologies. Further, both books reflect the spectrum of genetic analyses, from classical genetic analysis of agricultural crops to analysis of a variety of different invertebrate and vertebrate organisms, including the dauntingly complex field of human population genetics. We will first distill some of the theoretical approaches to QTL analysis that the reader is most likely to encounter. We will then offer practical advice and highlight some of the more common software packages that implement these tools.

The simplest type of QTL analysis considers the phenotype of interest and each individual marker genotype in isolation, without regard to information about adjacent markers on the same chromosome. This type of analysis can be accomplished using Pearson product–moment correlation. Non-parametric analysis can be used for non-normally distributed phenotypes. Each genotype is coded, for example A/A = 0, A/B = 1, and B/B = 2, and the correlation between the trait and the genotype at each marker is examined. The hypothesis being tested is that a QTL resides at the location of the marker. Low p values suggest a relationship between genotype and phenotype. This type of screen is especially useful for preliminary analysis of data, or as a tool to examine markers while an experiment is still in progress.

Interval mapping (IM) improves on the above approach (Lander and Botstein, 1989; Jansen, 1993). IM requires knowledge of the order of genetic markers on individual chromosomes, and estimates the probability that a QTL resides between each pair of adjacent markers. Estimations of QTL location along the length of an entire chromosome are used to create a two-dimensional probability map of the QTL location on a given chromosome. There is an increase in power associated with IM when compared with single locus mapping strategies (Lander and Botstein, 1989). Because the phenotype/genotype relationship will be affected by the presence of other QTL elsewhere in the genome, IM can be further refined to composite interval mapping (CIM), which allows information about other background markers to be taken into account, thus further refining the analysis (Zeng, 1994). Additional modifications to CIM approaches have to do with the method for selecting background markers (Manly and Olson, 1999).

The statistical result required to conclude that a QTL has been mapped has been a topic of scientific debate. In general, QTL mapping studies involve large numbers of multiple comparisons, and therefore, stringent criteria for significance must be set to avoid false positive errors. Statistical significance of a particular QTL is difficult to determine and is complicated by the use of a variety of different probability statistics employed by different analytical procedures. The most prominent controversy is between a set of standards developed by Lander and Kruglyak (1995) that are independent of the particular data set under analysis, and the use of the permutation test that randomly rearranges the particular phenotype and genotype data under analysis to determine the probability statistics empirically required to achieve the desired likelihood of false positive errors (Churchill and Doerge, 1994; Doerge and Churchill, 1996). The significance levels required by permutation tests are sometimes (but not always) less stringent.

The availability of the above-mentioned QTL analyses varies across the numerous statistical packages that are available. New versions are continually being developed and released. Some of them are made available to the scientific community free of charge, whereas others are commercially distributed. A good starting point for selection of an appropriate package is a recent article by Manly and Olson (1999), which describes all programs briefly, and gives Web addresses to obtain software downloads and further details. This paper also describes in more detail a former version of Map Manager QT, now superseded by Map Manager QTX, which is our preference (http://mapmgr.roswellpark.org/mapmgr.html).

Genomic maps and genotypes of particular inbred strains and RI lines, details about individual markers, and additional information can be found at the Mouse Genome Informatics web site (http://www.informatics.jax.org/). Related databases that are somewhat easier to utilize are maintained at (http://www.nervenet.org/MMfiles/MMlist.html).

1.8   OTHER TOOLS

Once a QTL has been localized to a small segment of a chromosome, other tools can be used to accumulate evidence that a particular gene, termed a "candidate gene," is the QTL. Below we describe some of the available strategies.

1.8.1   TRANSGENICS, KNOCKOUTS AND KNOCKINS

An extremely large and ever expanding arsenal of naturally occurring and induced single gene mutants are available. The technologies associated with production of knockout mice, along with some examples of their utilization, are described in Chapter 2. Knockout mice can be used to identify a phenotypic change that would suggest an impact of the deleted gene on that trait. The principles underlying the use of knockin and transgenic mice are largely the same as those for knockout mice; however, in these animals, a single gene's function is altered or the gene is over- or under-expressed, rather than deleted. Methods for the alteration of a gene's expression in specific cell types of an intact animal have also been developed (Bader, 2000). Excellent reviews describing the details of transgenic development and their usefulness for the study of complex diseases are available (e.g., Landel, 1991; Beal, 2001).

Sometimes, QTL are mapped in the vicinity of candidate genes for which single gene mutants exist. In this case, it makes sense to test these animals for an alteration in the mapped trait. Although identification of phenotypic differences between a single gene mutant and a wild type does not prove that the mutated gene is the QTL, it provides supportive evidence and encourages further study. For example, a QTL on mouse chromosome 9 near the dopamine D2 receptor gene was identified for voluntary ethanol drinking (Phillips et al., 1994). Subsequently, dopamine D2 receptor knockout mice were found to voluntarily consume about half as much alcohol as their wild type litter mates (Phillips et al., 1998b).

1.8.2   GENE EXPRESSION ASSAYS

Differences in gene expression can provide evidence for the involvement of candidate genes known to occupy a QTL region. Expression assays may be particularly useful in comparisons of CS or SDS strains to their background counterparts. Expression of candidate genes between genotypes can be compared by *in situ* hybridization, Western Analysis, Northern Analysis, or by other analogous methods, including microarray analysis (Lander, 1999). However, microarray methods are more likely to be used in a global analysis of thousands of genes or genetic sequences across the genome (Lewohl et al., 2001). Data from such analyses may indicate that genes in the QTL region are differentially regulated either basally, or in response to a specific event (such as administration of ethanol).

1.8.3   INDUCED MUTAGENESIS

One other strategy receiving close scrutiny as a method for identifying quantitative trait genes is mutagenesis induced by a chemical, such as ethylnitrosourea (ENU). The advantages of this method have been recently discussed (Nadeau and Frankel, 2000), and carefully contrasted with the QTL mapping strategies described in this chapter (Belknap et al., 2001). ENU mutagenesis is more likely to be used in a global search for phenotypic differences, followed by determination of the location of the mutation(s) resulting in the detected phenotypic alterations. A variant of the ENU approach disrupts the function of unknown genes by using a reporter vector that randomly integrates into the genome of an embryonic stem cell. Integration of these vectors into a locus downstream from a promoter site drives the production of the reporter gene in cells where the disrupted gene would normally be expressed. Furthermore, the function of the endogenous gene is typically disrupted, resulting in mutants similar to those induced by ENU. In contrast to ENU, the disrupted gene may be identified directly due to the presence of the reporter vector, whose sequence is known. This approach has already been used in mice to identify novel genes (Cecconi and Meyer, 2000). Mutagenesis projects based on these and similar approaches could produce strains with mutations in already mapped QTL regions. Such strains could be obtained and tested for the mapped traits. This may prove to be another valuable source of single gene mutants.

1.8.4   SEQUENCE ANALYSES

Full genome sequencing of several mouse strains will be complete in the near future. Until then, sequence analysis of candidate genes may be pursued to identify the polymorphisms with functional consequences. For example, a QTL for severity of acute ethanol withdrawal, provisionally mapped in BXD RI mice to chromosome 11, was near genes coding for several GABA$_A$ receptor subunits (Buck et al., 1997). Subsequent analyses identified a functionally relevant coding sequence polymorphism between the C57BL/6J and DBA/2J progenitor strains for the $\gamma2$ subunit gene of the GABA$_A$ receptor (Hood and Buck, 2000). Additional work will be needed to conclusively establish whether this polymorphism is actually the source of the phenotypic association in this chromosomal region.

1.9 FUTURE DIRECTIONS

This chapter has described methods for mapping QTL to progressively smaller regions of individual chromosomes. To date, identification of the precise polymorphisms that underlie QTL in murine populations has been difficult (Nadeau and Frankel, 2000). This is because QTL mapping methods alone do not have enough precision to isolate a single gene. Instead, after a QTL is mapped to a resolution of <1 cM, other approaches must be used for identification of the sequence difference that modifies the mapped trait. In the past, gene identification required sequencing vast chromosomal regions, often containing >1,000,000 base pairs. We have now entered a new era, where multiple commonly used laboratory strains (including C57BL/6J and DBA/2J) have been, or will soon be, fully sequenced, so that, once a QTL is mapped with high precision, the exact sequence of both strains can be obtained from existing databases and scanned for logical candidates based on known and predicted gene function, tissue expression patterns, and the identification of functional polymorphisms. As candidates for the source of a QTL are identified, multiple methods can be used to evaluate whether a particular gene is the QTL (Belknap et al., 2001). These approaches will vary based on both the nature of the suspected gene and knowledge of the gene's function. Approaches may include:

- Pharmacological or physiological manipulations
- Sense or anti-sense to alter gene expression
- Viral vectors to introduce allelic variants
- Transgenic, knockout or knockin mice to test effects of altered gene function
- Microarrays to identify expression differences, and perhaps others

It has been argued that the ideal experiment would be an "allele swap" in which only the candidate polymorphism was placed onto an opposing background. We have explored the state of the art of gene identification for finely mapped QTL in a recent review (Belknap et al., 2001).

1.10 CONCLUSIONS

QTL mapping experiments are not driven by existing hypotheses about which genes might be involved in a particular phenotype. This is an advantage, because novel systems may be implicated, thus expanding the understanding of mechanisms involved in the trait of interest. This approach has been termed "phenotype driven," to contrast it with, for example, knockout mouse studies, which are "genotype driven," that is, driven by the hypothesis that a particular gene is involved in governing a trait of interest. QTL mapping is especially valuable for the study of ethanol-related traits because the pharmacological targets of ethanol and their downstream effects are numerous and difficult to define. We have tried to convey that one need not be an expert in the field of genetics to accomplish a QTL mapping study.

However, it would be prudent to seek appropriate advice from experts in the phenotype of interest, quantitative and molecular genetics.

Alcoholism research has much to gain from the identification of novel genetic targets by QTL mapping methods. Just as our knowledge of the genome is ever changing, QTL mapping methods are in a continuing state of evolution. This has been, and will continue to be, of considerable advantage to the dedicated practitioners of these methods.

ACKNOWLEDGMENTS

The authors would like to acknowledge Drs. John C. Crabbe, John K. Belknap and Deborah A. Finn for their helpful comments and insights on all aspects of this manuscript. Additionally, we thank Drs. Kari J. Buck, Kenneth F. Manley and Robert W. Williams for helpful correspondence relating to specific issues discussed herein. Finally, we wish to thank Peter F. Kellers for his artistic assistance with the production of the figures that accompany this manuscript. This work was supported by National Institute on Alcohol Abuse and Alcoholism grants P50AA10760, F32AA05600 and T32AA07468, as well as a grant from the Department of Veterans Affairs.

REFERENCES

Aylsworth AS (1998) Defining disease phenotypes. In: *Approaches To Gene Mapping In Complex Human Diseases* (Haines JL and Pericak-Vance MA, Eds.), pp 53-76. New York, NY: Wiley-Liss.

Bader M (2000) Transgenic animal models for neuropharmacology. *Rev Neurosci* 11:27-36.

Beal MF (2001) Experimental models of Parkinson's disease. *Nat Rev Neurosci* 2:325-334.

Belknap JK (1998) Effect of within-strain sample size on QTL detection and mapping using recombinant inbred mouse strains. *Behav Genet* 28:29-38.

Belknap JK et al. (2001) QTL analysis and genome-wide mutagenesis in mice: complementary genetic approaches to the dissection of complex traits. *Behav Genet* 31:5-15.

Belknap JK et al. (1996) Type I and type II error rates for quantitative trait loci (QTL) mapping studies using recombinant inbred mouse strains. *Behav Genet* 26:149-160.

Belknap JK et al. (1997) Short-term selective breeding as a tool for QTL mapping: ethanol preference drinking in mice. *Behav Genet* 27:55-66.

Bennett B (2000) Congenic strains developed for alcohol- and drug-related phenotypes. *Pharmacol Biochem Behav* 67:671-681.

Bennett B and Johnson TE (1998) Development of congenics for hypnotic sensitivity to ethanol by QTL-marker-assisted counter selection. *Mamm Genome* 9:969-974.

Browman KE and Crabbe JC (2000) Quantitative trait loci affecting ethanol sensitivity in BXD recombinant inbred mice. *Alcohol Clin Exp Res* 24:17-23.

Buck KJ et al. (1997) Quantitative trait loci involved in genetic predisposition to acute alcohol withdrawal in mice. *J Neurosci* 17:3946-3955.

Cecconi F and Meyer BI (2000) Gene trap: a way to identify novel genes and unravel their biological function. *FEBS Letters* 480:63-71.

Churchill GA and Doerge RW (1994) Empirical threshold values for quantitative trait mapping. *Genetics* 138:963-971.

Crabbe JC (1998) Provisional mapping of quantitative trait loci for chronic ethanol withdrawal severity in BXD recombinant inbred mice. *J Pharmacol Exp Ther* 286:263-271.

Crabbe JC et al. (1994) Quantitative trait loci mapping of genes that influence the sensitivity and tolerance to ethanol-induced hypothermia in BXD recombinant inbred mice. *J Pharmacol Exp Ther* 269:184-192.

Crabbe JC et al. (1999a) Identifying genes for alcohol and drug sensitivity: recent progress and future directions. *Trends Neurosci* 22:173-179.

Crabbe JC et al. (1990) Estimation of genetic correlation: interpretation of experiments using selectively bred and inbred animals. *Alcohol Clin Exp Res* 14:141-151.

Crabbe JC, Wahlsten D and Dudek BC (1999b) Genetics of mouse behavior: interactions with laboratory environment. *Science* 284:1670-1672.

Crawley JN (2000) What's wrong with my mouse? In *Behavioral Phenotyping of Transgenic and Knockout Mice*. New York, NY: Wiley-Liss.

Cunningham CL (1995) Localization of genes influencing ethanol-induced conditioned place preference and locomotor activity in BXD recombinant inbred mice. *Psychopharmacology* 120:28-41.

Davarsi A (1998) Experimental strategies for the genetic dissection of complex traits in animal models. *Nat Genet* 18:19-24.

Davarsi A and Soller M (1994) Selective DNA pooling for determination of linkage between a molecular marker and a quantitative trait locus. *Genetics* 138:1365-1373.

Davarsi A and Soller M (1995) Advanced intercross lines, an experimental line for fine genetic mapping. *Genetics* 141:1199-1207.

DeFries JC et al. (1989) LS X SS recombinant inbred strains of mice: initial characterization. *Alcohol Clin Exp Res* 13:196-200.

Demarest K et al. (2001) Further characterization and high resolution mapping of quantitative trait loci for ethanol-induced locomotor activity. *Behav Genet* 31:79-71.

Demarest K et al. (1999) Identification of an acute ethanol response quantitative trait locus on mouse chromosome 2. *J Neurosci* 19:549-561.

Doerge RW and Churchill GA (1996) Permutation tests for multiple loci affecting a quantitative character. *Genetics* 142:285-294.

Erwin VG et al. (1997) Common quantitative trait loci for alcohol-related behaviors and central nervous system neurotensin measures: locomotor activation. *J Pharmacol Exp Ther* 280:919-926

Falconer DS and Mackay FC (1996) *Introduction to Quantitative Genetics, Fourth Ed.* Essex, England: Prentice Hall.

Germer S, Holland MJ and Higuchi R (2000) High-throughput SNP allele-frequency determination in pooled DNA samples by kinetic PCR. *Genome Res* 10:258-266.

Grisel JE (2000) Quantitative trait locus analysis. *Alcohol Res Health* 24:169-174.

Grisel JE and Crabbe JC (1995) Quantitative trait loci mapping. *Alcohol Health & Res World* 19:220-227.

Groot PC et al. (1992) The recombinant congenic strains for analysis of multigenic traits: genetic composition. *FASEB J* 6:2826-2835.

Hitzemann R (2000) Animal models of psychiatric disorders and their relevance to alcoholism. *Alcohol Res Health* 24:149-158.

Hitzemann R et al. (1998) Genetics of ethanol-induced locomotor activation: detection of QTLs in a C57BL/6J x DBA/2J F_2 intercross. *Mamm Genome* 9:956-962.

Hood HM and Buck KJ (2000) Allelic variation in the GABA A receptor gamma2 subunit is associated with genetic susceptibility to ethanol-induced motor incoordination and hypothermia, conditioned taste aversion, and withdrawal in BXD/Ty recombinant inbred mice. *Alcohol Clin Exp Res* 24:1327-1334.

Iakoubova OA et al. (2000) Microsatellite marker panels for use in high-throughput genotyping of mouse crosses. *Physiol Genomics* 3:145-148.

Janowsky A et al. (2001) Mapping genes that regulate the density of dopamine transporters and correlated behaviors in recombinant inbred mice. *J Pharmacol Exp Ther* 298:634-643.

Jansen RC (1993) Interval mapping of multiple quantitative trait loci. *Genetics* 135:205-211.

Landel CP (1991) The production of transgenic mice by embryo microinjection. *Genet Anal Tech Appl* 8:83-94.

Lander ES (1999) Array of hope. *Nature* 21:3-4.

Lander ES and Botstein D (1989) Mapping Mendelian factors underlying quantitative traits using RFLP linkage maps. *Genetics* 121:185-199.

Lander ES and Kruglyak L (1995) Genetic dissection of complex traits: guidelines for interpreting and reporting linkage results. *Nat Genet* 11:241-247.

Le Roy I (1999) Quantitative Trait Loci (QTL) Mapping. In: *Neurobehavioral Genetics Methods And Applications* (Jones BC and Mormede P Eds.), pp 69-76. Boca Raton, FL: CRC.

Lederhendler I and Schulkin J (2000) Behavioral neuroscience: challenges for the era of molecular biology. *Trends Neurosci* 23:451-454.

Lewohl JM et al. (2001) Application of DNA microarrays to study human alcoholism. *J Biomed Sci* 8:28-36.

Li T-K (2000) Clinical perspectives for the study of craving and relapse in animal models. *Addiction* 95:S55-S60.

Lindblad-Toh K et al. (2000) Large-scale discovery and genotyping of single-nucleotide polymorphisms in the mouse. *Nat Genet* 24:381-386.

Lipkin E et al. (1998) Quantitative trait locus mapping in dairy cattle by means of selective milk DNA pooling using dinucleotide microsatellite markers: analysis of milk protein percentage. *Genetics* 149:1557-1567.

Lynch M and Walsh B (1998) Genetics and analysis of quantitative traits. Sunderland, MA: Sinauer.

Manly KF and Olson JM (1999) Overview of QTL mapping software and introduction to map manager QT. *Mamm Genome* 10:327-334.

Markel PD et al. (1997a) Confirmation of quantitative trait loci for ethanol sensitivity in long-sleep and short-sleep mice. *Genome Res* 7:92-99.

Markel PD et al. (1996) Strain distribution patterns for genetic markers in the LSXSS recombinant inbred series. *Mamm Genome* 7:408-412.

Markel PD, DeFries JC and Johnson TE (1995) Use of repeated measures in an analysis of ethanol-induced loss of righting reflex in inbred long-sleep and short-sleep mice. *Alcohol Clin Exp Res* 19:299-304.

Markel PD et al. (1998) Quantitative trait loci for ethanol sensitivity in the LS X SS recombinant inbred strains: interval mapping. *Behav Gen* 26:447-458.

Markel P et al. (1997b) Theoretical and empirical issues for marker-assisted breeding of congenic mouse strains. *Nat Genet* 17:280-284.

McClearn GE and Kakihana R (1981) Selective breeding for ethanol sensitivity: Short-Sleep and Long-Sleep mice. In: *Development of Animal Models as Pharmacogenetic Tools* (McClearn GE, Deitrich RA and Erwin VG, Eds.), pp 147-159. Rockville, MD: U.S. Dept. Health and Human Services.

McClearn GE, Wilson JR and Meredith W (1970) The use of isogonic and heterogenic mouse stocks in behavioral research. In: *Contributions to Behavior-Genetic Analysis: The Mouse as A Prototype.* (Lindzey G and Thiessen, DD Eds.), pp 3-22. New York, NY: Appleton-Century-Crofts.

McPeek MS (2000) From mouse to human: fine mapping of quantitative trait loci in a model organism. *Proc Natl Acad Sci USA* 97:12389-12390.

Melo JA et al. (1996) Identification of sex-specific quantitative trait loci controlling alcohol preference in C57BL/6 mice. *Nat Genet* 13:137-138.

Metten P et al. (1998) High genetic susceptibility to ethanol withdrawal predicts low ethanol consumption. *Mamm Genome* 9:983-990.

Moen CJ et al. (1991) The recombinant congenic strains — a novel genetic tool applied to the study of colon tumor development in the mouse. *Mamm Genome* 1:217-227.

Moore KJ and Nagle DL (2000) Complex trait analysis in the mouse: the strengths, the limitations and the promise yet to come. *Annual Rev Genet* 34:653-686.

Mott R et al. (2000) A method for fine mapping quantitative trait loci in outbred animal stocks. *Proc Natl Acad Sci USA* 97:12649-12654.

Nadeau JH and Frankel WN (2000) The roads from phenotypic variation to gene discovery: mutagenesis vs. QTLs. *Nat Genet* 25:381-384.

Nadeau JH et al. (2000) Analyzing complex genetic traits with chromosome substitution strains. *Nat Genet* 24:221-225.

Paigen K and Eppig JT (2000) A mouse phenome project. *Mamm Genome* 11:715-717.

Palmer AA et al. (2000) Prepulse startle deficit in the Brown Norway rat: a potential genetic model. *Behav Neurosci* 114:374-388.

Peltonen L and McKusick VA (2001) Dissecting human disease in the postgenomic era. *Science* 291:1224-1229.

Phillips TJ et al. (1998a) Genes on mouse chromosomes 2 and 9 determine variation in ethanol consumption. *Mamm Genome* 9:936-941.

Phillips TJ et al. (1998b) Alcohol preference and sensitivity are markedly reduced in mice lacking dopamine D2 receptors. *Nature Neurosci* 1:610-615.

Phillips TJ et al. (1994) Localization of genes affecting alcohol drinking in mice. *Alcohol Clin Exp Res* 18:931-941.

Phillips TJ et al. (1995) Effects of acute and repeated ethanol exposures on the locomotor activity of BXD recombinant inbred mice. *Alcohol Clin Exp Res* 19:269-278.

Phillips TJ et al. (1996) Evaluation of potential genetic associations between ethanol tolerance and sensitization in BXD/Ty recombinant inbred mice. *J Pharmacol Exp Ther* 227:613-623.

Risinger FO and Cunningham CL (1998) Ethanol-induced conditioned taste aversion in BXD recombinant inbred mice. *Alcohol Clin Exp Res* 22:1234-1244.

Roberts AJ et al. (1995) Genetic analysis of the corticosterone response to ethanol in BXD recombinant inbred mice. *Behav Neurosci* 109:1199-1208.

Rodriguez LA et al. (1995) Alcohol acceptance, preference, and sensitivity in mice. II. Quantitative trait loci mapping analysis using BXD recombinant inbred strains. *Alcohol Clin Exp Res* 19:367-373.

Silver L (1995) *Mouse Genetics: Concepts and Applications.* London: Oxford University Press.

Talbot JT et al. (1999) High-resolution mapping of quantitative trait loci in outbred mice. *Nat Genet* 21:305-308.

Taylor BA (1978) Recombinant inbred strains: use in gene mapping. In: *Origins of Inbred Mice* (Morse HC, Ed.), pp 423-438. New York: Academic Press.

Vadasz C et al. (2000) Mapping of quantitative trait loci for ethanol preference in quasi-congenic strains. *Alcohol* 20:161-171.

Wakeland E et al. (1997) Speed congenics: a classic technique in the fast lane (relatively speaking). *Immunol Today* 18:472-477.

Williams RW et al. (2001) The genetic structure of recombinant inbred mice: high-resolution consensus maps for complex trait analysis. Release 1, January 15, 2001 at www.nervenet.org/papers/bxn.html or mickey.utmem.edu/papers/bxn.html.

Williams RW et al. (1996) Genetic and environmental control of variation in retinal ganglion cell number in mice. *J Neurosci* 16:7193-7205.

Zeng Z-B (1994) Precision mapping of quantitative trait loci. *Genetics* 136:1457-1468.

2 Knockout and Knockin Mice

Gregg E. Homanics

CONTENTS

2.1   INTRODUCTION

The ability to create custom designed mice with precise mutations in genes putatively involved in mediating or modulating alcohol-induced responses has ushered in a new era of alcohol research. Recent technological advances have provided us with the tools to create mice that harbor virtually any genetic alteration imaginable. The production of these designer mice has allowed for the direct testing of the relevance of specific gene products on alcohol-induced responses at the whole animal level. Because we are ultimately interested in understanding how alcohol ingestion leads to changes in behavioral responses, testing responses *in vivo* at the whole animal level is absolutely critical. Only by combining our understanding of alcohol's effects at the organism level with our understanding at the molecular, cellular, and systems levels will we fully understand how ethanol exerts its effects and how we can interrupt the process to negate the many adverse effects of this commonly used and abused drug.

The most important and widely used genetically engineered animal in alcohol research has been the global gene knockout mouse. With the traditional global knockout approach, the gene of interest is inactive in all cells of the body from the time the animal is conceived until the day that animal dies. To produce a global knockout with traditional technology, one simply either inserts a selectable marker gene into an important exon of the gene of interest or one replaces a portion of the gene that includes an important exon(s) with a marker gene (Homanics et al., 1998a).

While the traditional global knockout approach is the least technically demanding and fastest method to creating a knockout, this approach is subject to limitations that restrict its usefulness (see below). Instead, I advocate that second generation knockout technology be implemented to not only minimize or even overcome these limitations, but also to produce genetically altered mice that are vastly more useful. The most efficient strategy available for creating second generation knockouts relies on both the Cre/loxP (Sauer and Henderson, 1988; Sauer, 1993) and FLP/frt (O'Gorman et al., 1991; Dymecki, 1996) site-specific recombination systems. This combined approach was pioneered by Gail Martin's lab (Lewandoski et al., 1997a; Meyers et al., 1998).

With this approach, one first uses gene targeting in embryonic stem (ES) cells to create a floxed (flanked by loxP sites) gene of interest as illustrated in Figure 2.1. Note that, with this strategy, a selectable marker gene flanked by frt sites is located in intronic DNA. Following the specific removal of the marker gene by FLP mediated

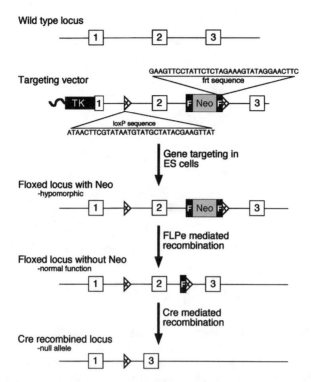

FIGURE 2.1 Overview of second generation gene knockout strategy. Gene targeting in mouse ES cells is used to flox (flank with loxP sites) a portion of the gene of interest. Note that floxed locus also contains selectable marker gene (neo) flanked by frt sites. FLPe mediated recombination is used to remove the neo gene from the locus, accomplished either by transient FLPe expression in ES cells *in vitro* or by mating to a FLPe general deleter mouse line *in vivo* (Rodriguez et al., 2000). Cre-mediated recombination is used to convert either floxed locus to a null allele by deleting an exon that is critical for gene function. Triangle with X = loxP site; black box with F = frt site; Neo = neomycin resistance gene; TK = herpes simplex virus thymidine kinase gene; numbered white boxes = exons; thin black line = intronic DNA; thick black line = plasmid backbone.

recombination, the floxed locus now contains only two loxP sites that flank an important part of the gene of interest, and a single frt site. Because these three recombination recognition sites are small and located in intronic DNA, by themselves they are innocuous and the gene of interest should function normally.

An animal with this type of floxed allele is extremely useful, as it can be used for a multitude of experiments. For example, as shown in Figure 2.2, by mating this floxed mouse line to a general deleter Cre-expressing transgenic strain, a global gene knockout can readily be produced. By mating the floxed mouse line to a transgenic mouse that expresses Cre recombinase in a tissue-specific or temporally regulated manner, conditional knockouts can be produced. By mating the floxed mouse line to transgenic mice that express Cre recombinase in an inducible manner, the timing of the knockout can be placed under investigator control. The floxed mouse line can even be used for experiments in which Cre recombinase is expressed from various

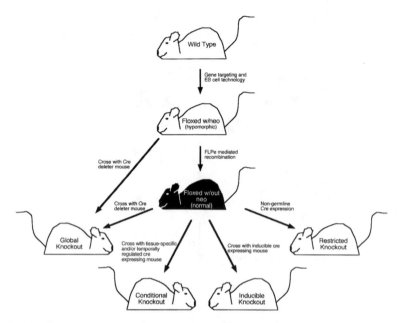

FIGURE 2.2 Summary of the many potential uses of a single floxed mouse line. Gene targeting and ES cell technology are used to create a floxed mouse that also contains a neomycin (neo) gene. The presence of neo may adversely affect expression of the gene of interest, thereby creating a hypomorphic allele, which can be converted to a global null allele by crossing to a Cre general deleter mouse line. Alternatively, the hypomorphic allele can be converted to a floxed allele that lacks neo and functions like a wild-type gene by crossing to an FLPe general deleter mouse line. The floxed mouse line without neo can be used for a wide variety of genetic experiments. This floxed mouse line can be crossed to a Cre general deleter mouse line to produce a global knockout. This floxed mouse line can be mated to various transgenic mouse lines that express Cre in a tissue-specific or temporally regulated manner to produce conditional knockouts. This floxed mouse line can be mated to various transgenic animals that express Cre in an inducible manner to produce knockouts that are temporally regulated by the investigator. Numerous vectors are available for the non-germline expression of Cre recombinase in virtually any tissue at any time to produce knockouts that are tissue-specific and temporally regulated.

viral and nonviral vectors (e.g., adeno-associated virus, liposomes, etc.). This affords the opportunity for the investigator to precisely control the location and the timing of gene inactivation. It is anticipated that stereotactic injection of Cre expression vectors into precise locations in the brain will be immensely useful to the alcohol research field.

In summary, using this second generation approach to gene knockouts has many advantages over traditional knockout technology. Creating a floxed mouse involves only slightly more time and effort than creating a traditional global knockout, but the usefulness of the floxed mouse is tremendous. Simply by mating a floxed mouse to any one of hundreds of Cre transgenic mouse lines available, or by injecting Cre-expressing vectors into various sites in the body, literally thousands of different genetic experiments can be conducted using a single floxed mouse line. The mix-and-match mating approach should be especially appealing to investigators who do not have the resources to create their own genetically engineered mice *de novo*. Already, various floxed mice and Cre-expressing mice are being collected for distribution by the Jackson Laboratory's Induced Mutant Repository (www.jax.org/resources/documents/imr/). Thus, as this collection expands, many knockout experiments of the future will consist of simply purchasing mice "off the shelf" and mating them together to produce the desired global or conditional knockout. A Web site that is attempting to keep track of the rapidly expanding number of Cre-expressing and floxed mouse lines that have been produced is being maintained by Andras Nagy (www.mshri.on.ca/nagy/cre.htm) Nagy, 2000).

2.2 TECHNICAL DESCRIPTIONS OF KNOCKOUT MOUSE PRODUCTION

Producing a gene knockout animal is a very long and expensive process. In a laboratory that specializes in producing gene knockout animals, the process takes, on average, ~1–1.5 years from the time the project is initiated until animals are ready for analysis. Table 2.1 lists the various steps involved and time estimates for each step. One way to minimize the hands-on work and expedite progress is to outsource to a commercial laboratory with knockout capabilities. For example, Incyte Genomics, Inc. (www.incyte.com) or Cell and Molecular Technologies, Inc. (www.cmt-inc.net) will perform all or part of the procedures required. However, outsourcing can be quite expensive. A complete knockout project may cost upwards of $65,000 just to produce the mouse line (i.e., not including subsequent analysis and per diem costs). An alternative is to take advantage of core facilities that have been set up at many institutions around the world. These facilities offer a variety of services and are often subsidized by the institution to keep costs at a minimum for members of the institution. Finally, although the entire procedure is quite challenging, it is possible for someone to create a gene knockout animal in his or her own laboratory. However, to enhance the chances of success, it is highly advisable to consult, or even visit, a knockout lab.

2.2.1 GENOMIC CLONE ISOLATION AND CHARACTERIZATION

2.2.1.1 Library Screening

The first step in any gene targeting procedure is to create a DNA construct that can be used to modify the gene of interest. This process begins by cloning a portion of the genomic locus that harbors the gene of interest. We have found that hybridization of commercially available high density arrays of bacterial artificial chromosome (BAC) libraries that contain mouse genomic DNA is a rapid and relatively inexpensive method for obtaining a genomic clone. We have used the BAC Mouse ES-129/SvJ Release 1 Easy to Screen Hybridization Filters (cat. No. FBAC-4431) from Incyte Genomics, Inc. (Palo Alto, CA) for all of our recent library screenings. We have reused this single filter set to isolate more than 10 different genes. It is important

TABLE 2.1
Estimated Time Required for Each Step of Knockout Mouse Production

Step	Estimated time required (months)
Clone and characterize genomic locus	1-4
Assemble targeting vector	1-4
Target locus in ES cells	1-6
Produce chimeras	2-6
Breed mice	6-8
Analyze mice	6 months-several years

to note that one should select a mouse genomic DNA library that is isogenic to the ES cell line that will eventually be targeted, as it has been demonstrated that nonisogenic DNA often reduces the frequency of gene targeting (teRiele et al., 1992; vanDeursen and Wieringa, 1992).

To isolate a genomic clone for the gene of interest, the high-density filter array is hybridized using a probe for the gene of interest. Typically, a partial cDNA probe from mouse, rat, or human is radio labeled and used to screen the filter set. Once the grid addresses of hybridizing clones are obtained, one simply orders those BAC clones from the commercial source that supplied the library.

2.2.1.2   BAC Clone Mapping and Characterization

BAC clones suspected of harboring the gene of interest are next analyzed to verify the integrity of the locus, i.e., that no rearrangements or deletions have occurred during construction of the library. This is easily accomplished by Southern blot hybridization of BAC DNA following digestion with several different restriction enzymes. The size of fragments that hybridize to the probe is compared with a similar Southern blot created with mouse genomic DNA. If the restriction pattern of the BAC clones is identical to that observed for the mouse genomic DNA, then no major alterations in the cloned gene have occurred.

To further characterize and make the cloned gene more manageable, small fragments of the BAC insert are subcloned into plasmid vectors. It is important to realize that the entire gene does not have to be subcloned. All that is needed is an ~7–10kb genomic DNA fragment that harbors one or more exons that encode a part of the protein that is critical for gene function. Typically, exons near the 5' part of the gene are favored for vector construction.

For subcloning, select a restriction enzyme that produces an ~7–10kb fragment on Southern blot analysis that hybridizes to the probe. We have found that the most efficient method for subcloning is to digest a small amount of the BAC clone with the selected restriction enzyme and shotgun subclone all resulting fragments into a similarly digested plasmid vector. There is no need for gel purification of the BAC fragments as long as the restriction enzyme can be heat inactivated prior to use in the ligation reaction.

To identify those recombinant plasmids that harbor the subclone of interest, colony lifts (Ausubel et al., 1993; Sambrook and Russell, 2001) are performed and subsequently hybridized with the probe. Hybridizing bacterial colonies are then picked, liquid cultures grown up, and plasmid DNA isolated. Plasmid DNA is further characterized by extensive restriction mapping and Southern blot analysis. The goal of this characterization is to identify the location of one or more exons within the genomic DNA subclone. Based on the restriction map created, smaller portions of the genomic subclone that hybridize to the probe can be further subcloned and characterized. Ultimately, one must sequence through at least one exon to (a) definitively establish that the correct gene has been cloned, and (b) to establish intron/exon boundaries.

While the procedures described for cloning and characterization of a genomic locus are quite efficient by today's standards, it is certain that, with the completion of the mouse genome sequencing project, that even more efficient approaches will replace this procedure in the near future. It is foreseeable that simple genomic database searches will allow for BAC clone identification and mapping information without the need for library screening and restriction mapping.

2.2.2 TARGETING VECTOR PRODUCTION

2.2.2.1 Vector Assembly

The design and construction of a DNA vector for targeting the locus of interest is a critical step in the gene knockout procedure that requires much forethought and planning. Of utmost importance in design is to ensure that, following introduction of the vector into ES cells, that homologous and nonhomologous gene insertion events can be readily distinguished. There is nothing worse than hastily performing a gene targeting experiment only to find that targeted and random integrants cannot be readily distinguished.

An optimal targeting vector will contain ~7–10kb of isogenic homologous genomic DNA, selectable marker genes, and recombination recognition sites, as illustrated in Figure 2.1. While many different strategies are possible for creating targeting constructs, the following has been found to work quite well. First, complementary loxP oligonucleotides are ligated into a suitable restriction enzyme site in an intron of the gene of interest. An insertion site should be at least 200 basepairs from an exon to avoid interference with processing of the RNA following gene expression. If a convenient restriction site is not available, one can efficiently be created using site-directed mutagenesis with the QuikChange Site Directed Mutagenesis Kit (Stratagene). The loxP oligonucleotides are synthesized commercially (Life Technologies, Inc.) and are usually designed to have ends that are compatible (sticky) with the cloning site that will be used. It is also often worthwhile to include in the oligo a restriction enzyme recognition site that will be useful for screening for the presence of the loxP site following homologous recombination in ES cells. Once the loxP oligos are inserted, the DNA is sequenced in that region to (a) definitively establish that only a single copy of the loxP oligo is inserted, (b) to establish the orientation of the loxP oligo, and (c) to verify the fidelity of the ligation reaction.

Next, a positive selection marker gene, typically PGKneo, is inserted into a different intron in a convenient restriction site. Note that, for the strategy described here, a PGKneo gene that is flanked on both sides by frt sites and flanked on one side by a single loxP site as originally described (Lewandoski et al., 1997a; Meyers et al., 1998), and as illustrated in Figure 2.1, must be used. Following insertion of the marker gene, it is important to definitively establish that the loxP site is in the same orientation as the loxP site that was previously inserted into a different intron. This is usually accomplished simply by restriction mapping.

Finally, a negative selection marker such as thymidine kinase (TK), is cloned into the vector such that it is at one end of the arms of the targeting vector (see Figure 2.1). The orientation of the TK gene does not seem to be important. If possible, the TK cassette should be cloned onto the homologous arm that is opposite from the restriction site that will ultimately be used to linearize the vector prior to introducing it into ES cells. Several different TK selectable cassettes have been described (Mansour et al., 1988; Tybulewicz et al., 1991).

As a final step, it is advisable to sequence through as much of the exonic DNA in the construct as possible to ensure that no unwanted mutations have been introduced into the coding sequence for the gene of interest.

2.2.2.2   Vector Preparation

The vector DNA that will be introduced into ES cells should be of the highest quality possible. DNA that has been purified by cesium chloride ultracentrifugation is preferred, but other methods such as QUIAGEN Maxiprep Kits (QUIAGEN Inc., Valencia, CA) have also been successfully utilized by other labs.

Prior to introduction into ES cells, the DNA targeting vector must be linearized by restriction enzyme digestion in the vector backbone. This is critical for the stable incorporation of the construct into the genome of the ES cells. Following complete digestion (verified by comparison to undigested DNA using gel electrophoresis) of a large quantity ($\sim$500µg) of the targeting vector, the DNA is ethanol precipitated, washed with 70% ethanol, briefly air dried, and resuspended in TE buffer (10mM Tris-HCl, 1.0mM EDTA, pH = 7.5). It is very important that the DNA remain sterile following the ethanol wash. To accomplish this, perform all steps after the ethanol wash in a tissue culture hood.

After the DNA has been completely resuspended in TE buffer, the concentration of DNA is checked by spectrophotometry. DNA is subsequently diluted with TE to 1.0µg/µl, dispensed into 20µl aliquots, and stored at −20ºC until needed.

2.2.3   GENE TARGETING IN EMBRYONIC STEM (ES) CELLS

2.2.3.1   Mouse Embryonic Fibroblast Preparation

Primary mouse embryonic fibroblasts (MEFs) are required as a feeder layer for culturing ES cells in an undifferentiated state. When isolating embryos for production of MEFs, it is important to maintain sterility. Surgical tools are autoclaved prior to use and are frequently dipped in ethanol and flamed during the procedure.

MEF culture medium consists of: 500ml Knockout D-MEM (Life Technologies, Inc., Cat. # 10829-018); 5ml glutamine (200mM stock; Life Technologies, Inc., Cat. # 25030-081); 25-50ml ES qualified fetal bovine serum (FBS; Life Technologies, Inc., Cat. # 16141-079), and 4μl β-mercaptoethanol (Sigma, Cat. # M-7522).

MEFs are isolated from late-gestation mouse embryos derived from animals that harbor a neomycin resistance expression cassette. Technical details of MEF isolation, culture, and freezing are provided in Appendix 2.1. Before using MEFs as a feeder layer for ES cell culture, they must be mitotically inactivated either by treating with mitomycin-C (see Appendix 2.1) or by irradiation (Abbondanzo et al., 1993; Hogan et al., 1994). To facilitate the mitomycin-C treatment process, allow MEFs to grow to a high density (see Figure 2.3A) before treatment. Once cells are treated, they can be trypsinized and plated at a density that is suitable for culturing ES cells; an appropriate density is one in which the entire plate is covered by a single layer of MEFs (see Figure 2.3B).

2.2.3.2 ES Cell Culture

This phase of the work is absolutely critical to the success of a gene targeting experiment. Thus, it is highly recommended that this phase be conducted under the direct supervision of someone with experience working with ES cells, be outsourced, or be conducted by a core facility. Failure to properly handle the ES cells will ultimately result in failure, i.e., chimeric animals that are incapable of transmitting the genetic change through the germline.

The choice of ES cell line is critical to the success of knockout experiments. The vast majority of ES cell lines available are derived from various Strain 129 substrains. A critical aspect to consider when selecting an ES cell line for use in a knockout experiment is to select one that has been proven to produce germline-competent chimeras in your lab under your specific culture conditions. It is highly recommended to obtain and test several different ES cell lines for germline competency before starting any targeting project in a new lab. In our labs, the R1 ES cell line (Nagy et al., 1990) has proven extremely reliable although we have also used a commercially available Strain 129/SvJ cell line from Incyte Genomics with good success. Several other cell lines that have worked in a number of labs have produced less than satisfactory results in our labs.

ES cell culture and passage:
1. Plate ES cells onto inactivated MEFs at a moderate density (see Figure 2.3C) in ES cell culture medium that is appropriate for the cell line in use. For the R1 ES cell line, culture medium consists of: 500ml Knockout D-MEM, 500 units/ml leukemia inhibitory factor (Chemicon Inc., Cat. # ESG 1107), 5ml glutamine (final concentration = 0.2M), 75ml ES qualified FBS, 4μl β-mercaptoethanol.
2. Cells are cultured at 37° C in a humidified atmosphere of 95% air/5% CO_2.

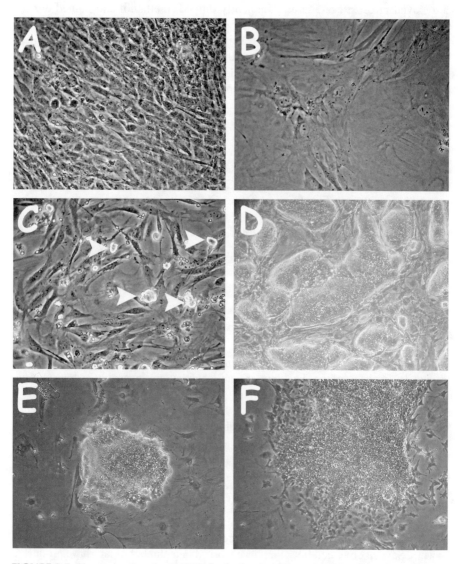

FIGURE 2.3 Photomicrographs of MEF feeder layers and murine ES cells. (A) MEFs that have been cultured to a high density and are ready to be mitotically inactivated. (B) MEFs that have been inactivated and plated at a low density suitable for serving as a feeder layer for ES cells. (C) Small ES cell colonies visible the day following passage. (D) ES cell colonies that have grown to such a density that they require passage. Note that several of the colonies are about to grow into each other. (E) A healthy, undifferentiated G418-resistant ES cell colony suitable for isolation and expansion. Note smooth, highly refractive border of the ES cell colony. (F) A G418-resistant ES cell colony that has overgrown and begun to differentiate. Note the appearance of individual cells at edge of colony and lack of smooth, highly refractive border.

3. Cells should be visually monitored daily and passaged when ~75% confluent (see Figure 2.3D). Continued culture of ES cells at appropriate densities should require passage every 2–3 days to maintain the cells in an undifferentiated state. Cells that are allowed to grow to too high a density quickly exhaust the buffering capacity of the culture medium and begin to differentiate and die.

4. To pass cells, gently wash dish twice with PBS and subsequently add 2ml trypsin and incubate at 37° C for 3–5 min. When cells have detached from dish, add 8ml culture medium and gently pipette twice to reduce the cells to a suspension composed of single cells and small clumps. Pass to freshly fed MEF feeder layers at 1:5 to 1:10 dilutions as appropriate. It is wise to pass at several different densities and use those that look the least differentiated.

2.2.3.3   Electroporation of ES Cells and Isolation of Clones

The most commonly used method for introducing targeting vectors into ES cells is electroporation. This procedure is simple, fast, and proven effective. Details of ES cell electroporation are provided in Appendix 2.2. Following electroporation, ES cells are cultured in selection medium and when surviving ES cell clones are apparent, they are picked, expanded, and split for freezing and DNA preparation as described (see Appendix 2.2).

2.2.3.4   Screen for Gene Targeting

The isolated DNA must be screened to identify those ES cell clones in which the targeting vector has inserted at the target locus by homologous recombination. While many labs have utilized various PCR screening strategies, we prefer to directly screen for targeting by Southern blot analysis. Southern blot analysis provides for definitive results and is not subject to the frequently encountered problems associated with PCR, e.g., false positive and false negative results.

DNA is digested in reactions that include: 10.0µl DNA (~5–10µg), 2.5µl appropriate buffer, 1.5µl restriction enzyme, 1.0µl spermidine (100mM stock; Sigma, Cat. # S0266), and 10.0µl water. Digestion is allowed to proceed at 37°C overnight with gentle rocking. Following digestion, DNA is subjected to agarose gel electrophoresis, Southern blotting, hybridization, and autoradiography using standard techniques (Ausubel et al., 1993; Sambrook and Russell, 2001).

A critical point that cannot be overlooked is that it is important to characterize putatively targeted loci in a detailed manner because unexpected recombination events are often observed (Hasty et al., 1991; Moens et al., 1992; Homanics et al., 1993) including rearrangements, deletions, and insertion of multiple copies of the targeting vector. Thus, all putatively targeted clones should be analyzed with several different restriction enzymes and several different probes, including probes that are either internal or external to the targeting construct. Also, putatively targeted clones should be analyzed with a probe that recognizes the selectable marker gene. This allows one to determine if additional copies of the targeting construct have been inserted randomly in the genome.

Once all (or most) clones derived from a 96-well plate have been screened for targeting, the master plate is thawed and targeted clones are recovered. To thaw the 96-well plate, remove from freezer, rinse plate with ethanol, incubate at 37°C until all ice has melted (~10 min), remove medium, add 200μl of warm ES cell culture medium, and culture overnight. Wells containing targeted clones are subsequently passed as needed to expand the culture to a point so that 5–10 vials of each clone can be frozen in cryovials as described above for MEFs. At this point, it is also advisable to prepare DNA from each clone and verify by Southern blot analysis that the correct clones have been identified.

2.2.4 ANIMAL PRODUCTION

2.2.4.1 Embryo Microinjection and Chimera Production

Chimeric mice can be produced from ES cells by either embryo aggregation (Wood et al., 1993a; Wood et al., 1993b) or microinjection techniques (Stewart, 1993; Hogan et al., 1994). We routinely use embryo microinjection. Methodological details of pipette production, embryo production, culture, and surgical transfer are found elsewhere (Wasserman and DePamphilis, 1993; Hogan et al., 1994).

Briefly, ES cells are prepared for embryo microinjection by trypsinization, followed by resuspension in ES cell culture medium that contains 20mM HEPES (Life Technologies, Inc., Cat. # 15630-080). Approximately 10–20 correctly targeted ES cells are microsurgically injected into the blastocoel of day 3.5 postcoitum C57BL/6J embryos produced from natural matings (i.e., not superovulated). Following ES cell injection, 10 to 15 blastocysts are surgically transferred to the uterine horns of day 2.5 postcoitum pseudopregnant females (CD1). Chimeric animals produced from manipulated embryos are identified on the basis of coat color. Black coat color indicates a C57BL/6J embryo contribution, whereas an agouti coat color indicates an ES cell (Strain 129) contribution. Visibly chimeric animals are mated to C57BL/6J mice. Offspring that receive the ES cell genome are identified by agouti coat color. Approximately 50% of these agouti mice are heterozygous for the modified gene. All ES cell derived mice are genotyped by genomic Southern analysis, just as was done to identify ES cells that contained the targeted locus.

2.2.4.2 Removal of Neomycin Gene and Breeding Strategy

The presence of the selectable marker gene within the target locus often reduces expression of the gene of interest. While such hypomorphic alleles can be exploited to gain insight into the function of the gene of interest (Meyers et al., 1998; Nagy et al., 1998), it is usually more desirable to remove the marker from the locus. With the strategy outlined here, this is easily accomplished by simply crossing mice that harbor the targeted locus with an FLPe general deleter strain of transgenic mouse (Rodriguez et al., 2000). This transgenic mouse line is available from the Jackson Laboratory's Induced Mutant Resource (stock # 003800). This strain of transgenic mouse ubiquitously expresses an enhanced FLP recombinase gene that very efficiently catalyzes recombination specifically between frt sites such as those that flank the neomycin gene in the targeted locus. Following FLPe mediated recombination, the neomycin gene is irreversibly deleted from the locus. The presence or absence of the FLPe transgene and the neomycin cassette must be monitored in the mice by analyzing tail DNA by Southern blot or PCR analysis. In subsequent generations, the FLPe transgene can be bred out of the pedigree.

To convert the floxed targeted locus to a global knockout, one simply has to perform a similar mating of the mice, but this time a Cre general deleter mouse line is used (Lakso et al., 1996; Meyers et al., 1998). Cre mediated recombination will irreversibly delete all DNA located between the loxP sites. If the floxing event was properly designed, deletion of the intervening DNA will eliminate a portion of the gene of interest that is critical for function and thus inactivate the gene. In subsequent generations, the Cre transgene can be bred out of the pedigree. Interbreeding of mice that are heterozygous for the recombined allele will result in mice that are wild type, heterozygous, or homozygous.

As described in the introduction, the floxed mouse line produced with this type of gene-targeting approach can also be used to create tissue-specific, temporally regulated, and inducible knockouts simply by mating to appropriate transgenic animals as described above. To date, over 100 different Cre expressing transgenic mouse lines have been successfully produced. Many of these mice are currently available from the Jackson Laboratory's Induced Mutant Resource.

2.2.4.3   Confirmation of Functional Change

Another critical part of any gene-targeting experiment is to definitively establish that the intended effect of the genetic alteration is actually realized. For example, when attempting to create a null allele it is important to establish that the modified gene is incapable of producing functional protein product. This is most often assayed using an antibody that is specific for the protein product of interest on a Western blot assay or with immunohistochemistry. If a specific antibody is not available, the effects of the targeting event can also be investigated at the RNA level using RT-PCR, Northern blot hybridization, or RNase protection assays (Ausubel et al., 1993; Sambrook and Russell, 2001), or at the protein level by assaying for function of the protein of interest. One should also be cognizant that some genetic alterations can result in truncated gene products that may have partial or novel functions that should be considered in the analysis.

2.3   EXAMPLES OF GENE KNOCKOUT ANIMALS IN ALCOHOL RESEARCH

In 1995, the first publication of gene knockout animals in the alcohol research field appeared (Harris et al., 1995). Despite being only recently applied to address ethanol's mechanism of action, numerous studies have been published to date. Several recent reviews have appeared describing some of these knockouts and alcohol action (Homanics et al., 1998b; Homanics, 2000). In the sections that follow, a few knockouts will be briefly discussed to illustrate some of the exciting results that can be obtained from knockout animals.

2.3.1 PKC KNOCKOUTS

2.3.1.1 PKC-gamma (PKC-γ) Knockout

The first knockout mouse to display an altered behavioral response to ethanol was the protein kinase C-gamma (PKC-γ) knockout mouse (Harris et al., 1995). Deletion of this brain specific protein kinase C isozyme resulted in mice that were less sensitive to the sedative/hypnotic and hypothermic effects of ethanol but were normally sensitive to other γ-aminobutyric acid type A (GABA$_A$) modulators on these same behavioral endpoints. In addition, GABA$_A$ receptors from brain membranes from these mice were also less sensitive to ethanol but not to flunitrazepam or pentobarbital. More recently, it has been demonstrated that the decreased ethanol sensitivity and tolerance development can be modulated by various genetic backgrounds (Bowers et al., 1999). Together, these results suggest that PKC-γ is a key modulator of behavioral effects of ethanol via the GABA$_A$ receptor and that the effect of PKC-γ is influenced by components of the genetic background. Phenotypic modulation by genetic background is a very important aspect of knockout experiments and is discussed in great detail elsewhere (Gerlai, 1996; Lathe, 1996; Crawley et al., 1997; Banbury Conference, 1997; Phillips et al., 1999; Homanics, 2000) and below.

2.3.1.2 PKC-Epsilon (PKC-ε) Knockout

The epsilon subunit of protein kinase C (PKC-ε) had been implicated in ethanol action from various *in vitro* studies (Messing et al., 1991; Gordon et al., 1997). However, because of obvious limitations of cell culture systems, involvement of PKC-ε in behavioral responses to ethanol was unapproachable with these experimental systems. To get at the importance of this subunit, Messing and colleagues at the Gallo Research Center created and analyzed PKC-ε gene knockout mice. They reported that the knockouts were much more sensitive to the acute behavioral effects of ethanol and other allosteric activators of the GABA$_A$ receptor (Hodge et al., 1999). Perhaps most interestingly, the PKC-ε knockout mice voluntarily consumed significantly less ethanol, but not saccharin or quinine, than control littermates. The reinforcing properties of ethanol were also shown to be dramatically reduced in the knockout animals and that this change in behavior may be related to changes in the mesolimbic dopamine system (Olive et al., 2000). Most recently, it has been demonstrated that severity of ethanol-induced withdrawal is reduced in the knockout mice (Olive et al., 2001). Together, these PKC-ε knockout studies suggest that inhibition of this enzyme reduces voluntary ethanol consumption and attenuates the withdrawal response in mice. This leads to the tantalizing hypothesis that specific inhibitors of PKC-ε may be useful therapeutic agents for treating excessive alcohol intake and alcoholism.

2.3.2 GABA$_A$ RECEPTOR SUBUNIT KNOCKOUTS

The GABA$_A$ receptor has been implicated as a key component of ethanol's mechanism of action. However, a lack of subunit-specific drugs has prevented elucidation of the involvement of individual subunits. At least 19 different subunits have been identified

(Macdonald and Olsen, 1994; Sieghart, 1995; Olsen and Homanics, 2000). Gene knockout technology has enabled a genetic dissection of the contribution of individual subunits to ethanol action. To date, numerous subunits have been knocked out including the α1 (Sur et al., 2001; Vicini et al., 2001), α6 (Homanics et al., 1997a; Jones et al., 1997), β2 (Sur et al., 2001), β3 (Homanics et al., 1997b), γ_2 (Gunther et al., 1995), γL (Homanics et al., 1999), and δ (Mihalek et al., 1999) subunits.

2.3.2.1 Gamma 2 Long (γ_2L) Knockout

The gamma 2 (γ_2) subunit of the GABA$_A$ receptor exists in two splice variants that differ by the inclusion (γ_2L) or exclusion (γ_2 short) of 8 amino acids derived from exon 9 of the γ_2 gene (Whiting et al., 1990; Kofuji et al., 1991). These eight amino acids have been implicated from various *in vitro* studies as being critical for ethanol potentiation of the effects of GABA at the GABA$_A$ receptor (Wafford et al., 1991; Wafford and Whiting, 1992) although this effect has not been universally observed (Sigel et al., 1993; Marszalec et al., 1994; Mihic et al., 1994; Sapp and Yeh, 1998).

To resolve the *in vivo* importance of the γ_2L splice variant at the whole animal level, we used traditional knockout technology to create mice in which exon 9 had been deleted from the genome, i.e., a γ_2L gene knockout mouse (Homanics et al., 1999). It was determined at the cellular level and at the whole animal level that the γ_2L subunit was not critical for various ethanol induced behavioral responses, including sedation, anxiolysis, tolerance, withdrawal hyperexcitability (Homanics et al., 1999), or consumption (unpublished observations, G. Homanics, B. Bowers, J. Wehner). This lack of a specific requirement for the γ_2L subunit has been confirmed in other complementary studies (Wick et al., 2000).

2.4 EVALUATION OF TRADITIONAL VS. SECOND GENERATION KNOCKOUT TECHNOLOGY

2.4.1 LIMITATIONS OF TRADITIONAL GENE KNOCKOUTS

Many very informative and exciting studies have utilized global gene knockouts produced by traditional methodology. However, such an approach is often less than optimal. For example, many global knockouts are embryonic- or neonatal-lethal. Even though much can often be learned before the animals expire, a lethal phenotype is catastrophic for experiments that rely on behaviors in adult animals (e.g., most alcohol-related behaviors).

Many global gene knockouts also display developmental compensation that confounds the interpretation of the results obtained. This is especially important when studying genes expressed in the developing nervous system, such as those that mediate or modulate ethanol-related behaviors. Oftentimes, it is possible to identify and quantify the extent of compensatory changes in genes that are obvious (Tretter et al., 2001). However, the most troubling aspect of compensation is that changes may occur in genes that are not obvious and, thus, are never analyzed.

The effects of polymorphic genes linked to the targeted locus also confound traditional global knockouts (Gerlai, 1996; Lathe, 1996; Crawley et al., 1997;

Banbury Conference, 1997; Phillips et al., 1999; Homanics, 2000). Most often, global knockouts are produced on a mixed genetic background, usually of C57BL/6J and Strain 129. The problem is that, in the knockout animals, genes physically linked to the knocked out allele are Strain 129 derived. In contrast, in the wild-type control animals, genes linked to the wild-type allele are C57BL/6J derived. If there are functional polymorphisms (which usually are unknown) linked to the gene of interest that impact on the phenotype being analyzed, then one might mistakenly attribute phenotypic differences to the knocked out gene when, in fact, it is the linked polymorphism that is responsible for the differences.

With global knockouts, even if an interesting behavioral phenotype is identified, often one cannot determine which cell types or which neuronal circuits are responsible for the phenotype because the gene was inactivated in all cells of the animal. While none of these caveats discredit traditional knockout studies, they force one to limit the conclusions that can be decisively drawn.

2.4.2 ADVANTAGES OF CRE/LOX KNOCKOUT TECHNOLOGY

Use of second-generation knockout technology as described above allows one to create a single floxed mouse that can be used for a multitude of experiments simply by crossing to Cre transgenic animals or by transiently expressing Cre recombinase in a tissue of interest. With a bit of forethought, experiments can be designed so that all of the limitations of traditional global knockouts are circumvented. For example, if the global null allele is neonatal lethal, one can use the same floxed mouse line to produce a postnatal knockout. This would also avoid developmental compensation. In addition, with this second-generation approach, one type of control animal will be floxed animals that do not have the Cre transgene and therefore linked genes will be identical between these controls and knockouts. Thus, linked polymorphisms will not confound the experiments. By restricting the knockout to specific cell types or circuits, one can efficiently dissect out the components of a pathway that are critical for a behavioral phenotype.

2.5 KNOCKIN MICE

While gene knockout animals have already revealed numerous novel insights into alcohol action, gene knockin animals have the potential to provide even greater advances into understanding how alcohol exerts its many effects. A knockin animal is an animal in which an endogenous gene has been modified so that, instead of producing its normal protein product, it produces a mutant protein. It is important to appreciate the fact that a knockin animal will express the gene of interest at the same amount, times, and tissues as the wild-type gene. The only difference will be that the function of the expressed gene product is altered.

2.5.1 KNOCKIN VECTOR PRODUCTION

Production of a gene-targeting vector for creating a gene knockin is similar to that described above for creating a gene knockout. One begins with a 7–10kb mouse genomic DNA fragment in a plasmid vector. The genomic fragment is then modified using site-directed mutagenesis to make the desired change in the coding region of the gene of interest. Typically, the desired change will involve the substitution of 1–3 nucleotides to change a single codon for an amino acid. When designing the mutation that will be introduced, it is important to carefully consider how that mutation will ultimately be screened for in ES cells. If the nucleotide substitution either eliminates or introduces a restriction enzyme site that can be used for screening, then that is sufficient. If the nucleotide substitution does not change a useful restriction site, then it is advisable to also introduce a "silent" mutation within a few basepairs of the knockin mutation. Because of redundancy in the genetic code, a silent mutation can be engineered that changes a useful restriction recognition site but does not change the codon.

These minor genetic modifications can be accomplished in the plasmid containing the entire 7–10kb genomic DNA insert with Stratagene's QuikChange XL site-directed mutagenesis kit. This kit was specifically designed for mutating large plasmids. Alternatively, a smaller fragment of the genomic DNA insert can be subcloned, mutated using any one of a large variety of commercially available kits for site-directed mutagenesis, and then recloned back into the parental vector. Following any mutagenesis procedure, all exonic DNA present in the vector should be sequenced (if possible) to ensure that no unwanted mutations are present and to verify the integrity of the desired mutation.

Following the mutagenesis, a positive selectable marker gene, typically PGKneo, that is flanked by recombination recognition sites is cloned into a convenient restriction site located in an intron that is adjacent to the mutated exon (see Figure 2.4).

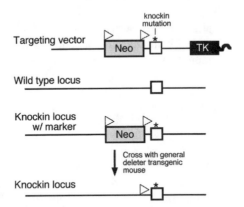

FIGURE 2.4 Overview of knockin gene targeting strategy. Gene targeting in mouse ES cells is used to introduce a specific mutation into an exon of the gene of interest and to insert a selectable marker gene into an adjacent intron. Note that the marker gene is flanked by recombinase recognition sequences. Either loxP or frt sites can be used. Site-specific recombination using either Cre or FLPe recombinase, respectively, can be used to remove the marker from the targeted locus. This can be accomplished either by transient recombinase expression in ES cells *in vitro* or by mating to a general deleter mouse line *in vivo*. Triangle = recombinase recognition sequences; Neo = neomycin resistance gene; TK = herpes simplex virus thymidine kinase gene; white box = exon; thin black line = intronic DNA; thick black line = plasmid backbone, * = knockin mutation.

Either the upstream or downstream intron will work. It is important to place the marker gene at least 200 basepairs from the intron/exon boundary to avoid interference with processing of the RNA following gene expression. A negative selectable marker gene such as TK can be added to the vector as illustrated, although this is not absolutely necessary.

2.5.2 GENE TARGETING AND MOUSE PRODUCTION

The procedures for vector preparation, electroporation, identification of targeted clones, and animal production are identical to that described above for gene knockouts. Following animal production, the positive selectable marker gene must be removed from the targeted locus to avoid adverse effects of the marker gene on expression of the gene of interest. This is easily accomplished simply by mating the marker-containing knockin animal to a general deleter mouse line that expresses the appropriate recombinase in all tissues (Lakso et al., 1996; Meyers et al., 1998; Rodriguez et al., 2000) or specifically in the germ line (Lewandoski et al., 1997b; O'Gorman et al., 1997). Because the recombination event induced is permanent and irreversible, the recombinase can be bred out of the pedigree in subsequent generations.

2.5.3 KNOCKIN MOUSE CHARACTERIZATION

Several important analyses must be performed to ensure that the knockin is indeed present and functioning as expected. To verify that the intended mutation is present, DNA sequence analysis of a RT-PCR product derived from the gene of interest should be performed. One should also verify that no other mutations are present in the gene of interest that could confound interpretation of the results. In addition, one should analyze the amount and tissue distribution of the gene product to ensure that any observed phenotype is not due to an unintended change in one of these parameters. This can be accomplished using *in situ* hybridization, immunohistochemistry, or semi-quantitative Western blot analysis.

2.5.4 EXAMPLES OF KNOCKIN ANIMALS

Recently, a highly relevant series of studies using several related $GABA_A$ receptor knockins have been published. These studies were conducted to dissect the role of the various alpha subunits of the $GABA_A$ receptor in mediating the various behavioral effects of benzodiazepines. This class of drugs, which includes valium, is of considerable clinical importance as they are some of the most widely prescribed drugs in the world. They induce sedation, amnesia, myorelaxation, anxiolysis, and impairment of cognition by modulating $GABA_A$ receptor function. $GABA_A$ receptors containing $\alpha 1$, $\alpha 2$, $\alpha 3$, or $\alpha 5$ have been demonstrated to be benzodiazepine sensitive, whereas those containing $\alpha 4$ or $\alpha 6$ are benzodiazepine insensitive (Mohler et al., 2000). The difference between sensitive and insensitive alpha subunits has been demonstrated to result from one key amino acid located at position 101 ($\alpha 1$ numbering). It is a histidine in sensitive receptors but an arginine in insensitive receptors. *In vitro* expression studies have demonstrated that sensitive receptors can be converted to benzodiazepine insensitive receptors simply by substitution of arginine for histidine at position 101 (Wieland et al., 1992; Benson et al., 1998). A very important additional point is that this amino acid substitution does not affect the receptor in any other way, e.g., the receptor responds normally to neurotransmitter. Therefore, knockin mice in which histidine 101 has been replaced by arginine individually in the various benzodiazepine sensitive alpha subunits of the $GABA_A$ receptor allows for the dissection of the importance of each sensitive subunit in the various behavioral responses that are induced by these drugs.

Mice with benzodiazepine insensitive $\alpha 1$ subunits demonstrate a lack of sensitivity to benzodiazepines on sedation and amnesia (Rudolph et al., 1999; McKernan et al., 2000). However, other benzodiazepine-induced responses, such as anxiolysis, myorelaxation, and ethanol potentiation, were intact. Thus, it was concluded that the sedative and amnestic effects of benzodiazepines are mediated by $\alpha 1$-containing $GABA_A$ receptors. In contrast, $\alpha 2$ knockin animals demonstrated that $\alpha 2$-containing $GABA_A$ receptors are selectively responsible for the anxiolytic (Low et al., 2000) and myorelaxant (Crestani et al., 2001) effects of benzodiazepines. These elegant studies have recently been reviewed (Rudolph et al., 2001).

These knockin studies are revolutionary in that they have established the functional role of specific subunits in specific drug induced behaviors. These studies open the door for developing subtype selective ligands that have the desirable effects of the drugs while being devoid of the deleterious side effects. For example, $\alpha 2$ selective compounds would be expected to display the beneficial anxiolytic effect and be devoid of the undesirable sedative and amnestic effects. Already, novel ligands have been developed and appear to display this profile (McKernan et al., 2000; Griebel et al., 2001).

The knockin approach is expected to provide equally remarkable insight into ethanol action. Specific point mutations in the $GABA_A$ and glycine receptors have been identified that eliminate the effects of ethanol at these receptors *in vitro* (Mihic et al., 1997). Production of knockin mice that harbor these ethanol insensitive mutations should reveal the functional role of specific receptor isoforms in the various behavioral responses to ethanol. In contrast to benzodiazepines that are known to function only at $GABA_A$ receptor, it is clear that ethanol has effects at multiple molecular targets. It is, therefore, unlikely that specific ethanol-induced behaviors will be totally ablated by a single knockin mutation. However, this awaits experimental demonstration.

2.6 FUTURE DIRECTIONS FOR GENETICALLY ALTERING ANIMALS

2.6.1 INDUCIBLE KNOCKOUTS

For many experiments, it is highly desirable for the investigator to control the timing of the gene knockout event. Temporal control allows for the knockout to occur only after critical periods of development have passed for example. Also, it is even possible to study the same animal before and after gene knockout. Such possibilities offer many powerful opportunities in the alcohol research field that are not possible with conventional approaches.

2.6.1.1 Tetracycline-Regulated Recombination

One promising method for creating inducible gene knockouts relies on the tetracycline-regulated system of gene expression in conjunction with the Cre/loxP system (Stonge et al., 1996; Holzenberger et al., 2000). This system works by combining a floxed mouse line with a transgenic mouse line that expresses a Cre gene from a construct regulated by the presence or absence of tetracycline. For example, with the tet-ON system, Cre will not be expressed in the absence of tetracycline and thus the floxed gene will continue to function as a wild-type gene. However, the simple addition of tetracycline to the drinking water of the animal will stimulate production of Cre and consequently inactivation of the floxed gene. This approach can even be combined with a tissue-specific promoter to produce a tissue-specific inducible knockout.

2.6.1.2 Hormone-Regulated Recombination

Another very promising technology for creating inducible knockouts also relies on Cre/loxP technology. To make Cre enzymatic activity inducible, Cre is expressed from a transgene as a fusion protein with a modified ligand-binding domain (LBD) of a steroid hormone receptor. For example, Cre has been fused to a modified LBD of the estrogen receptor. This LBD has been modified such that it no longer recognizes endogenous estrogens but instead has very high affinity for synthetic estrogen analogues such as tamoxifen. When the fusion protein is produced in a cell in the absence of ligand, the LBD (and Cre because it is attached) is sequestered in the cytoplasm. Because it is in the cytoplasm, it does not have access to the loxP sites in the DNA in the nucleus and consequently, the floxed gene is not recombined. However, in the presence of ligand, the LBD-Cre fusion is transported to the nucleus where Cre can function to inactivate the floxed gene. Thus, with this system, the investigator controls the timing of the knockout simply by injecting the animal with tamoxifen. Inducible systems such as this have been successfully used to induce recombination in many tissues including the central nervous system (Wang et al., 1997; Kellendonk et al., 1999; Vasioukhin et al., 1999; Imai et al., 2001).

2.6.2 Third Generation Knockouts

The most recent development for creating conditional gene knockout animals has been termed "third generation knockout technology" (Shimizu et al., 2000). This new development combines a tissue-specific knockout and tetracycline-regulated transgenic technologies to create a knockout that is inducible, reversible, and tissue-specific. With this system, Cre/loxP technology is used to create the tissue-specific knockout. Added onto this is a tetracycline-regulated transgene that is able to restore function of the ablated gene in a tissue-specific, inducible, and reversible manner. This system overcomes all of the limitations of conventional knockout technology and opens the door for numerous exciting experimental possibilities. To date, this system has not been employed to studies of alcohol action.

An equally exciting system that allows for reversible inactivation and overexpression of a gene of interest without disrupting normal tissue specificity relies solely on the tet-regulatable system (Bond et al., 2000). With this approach, a tet-based genetic switch is inserted into the 5' untranslated region of the gene of interest using homologous recombination. Insertion of this cassette places control of expression of the gene of interest under direct control of the investigator.

2.6.3 CLONING AND GENE KNOCKOUTS IN OTHER SPECIES

To date, the only species routinely employed for gene knockouts has been the mouse, as this is the only species from which pluripotent embryonic stem cells have been isolated. However, the advent of cloning technology is opening the door to the utilization of other species that may in some ways better serve the alcohol research community. It is now theoretically possible to take virtually any cell type, such as fibroblasts, genetically manipulate those cells *in vitro* using homologous recombination, and, with cloning technology, ultimately produce a gene-targeted animal. Already, cloned sheep (Wilmut et al., 1997), cows (Lanza et al., 2000), pigs (Betthauser et al., 2000; Onishi et al., 2000; Polejaeva et al., 2000), mice (Wakayama et al., 1998), and goats (Baguisi et al., 1999) have been produced. Some of these cloned animals have had transgenes added to them (Schnieke et al., 1997; Baguisi et al., 1999; Lanza et al., 2000). In one instance, even gene knockout sheep have been produced (McCreath et al., 2000). It is highly likely that, in the not-too-distant future, nonhuman primate cloning (and gene targeting) will also become a reality. In contrast to knockout mice, designer animals such as genetically manipulated primates may more closely model humans and, thus, may be of considerable utility in studies of alcohol action.

ACKNOWLEDGEMENTS

The author would like to thank all of the members of the Homanics laboratory, especially Carolyn Ferguson, Ed Malick, and Joanne Steinmiller, for their dedication and enthusiastic support. This work was supported by grants from the National Institute on Alcohol Abuse and Alcoholism (AA14022) and the National Institute of General Medical Sciences (GM47818 and GM52035).

REFERENCES

Abbondanzo SJ, Gadi I and Stewart CL (1993) Derivation of embryonic stem cell lines. *Meth Enzymol* 225:803-823.

Ausubel FM et al., Eds., (1993) *Current Protocols in Molecular Biology.* New York, John Wiley and Sons, Inc.

Baguisi A et al., (1999) Production of goats by somatic cell nuclear transfer. *Nat Biotechnol* 17:456-461.

Banbury Conference on Genetic Background in Mice (1997) Mutant mice and neuroscience: recommendations concerning genetic background. *Neuron* 19:755-759.

Benson JA et al., (1998) Pharmacology of recombinant gamma-aminobutyric acidA receptors rendered diazepam-insensitive by point-mutated alpha-subunits. *FEBS Lett* 431:400-404.

Betthauser J et al., (2000) Production of cloned pigs from *in vitro* systems. *Nat Biotechnol* 18:1055-1059.

Bond CT et al., (2000) Respiration and parturition affected by conditional overexpression of the Ca(2+)-activated K(+) channel subunit, SK3. *Science* 289:1942-1946.

Bowers BJ et al., (1999) Decreased ethanol sensitivity and tolerance development in gamma-protein kinase C null mutant mice is dependent on genetic background. *Alcohol Clin Exp Res* 23:387-397.

Crawley JN et al., (1997) Behavioral phenotypes of inbred mouse strains: implications and recommendations for molecular studies. *Psychopharmacology* 132:107-124.

Crestani F et al., (2001) Molecular targets for the myorelaxant action of diazepam. *Mol Pharmacol* 59:442-445.

Dymecki SM (1996) Flp recombinase promotes site-specific DNA recombination in embryonic stem cells and transgenic mice. *Proc Natl Acad Sci USA* 93:6191-6196.

Gerlai R (1996) Gene-targeting studies of mammalian behavior: is it the mutation or the background genotype? *Trends Neurosci* 19:177-181.

Gordon AS et al., (1997) Ethanol alters the subcellular localization of delta- and epsilon protein kinase C in NG108-15 cells. *Mol Pharmacol* 52:554-559.

Griebel G et al., (2001) SL651498: An anxioselective compound with functional selectivity for alpha 2- and alpha 3-containing gamma-aminobutyric AcidA (GABA$_A$) receptors. *J Pharmacol Exp Ther* 298:753-768.

Gunther U et al., (1995) Benzodiazepine-insensitive mice generated by targeted disruption of the gamma 2 subunit gene of gamma-aminobutryic acid type A receptors. *Proc Natl Acad Sci USA* 92:7749-7753.

Harris RA et al., (1995) Mutant mice lacking the gamma isoform of protein kinase C show decreased behavioral actions of ethanol and altered function of gamma-aminobutyrate type A receptors. *Proc Natl Acad Sci USA* 92:3658-3662.

Hasty P et al., (1991) Target frequency and integration pattern for insertion and replacement vectors in embryonic stem cells. *Mol Cell Biol* 11:4509-4517.

Hodge CW et al., (1999) Supersensitivity to allosteric GABA(A) receptor modulators and alcohol in mice lacking PKC epsilon. *Nat Neurosci* 2:997–1002.

Hogan B et al., (1994) *Manipulating the Mouse Embryo, 2nd ed.* Cold Spring Harbor, NY: Cold Spring Harbor Laboratory Press.

Holzenberger M et al., (2000) Ubiquitous postnatal LoxP recombination using a doxycycline auto-inducible Cre transgene (DAI-Cre). *Genesis* 26:157-159.

Homanics GE (2000) Gene targeting strategies in the analysis of alcohol and anesthetic mechanisms. In: *Genetic Manipulation of Receptor Expression and Function* (Accili D, Ed,), pp 93-110. New York, John Wiley & Sons, Inc.

Homanics GE et al., (1998a) Genetic dissection of the molecular target(s) of anesthetics with the gene knockout approach in mice. *Toxicol Lett* 100-101:301-307.

Homanics GE et al., (1998b) Alcohol and anesthetic mechanisms in genetically engineered mice. *Frontiers in Bioscience* 3:d548-558.

Homanics GE et al., (1993) Targeted modification of the apolipoprotein B gene results in hypobetalipoproteinemia and developmental abnormalities in mice. *Proc Natl Acad Sci USA* 90:2389-2393.

Homanics GE et al., (1997a) Gene knockout of the alpha-6 subunit of the gamma-aminobutyric acid type A receptor: Lack of effect on responses to ethanol, pentobarbital, and general anesthetics. *Mol Pharmacol* 51:588-596.

Homanics GE et al., (1999) Normal electrophysiological and behavioral responses to ethanol in mice lacking the long splice variant of the γ2 subunit of the γ-aminobutyrate type A receptor. *Neuropharmacology* 38:253-265.

Homanics GE et al., (1997b) Mice devoid of γ-aminobutyrate type A receptor β3 subunit have epilepsy, cleft palate, and hypersensitive behavior. *Proc Natl Acad Sci USA* 94:4143-4148.

Imai T et al., (2001) Impaired adipogenesis and lipolysis in the mouse upon selective ablation of the retinoid X receptor alpha mediated by a tamoxifen-inducible chimeric Cre recombinase (Cre-ERT2) in adipocytes. *Proc Natl Acad Sci USA* 98:224-228.

Jones A et al., (1997) Ligand-gated ion channel subunit partnerships — GABA(a) receptor alpha(6) subunit gene inactivation inhibits delta subunit expression. *J Neurosci* 17:1350-1362.

Kellendonk C et al., (1999) Inducible site-specific recombination in the brain. *J Mol Biol* 285:175-182.

Kofuji P et al., (1991) Generation of two forms of the γ-aminobutyric acid-A receptor γ 2 subunit in mice by alternative splicing. *J Neurochem* 56:713-715.

Koller BH et al., (1990) Normal development of mice deficient in beta 2M, MHC class 1 proteins, and CD8+ T cells. *Science* 248:1227-1230.

Laird PW et al., (1991) Simplified mammalian DNA isolation procedure. *Nucleic Acids Res* 19:4293.

Lakso M et al., (1996) Efficient *in vivo* manipulation of mouse genomic sequences at the zygote stage. *Proc Natl Acad Sci USA* 93:5860-5865.

Lanza RP et al., (2000) Extension of cell life-span and telomere length in animals cloned from senescent somatic cells. *Science* 288:665-669.

Lathe R (1996) Mice, gene targeting and behavior: more than just genetic background. *Trends Neurosci* 19:183-186.

Lewandoski M, Meyers EN and Martin GR (1997a) Analysis of Fgf8 gene function in vertebrate development. *Cold Spring Harb Symp Quant Biol* 62:159-168.

Lewandoski M, Wassarman KM and Martin GR (1997b) Zp3-cre, a transgenic mouse line for the activation or inactivation of loxP-flanked target genes specifically in the female germ line. *Curr Biol* 7:148-151.

Low K et al., (2000) Molecular and neuronal substrate for the selective attenuation of anxiety. *Science* 290:131-134.

Macdonald RL and Olsen RW (1994) GABA$_A$ receptor channels. *Ann Rev Neurosci* 17:569-602.

Mansour SL et al., (1988) Disruption of the proto-oncogene *int-2* in mouse embryo-derived stem cells: a general strategy for targeting mutations to non-selectable genes. *Nature* 336:348-353.

Marszalec W et al., (1994) Selective effects of alcohols on gamma-aminobutyric acid A receptor subunits expressed in human embryonic kidney cells. *J Pharmacol Exp Ther* 269:157-163.

McCreath KJ et al., (2000) Production of gene-targeted sheep by nuclear transfer from cultured somatic cells. *Nature* 405:1066-1069.

McKernan RM et al., (2000) Sedative but not anxiolytic properties of benzodiazepines are mediated by the GABA(A) receptor alpha1 subtype. *Nat Neurosci* 3:587-592.

Messing RO, Petersen PJ and Henrich CJ (1991) Chronic ethanol exposure increases levels of protein kinase C delta and epsilon and protein kinase C-mediated phosphorylation in cultured neural cells. *J Biol Chem* 266:23428-23432.

Meyers EN, Lewandoski M and Martin GR (1998) An Fgf8 mutant allelic series generated by Cre- and Flp-mediated recombination. *Nat Genet* 18:136-141.

Mihalek RM et al., (1999) Attenuated sensitivity to neuroactive steroids in GABA type A receptor delta subunit knockout mice. *Proc Natl Acad Sci USA* 96:12905-12910.

Mihic S et al., (1997) Molecular sites of volatile anesthetic action on GABA$_A$ and glycine receptors. *Nature* 389:385-389.

Mihic SJ, Whiting PJ and Harris RA (1994) Anaesthetic concentrations of alcohols potentiate GABA$_A$ receptor-mediated currents: lack of subunit specificity. *Eur J Pharmacol* 268:209-214.

Moens CB et al., (1992) A targeted mutation reveals a role for N-myc in branching morphogenesis in the embryonic mouse lung. *Genes Dev* 6:691-704.

Mohler H et al., (2000) The benzodiazepine site of GABA$_A$ receptors. In: *GABA in the Nervous System: The View at Fifty Years* (Martin DL, Olsen RW, Eds.), pp 97-112. Philadelphia, PA: Lippincott Williams & Wilkins.

Nagy A (2000) Cre recombinase: the universal reagent for genome tailoring. *Genesis* 26:99-109.

Nagy A et al., (1990) Embryonic stem cells alone are able to support fetal development in the mouse. *Development* 110:815-821.

Nagy A et al., (1998) Dissecting the role of N-Myc in development using a single targeting vector to generate a series of alleles. *Curr Biol* 8:661-664.

O'Gorman S, Fox DT and Wahl GM (1991) Recombinase-mediated gene activation and site-specific integration in mammalian cells. *Science* 251:1351-1355.

O'Gorman S et al., (1997) Protamine-Cre recombinase transgenes efficiently recombine target sequences in the male germ line of mice, but not in embryonic stem cells. *Proc Natl Acad Sci USA* 94:14602-14607.

Olive MF et al., (2000) Reduced operant ethanol self-administration and *in vivo* mesolimbic dopamine responses to ethanol in PKC epsilon-deficient mice. *Eur J Neurosci* 12:4131-4140.

Olive MF et al., (2001) Reduced ethanol withdrawal severity and altered withdrawal-induced c-fos expression in various brain regions of mice lacking protein kinase C-epsilon. *Neuroscience* 103:171-179.

Olsen RW and Homanics GE (2000) Structure and function of GABAa receptors: Insights from mutant and knockout mice. In: *GABA in the Nervous System: The View at Fifty Years* (Martin DL and Olsen RW, Eds.), pp 81-96. Philadelphia, PA: Lippincott, Williams, & Wilkins.

Onishi A et al., (2000) Pig cloning by microinjection of fetal fibroblast nuclei. *Science* 289:1188-1190.

Phillips TJ, Hen R and Crabbe JC (1999) Complications associated with genetic background effects in research using knockout mice. *Psychopharmacology* (Berl) 147:5-7.

Polejaeva IA et al., (2000) Cloned pigs produced by nuclear transfer from adult somatic cells. *Nature* 407:86-90.

Rodriguez CI et al., (2000) High-efficiency deleter mice show that FLPe is an alternative to Cre- loxP. *Nat Genet* 25:139-140.

Rudolph U, Crestani F and Mohler H (2001) GABA(A) receptor subtypes: dissecting their pharmacological functions. *Trends Pharmacol Sci* 22:188-194.

Rudolph U et al., (1999) Benzodiazepine actions mediated by specific gamma-aminobutyric acid(A) receptor subtypes. *Nature* 401:796-800.

Sambrook J and Russell DW (2001) *Molecular Cloning: a Laboratory Manual, 3rd ed.* Cold Spring Harbor, NY: Cold Spring Harbor Laboratory Press.

Sapp DW and Yeh HH (1998) Ethanol-GABA(a) receptor interactions — a comparison between cell lines and cerebellar purkinje cells. *J Pharmacol Exp Ther* 284:768-776.

Sauer B (1993) Manipulation of transgenes by site-specific recombination: Use of Cre recombinase. In: *Guides to Techniques in Mouse Development* (Wassarman PM, DePamphilis ML, Eds), pp 890-900. San Diego, CA: Academic Press, Inc.

Sauer B and Henderson N (1988) Site-specific DNA recombination in mammalian cells by the Cre recombinase of bacteriophage P1. *Proc Natl Acad Sci USA* 85:5166-5170.

Schnieke AE et al., (1997) Human factor IX transgenic sheep produced by transfer of nuclei from transfected fetal fibroblasts. *Science* 278:2130-2133.

Shimizu E et al., (2000) NMDA receptor-dependent synaptic reinforcement as a crucial process for memory consolidation. *Science* 290:1170-1174.

Sieghart W (1995) Structure and pharmacology of γ-aminobutyric acid$_a$ receptor subtypes. *Pharmacol Rev* 47:181-234.

Sigel E, Baur R and Malherbe P (1993) Recombinant GABA$_A$ receptor function and ethanol. *FEBS Lett* 324:140-142.

Stewart CL (1993) Production of chimeras between embryonic stem cells and embryos. *Methods Enzymol* 225:823-855.

Stonge L, Furth PA and Gruss P (1996) Temporal control of cre recombinase in transgenic mice by a tetracycline responsive promoter. *Nucleic Acids Res* 24:3875-3877.

Sur C et al., (2001) Loss of the major GABA$_A$ receptor subtype in the brain is not lethal in mice. *J Neurosci* 21:3409-3418.

teRiele H, Maandag ER and Berns A (1992) Highly efficient gene targeting in embryonic stem cells via homologous recombination with isogenic DNA constructs. *Proc Natl Acad Sci USA* 89:5128-5132.

Tretter V et al., (2001) Targeted disruption of the GABAa receptor delta subunit gene leads to an upregulation of gamma2 subunit-containing receptors in cerebellar granule cells. *J Biol Chem* 276:10532-10538.

Tybulewicz V et al., (1991) Neonatal lethality and lymphopenia in mice with a homozygous disruption of the c-*abl* proto-oncogene. Cell 65:1153-1163.

vanDeursen J and Wieringa D (1992) Targeting of the creatine kinase M gene in embryonic stem cells using isogenic and nonisogenic vectors. *Nucleic Acids Res* 20:3815-3820.

Vasioukhin V et al., (1999) The magical touch: genome targeting in epidermal stem cells induced by tamoxifen application to mouse skin. *Proc Natl Acad Sci* USA 96:8551-8556.

Vicini S et al., (2001) GABAa receptor α1 subunit deletion prevents developmental changes of inhibitory synaptic currents in cerebellar neurons. *J Neurosci* 21:3009-3016.

Wafford KA and Whiting PJ (1992) Ethanol potentiation of GABA-A receptors requires phosphorylation of the alternatively spliced variant of the γ2 subunit. *FEBS Lett* 313:113-117.

Wafford KA et al., (1991) Ethanol sensitivity of the GABA-A receptor expressed in Xenopus oocytes requires 8 amino acids contained in the γ2L subunit. *Neuron* 7:27-33.

Wakayama T et al., (1998) Full-term development of mice from enucleated oocytes injected with cumulus cell nuclei. *Nature* 394:369-374.

Wang YL et al., (1997) Ligand-inducible and liver-specific target gene expression in transgenic mice. *Nat Biotechnol* 15:239-243.

Wasserman PM and DePamphilis ML, Eds. (1993) *Guide to Techniques in Mouse Development*. New York, NY: Academic Press, Inc.

Whiting P, McKernan RM and Iversen LL (1990) Another mechanism for creating diversity in γ-aminobutyrate type A receptors: RNA splicing directs expression of two forms of γ2 subunit, one of which contains a protein kinase C phosphorylation site. *Proc Natl Acad Sci USA* 87:9966-9970.

Wick MJ et al., (2000) Behavioural changes produced by transgenic overexpression of gamma2L and gamma2S subunits of the GABA$_A$ receptor. *Eur J Neurosci* 12:2634-2638.

Wieland HA, Lüddens H and Seeburg PH (1992) A single histidine in GABA$_A$ receptors is essential for benzodiazepine agonist binding. *J Biol Chem* 267:1426-1429.

Wilmut I et al., (1997) Viable offspring derived from fetal and adult mammalian cells. *Nature* 385:810-813.

Wood SA et al., (1993a) Non-injection methods for the production of embryonic stem cell-embryo chimeras. *Nature* 365:87-89.

Wood SA et al., (1993b) Simple and efficient production of embryonic stem cell-embryo chimeras by coculture. *Proc Natl Acad Sci USA* 90:4582-4585.

Appendix 2.1
Preparation of Mouse Embryonic Fibroblast Feeder Cells

A. ISOLATION AND CULTURE OF FIBROBLASTS

1. Set up matings between mice homozygous for a functional neomycin resistance gene (most knockout mouse lines will work, e.g., Koller et al., 1990; Mihalek et al., 1999).
2. Sacrifice pregnant females on day 14–18 of pregnancy.
3. Disinfect mouse with 70% ethanol.
4. Remove skin from the abdomen with scissors and forceps. Wash exposed peritoneum with 70% ethanol.
5. Open abdominal cavity with scissors to expose uterus. Remove uterus and transfer to tube containing sterile phosphate buffered saline (PBS; Life Technologies, Inc., Cat. # 20012-027).
6. Transfer tube containing uterus to tissue culture hood.
7. Wash uterus in several changes of PBS until majority of blood has been removed.
8. Transfer uterus to 10cm^2 tissue culture dish (Becton Dickinson, Falcon #35-3003). Cut uterus between embryos. Remove embryos from uterus and transfer to fresh dish containing PBS.
9. Remove extra embryonic membranes that surround each embryo. Open up each embryo and eviscerate. If embryos are greater than 16 days postcoitus, remove the head also.
10. Transfer eviscerated embryos to a clean dish containing 10–15 ml trypsin (Life Technologies, Inc., Cat. # 25300-054).
11. Using two scalpels, mince the embryos until only small pieces remain.
12. Using a large-bore pipette, transfer embryo/trypsin mixture to a sterile 125ml Erlenmeyer flask. Add an additional 20–30 ml of trypsin.
13. Gently shake at 37°C for 25–30 min in an orbital shaker.
14. Pipette off as much trypsin as possible without removing large pieces of tissue. Place the trypsin/cell mixture in a 50 ml centrifuge tube and add an equal volume of MEF culture medium.
15. To the pieces of tissue that remain in the flask, add 20–30 ml trypsin and return to shaker for an additional 15–25 min.
16. Pipette off as much trypsin as possible without removing large pieces of tissue that remain. Place the trypsin/cell mixture in a 50 ml centrifuge tube and add an equal volume of MEF culture medium.
17. Centrifuge tubes containing cells at 300 xg for 10 min.
18. Aspirate off supernatant and resuspend cells in fresh MEF culture medium. Pool cells into one tube.

19. Plate cells in 10cm² tissue culture dishes at one embryo per dish. Culture in MEF culture medium at 37°C in a humidified atmosphere of 5% CO_2/95% air.
20. Change medium at 24 h post plating.
21. When nearly confluent, either freeze the cells for later use or expand by passing at 1:5–1:10.
22. To freeze cells, wash each plate with 10ml PBS, trypsinize with 2ml trypsin (37°C, ~5 min), add 8ml MEF culture medium to neutralize trypsin, and centrifuge as above. Resuspend cell pellet in freezing medium (11% DMSO/50% FCS/39% MEF culture medium) at 0.5 ml medium per dish. Transfer 0.5 ml of cell suspension to prelabeled cryovial. Place cryovials in Nalgene Cryo 1°C Freezing Container (Cat. # 5100–0001) and transfer to –80°C for 24 h. Subsequently, transfer to liquid nitrogen tank for long-term storage.

B.   MITOTIC INACTIVATION OF FIBROBLASTS

1. Prepare MEF culture medium containing 10µg/ml mitocycin-C (Sigma; Cat. # M-0503).
2. Feed MEFs with 6–10ml of mitomycin-C containing medium and culture for 2–2.75 h.
3. Wash cells three times with PBS containing 2.5% FCS.
4. Culture MEFs in 10 ml of MEF culture medium until needed. Treated cells should be used within 10 days of treatment.
5. Before using for ES cell culture, trypsinize and pass at appropriate density.

Appendix 2.2
ES Cell Transformation
by Electroporation

1. Grow ES cells to a moderately high density (i.e., just prior to the point that they require passage; see Figure 2.3D). Feed cells with fresh ES cell culture medium 2–4h prior to use. One 10cm^2 dish of ES cells is required for each electroporation that will be performed.
2. Trypsinize cells as described above and pool cells in centrifuge tube. Centrifuge at 320Xg for 5 min. Remove medium and resuspend cell pellet in 0.5ml of ES culture medium per each dish that was trypsinized.
3. Add 0.4ml of cell suspension to an aliquot of linearized targeting vector DNA and mix gently. Transfer to 2mm gap size electroporation cuvette (Genetronics, Inc., San Diego, CA, Model #620).
4. Using a BTX Electro Cell Manipulator 600 electroporation unit (Genetronics, Inc.), pulse at 250V, 50μF, 500V capacitance, and 350 ohm resistance.
5. Immediately transfer cell suspension to 12ml of ES cell culture medium.
6. Add 2ml of the diluted cell suspension to each of six MEF feeder plates and culture for 24h. Note: before using feeder plates, replace medium with ES cell culture medium.
7. Apply positive/negative selection (Mansour et al., 1988) to enrich for targeted cells. To do this, 24-h postelectroporation, feed cells with ES cell culture medium supplemented with 250μg/ml of geneticin (Life Technologies, Inc., Cat. # 10131-035) and 2μM ganciclovir (Syntex, Inc., Palo Alto, CA). Replace medium as needed (typically every 1–3 days). Continue selection for ~7–10 days. During this selection procedure, massive cell death will occur initially and viable colonies of ES cells will be difficult to detect. However, by ~7–10 days into the selection procedure, most of the dead cells will be washed away and small clumps of viable ES cells will begin to appear.
8. Pick healthy, undifferentiated ES cell clones that survive the selection procedure (Figure 2.3E,F). To accomplish this, use a mouth-controlled sterile pipette (Unopipette, Beckton Dickson and Co., Cat. # 5878) that has been pulled in a flame and cut so that the opening is ~50μM. To sterilize the pipette, rinse with ethanol followed by several rinses with culture medium. ES cell clones are picked while viewed under an inverted microscope at low power. This is usually done on the lab bench top. If care is taken to minimize exposure of the culture dish to sources of contamination, sterility can easily be maintained. ES cells are picked by scraping the clone from the surface of the dish, aspiration of the clone into the pipette, and transfer to a single well of a Microtest Tissue Culture Plate 96 Well (Beckton Dickinson, #35-3072) that has been preseeded

with inactivated MEFs. It is a good idea to wash the pipette with ethanol and medium frequently to reduce clone-to-clone cross contamination.

9. Culture recently picked clones for 24 h.

10. Trypsinize clones and replate in the same well. To do this, remove medium, wash with PBS, add 50μl of trypsin per well, wait 5 min, add 0.2ml ES cell culture medium, and titrate gently to dissociate cell clumps. Return to incubator and culture for 1–3 days. Processing cells in 96 well plates is facilitated by the use of a multichannel pipette.

11. Visually monitor cell growth daily. When most of the wells have ES cell colonies that are sufficiently grown up, i.e., they have grown to a point that they require passage, trypsinize the entire plate. Half of the trypsinized cell suspension is passed onto a new 96-well plate that has been preseeded with MEF feeder cells. The remaining half of the cell suspension is replated in the original 96-well plate. One of these plates will be a master plate that will be frozen. The ES cells on the other plate will be expanded and used for isolation of DNA.

12. Culture the master plate for 1 day then freeze all ES cell clones as follows. Remove medium from wells and replace with 100μl freezing medium. Seal edges of plate with parafilm, wrap plate in Saran Wrap, transfer plate into a styrofoam cooler and transfer to –80°C for 24 h. Subsequently, remove from Styrofoam cooler and store at –80°C for up to 4 months.

13. Culture the plate for DNA preparation until the culture medium in the majority of wells has turned yellow. When growing cells for DNA, it is not necessary to prevent overgrowth and differentiation.

14. Expand cultures for DNA preparation by trysinization and passage of each well into a single well of a 24-well plate (Becton Dickinson, Multiwell #35-3047) that has been pretreated with gelatin. Allow each well to grow until confluent. (To coat the plate with gelatin, add several drops of sterile 0.1% gelatin (Mallinckrodt Specialty Chemicals Co., Cat. # H219-59), allow to stand at room temperature for ~5 min, remove gelatin and add ES cell culture medium.)

15. Trypsinize and pass the contents of each well onto a single well of a 6-well plate (Becton Dickinson, Multiwell #35-3046) that has been pre-treated with gelatin. Allow each well to grow until confluent.

16. Lyse individual wells when confluent as described (Laird et al., 1991). Briefly, add 1.5ml of cell lysis solution and return to incubator. Lysis solution consists of: 100mM Tris-HCl pH 8.5, 5mM EDTA, 0.2% SDS, 200mM NaCl, and 100μg/ml Proteinase K (Life Technologies, Inc., Cat. #25530-015). When all cells of each six-well plate have been lysed and incubated at 37°C at least a few hours, DNA is precipitated by adding 1.5ml of isopropanol and shaking plate on a mechanical shaker at room temperature for at least 1 hour. After shaking, a web of white DNA will be visible. Using a 200μl plastic pipette tip, remove the DNA, dip into a tube of 70% ethanol, remove as much ethanol as

possible, and transfer DNA to a tube containing 50–150µl of TE. Allow DNA to dissolve at room temperature or 37°C for 48 h.

3 Molecular Studies by Chimeragenesis and Mutagenesis

*Michael J. Beckstead, Gregory F. Lopreato,
Kent E. Vrana and S. John Mihic**

CONTENTS

---

* The first two authors contributed equally to the writing of this chapter.

3.1 INTRODUCTION

The development of the current techniques for chimeragenesis and mutagenesis has greatly increased our understanding of the molecular mechanisms of action of a large number of proteins, including enzymes, transporters and ion channels. They have also been instrumental in elucidating how a wide variety of compounds interact with specific receptor sites on these proteins. In this chapter, we will describe the principles underlying the techniques of chimeragenesis and mutagenesis, provide examples of how they can be best implemented, and summarize how these techniques have been used in alcohol research. The limitations of these techniques, as well as possible future applications, are also discussed.

3.2 CHIMERAGENESIS

3.2.1 BACKGROUND

In the *Iliad,* Homer writes that the Chimera was a lion in its foreparts, a goat in the middle, and a serpent in its hindparts. The mythical creature's name is used to describe the products of recombinant cDNAs generated in the laboratory that, when expressed, produce novel composite proteins not derived from a single gene. These recombinant products may have a number of different uses, such as determining the roles that specific regions of proteins have on different protein functions. For example, the nicotinic acetylcholine (nACh) α7 and serotonin-3A ($5HT_{3A}$) receptor subunits are members of a superfamily of ligand-gated ion channels and show clear evidence of nucleotide and amino acid sequence homology. These receptors differ in the compounds that can be used as agonists to open their integral ion channels. A number of chimeric receptors containing portions of the nACh α7 and 5-HT_{3A} receptor subunits were constructed to determine the domain of these receptors governing agonist selectivity. One of these chimeras was functional when tested electrophysiologically, and demonstrated that the extracellular *N*-terminal portion of these subunits contains the agonist binding domain (Eisele et al., 1993).

To map the region of a protein that mediates a particular function, one begins by aligning and comparing the amino acid sequence of that protein with that of at least one other protein structurally related to the first, but differing from it in some way in its functioning. For example, one could align the related $GABA_A$ receptor α1 and α2 subunits to determine, by chimeragenesis, the amino acids responsible for differing benzodiazepine actions on receptors containing these subunits. Chimeric proteins can be generated either from equivalent proteins of different species (Chen and White, 2000; Adkins et al., 2001), or from different proteins within a gene family of a single species (Boileau and Czajkowski, 1999; Martinez-Torres et al., 2000). It is also possible to make chimeras from two different parental cDNAs. For example, a green fluorescent protein (GFP) tag can be added to the protein of interest, allowing monitoring of the intracellular trafficking of proteins (Fei et al., 2000; Kittler et al., 2000; Martinez-Torres and Miledi, 2001). Reviewed below are the two general approaches available for chimeragenesis: restriction site-dependent formation of chimeras (RSDFC) and restriction site-independent formation of chimeras (RSIFC).

3.2.2 Methodology

3.2.2.1 Restriction Site-Dependent Formation of Chimeras (RSDFC)

RSDFC construction relies on the existence or creation of restriction sites in the two parental clones that will be used to make the chimera. Each parental cDNA must have two unique restriction sites that are used to delineate the boundaries of domain swapping. It is best if the two restriction sites in each parental cDNA differ to ensure unidirectional ligation of the insert. It is necessary to create identical restriction enzyme sites at homologous positions of the two parental cDNAs to facilitate domain swapping. Introduction of restriction enzyme sites is accomplished by site-directed mutagenesis (discussed below), unless such a site already exists at the correct position. These restriction sites must be unique, to avoid cutting either the vector or the insert at positions other than those desired. Each of the parental clones is digested at these restriction sites such that compatible fragments are produced. These fragments are then gel-purified so that they can be recombined in a swapping fashion. Because the two parental cDNAs have identical pairs of restriction sites, it is easy to make not only the desired chimera but also its reciprocal. Unlike RSIFC, this technique has the advantage of producing chimeric constructs whose structure is predetermined; i.e., the chimera junction site is preselected by the experimenter. This technique is most useful to test the importance of a specific domain of a protein. A disadvantage is that a different construct must be prepared for each chimera that is made.

The first step in RSDFC chimeragenesis is the selection of the junction site or sites within the insert. The site(s) should ideally be positioned in a region of homology between the two parental cDNAs and this homology should be unambiguously demonstrable at the amino acid level. This increases the likelihood that the chimeric protein will be functional. The steps involved in creating a chimera via the generation of restriction enzyme sites is perhaps best explained by using an example, as described below.

The structures of ligand-gated ion channel subunits can basically be divided into a large extracellular amino terminus and a number of transmembrane domains with intracellular loops. Analogous to the Eisele et al. (1993) experiment with nACh $\alpha 7/5$-HT$_{3A}$ chimeras, a chimeric protein of two hypothetical related channels, A and B, with a junction site at the point where the extracellular domain connects to the first transmembrane domain (Figure 3.1A), could be created. This facilitates determination of the roles of the two halves of the parental proteins. In Figure 3.1A, the junction site region is shown to display a high degree of homology and is positioned roughly in the middle of the protein segments shown.

The parental cDNAs coding for A and B must be first subcloned into the same expression vector, at the same clonal sites (e.g., *Xho* I / *Not* I), and in the same orientation (Figure 3.2). Having selected the position of the junction site, a suitable restriction enzyme recognition sequence must then be found or created at the site. In our example, a *BamH* I recognition sequence is created in the two inserts by site-directed mutagenesis. A *BamH* I site was chosen because the native sequence at the

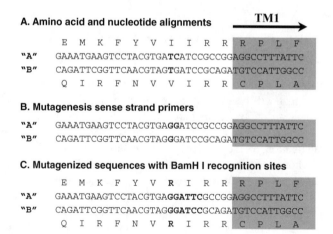

FIGURE 3.1 Creation of chimeras between two hypothetical ion channel subunits. (A) The wild-type sequence amino acid and nucleotide alignments of channels A and B are shown, with the proposed junction site in bold. (B) Mutagenesis of the A and B subunits is required to engineer *BamH* I restriction sites into both subunits. The nucleotides to be changed are shown in bold. (C) After mutagenesis, both subunits contain *BamH* I restriction sites (in bold).

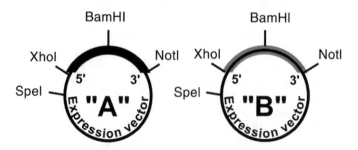

FIGURE 3.2 Restriction map of constructs A and B. The two parental cDNAs have been subcloned into the same expression vector, in the same orientation and into the same sites (*Xho* I / *Not* I). Note the location of the *BamH* I restriction site that will define the junction point in chimeragenesis.

desired junction point requires only two nucleotide changes for A and a single nucleotide change for B. The sense strand mutagenesis primer sequences are shown (Figure 3.1B), and the mutagenized sequences in Figure 3.1C. The restriction map of the constructs after the mutagenesis is shown in Figure 3.2. In our example, the introduction of a *BamH* I recognition site at the chimera junction caused an isoleucine to arginine mutation (Figure 3.1C). This mutation must be corrected by site-directed mutagenesis in the completed chimera.

The mutagenized parental constructs are digested in a manner that generates fragments that can only be religated unidirectionally. In our example, the constructs are digested with *Spe* I and *BamH* I. The digested product is then fractionated electrophoretically, and the fragments are isolated from 0.7% TAE gel (Figure 3.3). While commercial kits to isolate DNA from agarose gels are available, they are not

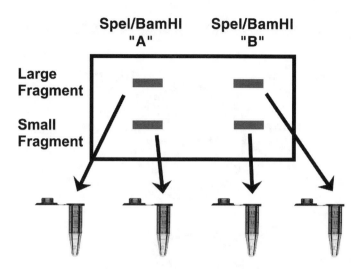

FIGURE 3.3 Digestion of constructs A and B with the enzymes *Spe* I and *BamH* I yields two linear fragments of each of the plasmids. These large and small fragments are isolated by gel electrophoresis, placed in different Eppendorf tubes for purification and recombined to make chimeras.

necessary. Instead, the bands can be excised from the gel with a sterile razor blade. The gel is visualized under low-intensity UV illumination during excision and the surface upon which the gel rests must be scrupulously cleaned. The liquid within each gel slice is separated by centrifugation. This is done as follows. A blue P1000 pipette tip is fitted with a bit of pillow stuffing or glass wool and autoclaved. The excised gel slice is placed atop the stuffing in the blue pipette tip, which is, in turn, placed in a 15 ml conical tube. This is centrifuged at $2500 \times g$ for 5 min. The liquid eluted from the slice contains the DNA fragment. The DNA is then ethanol precipitated, resuspended, and quantified by spectrophotometry. Including tRNA as a carrier during the ethanol precipitation step may improve the yield, although it complicates spectrophotometric quantification.

Two reciprocal ligation reactions (to make reciprocal chimeras) are set up to include 100 ng of the large isolated fragment from one parental cDNA, a threefold stoichiometric excess of the smaller fragment from the other parental cDNA, the appropriate buffer, and T4 DNA Ligase (4 units) in a final volume of 10 µl. The ligation proceeds overnight at 16°C.

Three transformations are performed for each chimera to be generated: the experimental transformation, and two controls. In the two controls, transformation is separately attempted on the unligated large and small fragments. The controls ensure that the fragments do not contain uncut parental plasmid. Transformants are plated on 1.5% LB/agar containing the appropriate antibiotic and allowed to grow overnight at 37°C. The next day, colonies are selected, grown up in LB media containing antibiotic, and, on the following day, the plasmid DNA is isolated by using a miniprep kit. Successful chimeragenesis is verified by restriction analysis (Figure 3.4). In our example, the A/B chimera has a *Pst* I restriction site within the

RSDFC-generated reciprocal chimeras

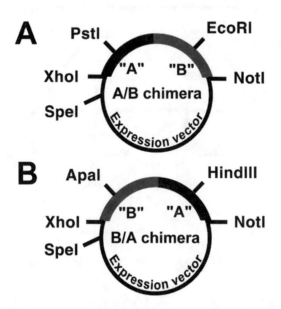

FIGURE 3.4 Verification of successful chimeragenesis by restriction analysis. Two reciprocal chimeras are shown. (A) This chimera contains the 5' region of construct A and the 3' region of construct B. (B) This chimera contains the 5' region of construct B and the 3' region of construct A.

5' portion of A that is in the chimera, as well as an *EcoR* I site in the 3' portion derived from B. If the location of the junction site in the chimera is known, the size of the fragment generated, if the chimera is digested with those two enzymes, can be calculated. In contrast, in the B/A chimera, in the 5' portion of B there is a restriction site for *Apa* I, and, in the 3' portion of A, there is a site for *Hind* III. Restriction digest patterns, such as these, allow for the unambiguous determination of successful chimeragenesis. Nevertheless, double-stranded sequencing of the entire chimera is still highly recommended.

3.2.2.2 Restriction Site-Independent Formation of Chimeras (RSIFC)

RSIFC was first described, but not published, by Dr. Randall R. Reed of the Johns Hopkins University School of Medicine, and later published by Moore and Blakey (1994) and Kim and Devreotes (1994). This technique has the advantage of being able to generate multiple chimeras from a single construct without the need to first introduce compatible restriction sites, as is required for RSDFC. This less targeted approach for chimeragenesis is the technique of choice when there is no clear justification for creating junction sites at specific amino acids, or when one desires to make a number of chimeras concurrently.

Flowchart of RSIFC chimera construction

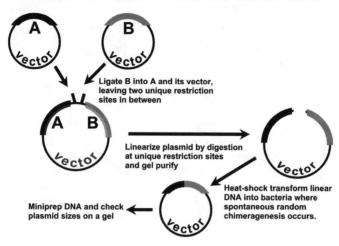

Ligate B into A and its vector, leaving two unique restriction sites in between

Linearize plasmid by digestion at unique restriction sites and gel purify

Heat-shock transform linear DNA into bacteria where spontaneous random chimeragenesis occurs.

Miniprep DNA and check plasmid sizes on a gel

FIGURE 3.5 Flowchart of the restriction site-independent formation of chimeras. Insert B is ligated into the plasmid-containing insert A such that A and B are cloned in a tandem head-to-tail configuration. (Adapted from Moore and Blakely, 1994.)

RSIFC requires the generation of a plasmid containing the cDNAs of two related subunits, A and B, cloned in a tandem head-to-tail configuration. One configuration is with inserts A then B in a 5'–3' to 5'–3' tandem configuration (Figure 3.5) and the converse with B then A in the 5'–3' to 5'–3' tandem configuration (not depicted). An A/B configuration will result in a chimera composed of the 5' region from subunit A and the 3' region from subunit B, while the B/A configuration will produce a chimera with the 5' region from subunit B and the 3' region from subunit A.

Several approaches can be taken in creating the tandem insert construct. If suitable unique restriction sites exist flanking one of the inserts (e.g., A), it can be isolated by restriction digestion and gel purification. If the same two restriction sites are not already present, one then introduces either 5' or 3' to the second insert (B) residing on another plasmid and cuts it with those restriction enzymes. These restriction sites must be unique, to ensure that the plasmid is linearized but not fragmented. Insert A can then be ligated into the plasmid containing insert B using T4 DNA ligase as described above. An alternative approach involves using a high-fidelity thermostable DNA polymerase to amplify one of the tandem inserts (e.g., insert A) by polymerase chain reaction (PCR), using specific primers that contain restriction sequences matching the position into which the insert will be cloned in the plasmid containing B. These restriction sites must not be found in the inserts and must be different from one another. The PCR product and the plasmid containing B are then digested using those enzymes. The digestion products are gel-purified, isolated by centrifugation and ligated. Regardless of the approach taken, one must ensure that a linker region exists between the two inserts and that it contains two additional unique restriction sites. The tandem insert plasmid is linearized using those two

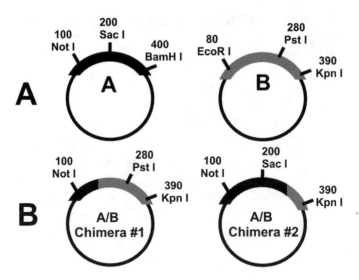

FIGURE 3.6 Identification of approximate junction points in chimeras. Restriction digestion patterns obtained after digestions with different pairs of restriction enzymes allow for the determination of the approximate site of the junction between the two parental sequences.

restriction enzymes to produce incompatible termini incapable of religating. This linearized construct is then transformed into bacteria using the heat shock method (Moore and Blakely, 1994).

The exact mechanism of rearrangement by bacteria is not understood, and the number of chimeras generated appears to be limited only by the number of homologous regions between the tandem inserts. Transformants are miniprepped by alkaline lysis. Gel electrophoresis is used to determine whether the plasmids contain a correctly sized chimeric insert, by comparison with the A and B construct standards.

Those plasmids that are determined to have correctly sized inserts are subjected to restriction analysis to determine the approximate junction point in each chimera. In our example, equally sized inserts A and B contain unique diagnostic restriction sites (Figure 3.6A). Two hypothetical chimeras are depicted in Figure 3.6B, one having a junction site between the *Not* I (from A) and *Pst* I (from B) sites (A/B Chimera #1), and one having a junction site between the *Sac* I (from A) and *Kpn* I (from B) sites (A/B Chimera #2). A double digestion with restriction enzymes *Not* I and *Pst* I would release a 180 base pair (bp) fragment from A/B Chimera #1. This means that the junction site in A/B Chimera #1 is positioned somewhere between base pairs 100 and 280. To determine if the *Sac* I site in insert A had been lost during chimeragenesis, one would test whether it could linearize the chimeric plasmid. If the *Sac* I site is present, it means that the chimera junction site must be between bp 200 and 280. If it is absent, the junction site is between bp 100 and 200, as depicted in Figure 3.6B. Digesting A/B Chimera #2 with *Sac* I and *Kpn* I would reveal a junction site somewhere between base pairs 200 and 390. If *Pst* I was unable to linearize the plasmid then one could further conclude that the junction site must be between base pairs 280 and 390. Information obtained from restriction digest patterns such as these is used to focus the sequencing efforts to the suspected chimeric junction region.

The disadvantages of the RSIFC technique include the inability to direct the positioning of the junction site and the difficulty of recovering reciprocal chimeras. Potential advantages include greater efficiency, reduced costs, and reduced experimenter bias in the selection of the chimera junction point. This random chimeragenesis technique has been successfully used by a number of laboratories studying HIV strains (Wang et al., 1999), cytochrome P450 enzymes (Pikuleva et al., 1996; Brock and Waterman, 2000), inorganic pyrophosphatase (Satoh et al., 1999), lipoxygenase (Jisaka et al., 2000), GABA/glycine receptors (Mihic et al., 1997) and flavin-containing monooxygenase (Wyatt et al., 1998).

Chimeragenesis is a powerful means of testing and mapping functional domains. Often, a combined strategy can be employed utilizing both RSIFC and RSDFC to initially home in on a relevant protein domain. Once the domain is roughly mapped by this approach, the amino acid sequence composing it can be studied by site-directed mutagenesis, as discussed in the next section.

3.3 MUTAGENESIS

3.3.1 BACKGROUND

Site-directed mutagenesis allows selective mutatation of one or more amino acid residues of the protein of interest with any other amino acid. Although the 20 naturally occurring amino acids are typically used in mutagenesis, it is also possible instead to incorporate unnatural amino acids, including those that contain fluorophors (Nowak et al., 1998). A number of different methodologies can be employed for mutagenesis, and many mutagenesis kits can be purchased from companies such as Promega (Altered Sites® II -Ex2 *in vitro* Mutagenesis System), Stratagene (QuikChange™ XL Site Directed Mutagenesis Kit), Clontech (Transformer™ Site-Directed Mutagenesis Kit), Pharmacia Biotech (Unique Site Elimination [U.S.E.] Kit) or Bio-Rad (Muta-Gene *in vitro* Mutagenesis Kit®). Most mutagenesis kits are based on the principle of oligonucleotide-directed mutagenesis, using mutagenesis oligos designed by the user and either synthesized locally or ordered from a commercial vendor. These mutagenesis oligos are complementary to, and hybridize with, a portion of the gene of interest that is to be mutated. The kits vary in the mechanisms by which mutations are introduced and also in their requirements for the use of unique restriction sites or specific vectors, enzymes or host strains of bacteria used in transformation.

3.3.2 METHODOLOGY

We have had considerable success using Stratagene's QuikChange™ kit and will describe how mutagenesis can be easily performed using this kit. Mutagenesis typically begins with a supercoiled double-stranded DNA vector containing the insert of interest, as well as two complementary synthetic oligonucleotide primers containing the desired mutation. A major advantage of using this mutagenesis technique is that it is vector-independent; virtually any double-stranded plasmid can be mutated without the need for subcloning. Just as for chimeragenesis, it is perhaps most effective to explain mutagenesis by means of an example. Let us assume that, of two related receptors, receptor X is sensitive to ethanol while receptor Y is not. These two receptors share significant sequence homology but differ in a number of amino acids; one of these differences is a tyrosine residue at a certain position in receptor X, while a leucine residue is found at the equivalent position in receptor Y. To test whether this amino acid difference is important for determining the alcohol sensitivity of receptor X, one should mutate the tyrosine to leucine in receptor X, and conversely also mutate leucine to tyrosine in receptor Y. If the mutation decreases ethanol sensitivity in X and produces some sensitivity in Y, it suggests that this residue is important for ethanol effects on this family of receptors. Note that this does not provide evidence that this residue is necessarily part of the binding site for ethanol, an issue that is addressed below. Mechanistic information regarding the physical properties of amino acids that are critical for ethanol actions can also be ascertained. For example, a tyrosine to phenylalanine mutation would address whether the hydroxyl group of tyrosine is important for the actions of ethanol; the structure of phenylalanine is identical to that of tyrosine except it lacks a hydroxyl group. Similarly, the mutation of tyrosine to a smaller amino acid also bearing a hydroxyl residue (serine or threonine) would address the importance of amino acid size.

3.3.2.1 Oligonucleotide Design

Mutagenesis begins with the cDNA of interest incorporated in a plasmid that is appropriate for the expression system that will later be used. A pair of perfectly complementary oligonucleotide primers 30–45 base pairs in length is synthesized; these hybridize to the portion of the cDNA that will be mutated. Both the sense and antisense primers should be an exact match to the wild-type sequence that is to be mutated, except for the nucleotides that will be changed to produce the desired mutation. The altered nucleotides should fall close to the middle of the primers and there should be 100% complementarity with the wild-type sequence for at least 12–15 base pairs on the primer on either side of the mutation. Ideally, the sequence should have a minimum GC base content of 40% and both ends of the primers should terminate with at least one G or C base.

Example: Let us assume that partial amino acid and double-stranded nucleotide sequences for a hypothetical gene are as follows:

Amino Acid	93	94	95	96	97	98	99	100	101	102	103	104	105	106
	Val	Gly	Gly	Asp	Pro	Ser	Thr	Tyr	Thr	Arg	Tyr	Asn	Val	Trp

5' GTG GGA GGA GAT CCC AGT ACC TAT ACA CGA TAC AAC GTT TGG 3'

3' CAC CCT CCT CTG GGG TCA TGG ATA TGT GCT ATG TTG CAA ACC 5'

We wish to mutate amino acid residue #100 from tyrosine (Tyr) to serine (Ser). The nucleotide triplet TAT codes for Tyr, while Ser can be encoded by TCT, TCC,

TCA, TCG, AGT or AGC. In mutating the wild-type sequence, it is preferable to change as few nucleotides as possible. Of the six possibilities, the best choice to encode Ser in the mutagenesis primers is TCT, because it involves only a single nucleotide change from TAT. Sometimes, all three nucleotides may need to be changed (e.g., TGG (tryptophan) to CCC (proline)), but this is not likely to be problematic if the mutagenesis primers are well designed. The following 35-mer, with the altered nucleotide enlarged, could be used as one of the pair of mutagenesis primers.

5'-GGA GAT CCC AGT ACC T**C**T ACA CGA TAC AAC GTT TG-3'

There may be times where there is an AT rich region 12–15 nucleotides removed from the mutation site, making it difficult to end the primer with a G or C base. This makes it necessary to either extend that portion of the primer to allow it to end with a G or C base, or instead to position the mutation off-center in the primer (e.g., 13 residues on the 5' side and 19 residues on the 3' side). While not optimal, adjustments such as this are usually not disruptive. It is also important that the GC percentage of the primer be over 40% (in our case it is 49%) and that the melting temperature of the primers (Tm) be at least 78°C. The Tm is calculated using the equation Tm = 81.5 + 0.41(%GC) − 675/N − % mismatch, where N is the primer length in base pairs, and % mismatch refers to the percentage of nucleotides in the primer that do not match the wild-type sequence. The Tm for our sample primer is approximately 79°C.

Increasing the length of the primer or increasing the GC ratio will increase the primer Tm if it is too low. A complementary mutagenesis primer encoding the mutation (in bold) also is required. It binds to the complementary strand of the plasmid. The sequence of this complementary primer is shown below.

5' CAA ACG TTG TAT CGT GTA **G**AG GTA CTG GGA TCT CC-3'

Note that the sequence of this primer is the converse reverse of the first; i.e., the first primer ended with CGT TTG and this one begins with CAA ACG. In designing primers, one needs to avoid the formation of hairpin loops, especially with G and C bases. If a pair of oligonucleotides fails to produce the desired mutation, it is possible that a primer hairpin loop (such as would be created by the sequence GGGGATCCCCC) is inhibiting the annealing of primers to the template plasmid. Before use, primers must be purified by either fast polynucleotide liquid chromatography (FPLC) or by polyacrylamide gel electrophoresis (PAGE).

3.3.2.2 Thermal Cycling

To perform the mutagenesis reaction, primers (125 ng each) and template DNA (5–50 ng) are combined with buffer, nucleotides (dNTPs) and *Pfu Turbo* DNA polymerase (provided in the QuikChange™ mutagenesis kit). The *Pfu Turbo* DNA polymerase is a high-fidelity proofreading polymerase that minimizes random mutations. Thermal cycling is performed as described in the instruction manual of the Stratagene QuikChange™ Site-Directed Mutagenesis Kit. Despite occurring in a thermal cycler, the temperature cycling involved in the mutagenesis step is not an example of PCR, as PCR involves the doubling of product with each temperature cycle. The product created using this kit is a single-stranded linear oligonucleotide that cannot serve as a template for further reactions. Thus, the amount of product produced is linearly proportional to the number of temperature cycles rather than exponentially related as in the case of PCR.

Each temperature cycle contains three temperature steps. First, the sample is heated to 95°C to melt all DNA into single strands. The temperature is then lowered to 55°C, whereupon mutagenesis primers anneal to complementary regions on the template. The temperature is then raised to 68°C and maintained there to allow the polymerase to extend the primer into a linear strand of DNA equal in length to the template. The new strand of DNA will be identical to the template, except for the mutation that was introduced in the primer. If the thermal cycling should fail to produce the desired mutant, one can try doubling the amounts of primers used, lower the annealing temperature (to 52°C), or increase the number of temperature cycles to 20–25.

3.3.2.3 Transformation and Amplification

After thermal cycling, the mutated linear DNA must be separated from the circular template wild-type DNA. This is done by incubating the mixture with the restriction enzyme *Dpn* I, which selectively breaks down methylated DNA. Previous passage of the wild-type plasmid through bacteria results in its methylation. The mutated linearized plasmids are then transformed into bacteria (XL-1 Blue supercompetent cells in the case of the QuikChange kit). These plasmids contain, in addition to the cDNA of interest, a gene coding for a resistance factor to antibiotics, such as ampicillin or kanamycin, allowing one to select for bacteria that have taken up the plasmid DNA and are expressing its products. After transformation, 50-200 µl volumes of product should be plated on 3.6% agar plates containing the appropriate antibiotic, and the plates inverted and incubated at 37°C for 15–17 h. Single colonies will be discernible on the plates and should be separated sufficiently from one another that they can be cleanly picked with an inoculation loop or a pulled glass pipette. Three or four colonies should be picked and placed into separate 15 ml Falcon tubes containing 3 ml of Luria's broth or similar media and the appropriate concentration of antibiotic. These tubes are then placed in a 37°C shaker at 225–250 rpm for 17–19 h. Each sample is then mini-prepped separately to isolate the amplified mutant DNA. Both the sense and antisense strands are sequenced to ensure that the desired mutation is present. Bacterial processes transform the previously linear DNA back into circular plasmids, which are then passed on as the cells divide. If mini-prepped carefully, this process should yield approximately 10–20 µg of mutated plasmid DNA, which can then be used to determine the effect of the point mutation.

3.3.2.4 Strategies for the Concurrent Multiple Mutation
of a Single Amino Acid

We have developed and use extensively a technique for the concurrent production of multiple mutations of a single amino acid. Like the technique described by Parikh and Guengerich (1998), our random mutagenesis approach employs a strategy using degenerate oligonucleotide primers and the Stratagene QuikChange™ mutagenesis kit, but is considerably easier to implement. Only one extra transformation and miniprep step are required to make multiple mutants at a specific site, compared with making a single mutant as described above. For example, if one hypothesized that either amino acid size or the presence of a hydroxyl group was the important characteristic of a particular tyrosine residue on a protein, one could test this by mutating that tyrosine residue to phenylalanine (similar size but lacking hydroxyl group), serine (smaller amino acid bearing a hydroxyl group), and perhaps also valine and alanine (small amino acids lacking hydroxyl groups). A degenerate primer encoding the nucleotide triplet KYT (where K = G or T; Y = C or T) replacing the wild-type tyrosine would yield the Ser (TCT), Phe (TTT), Ala (GCT) and Val (GTT) mutants.

Because degenerate complementary oligonucleotides are used in the mutagenesis reaction, one obtains, in our example, a population of four different sense and antisense strands of the plasmid, without any assurance that any particular double-stranded plasmid is completely complementary. That is, the two complementary strands of a particular plasmid could well code for different amino acids at that position. To correct for this, after digestion with *Dpn* I (as described above), plasmid DNA is transformed into XL1-Blue Supercompetent cells and incubated at 37°C (shaking at 225 rpm) for 17 h in LB media containing the antibiotic of choice. Bacterial replication of plasmids ensures that both strands of each plasmid would be entirely complementary. Plasmid DNA from these cells is then isolated and this pool of four different, but perfectly complementary, plasmids is again transformed into XL-1 Blue cells and plated on 3.6% agar plates. Single colonies are selected, cultured and subjected to miniprep isolation as described above. It is expected that each colony is derived from a single cell that took up a single copy of a plasmid. Successful mutagenesis is confirmed by double-stranded sequencing of miniprep DNA. This also assures that no unwanted mutations have been accidentally introduced. A schematic of the degenerate oligonucleotide mutagenesis protocol is shown in Figure 3.7.

Mutagenesis using degenerate oligonucleotides can be undertaken on a large or small scale. For example, the NNS (N = A, C, G or T; S = G or C) degeneracy series allows one to create all possible mutations at a particular site. However, one may instead be interested in producing only two mutants that differ in a specific physical property of interest. For example, if one wished to test the importance of amino acid charge at a particular residue, one could use this mutagenesis technique to concurrently introduce lysine (coded for by AAA) and glutamate (coded for by GAA) mutations at the site of interest. In this case, the degeneracy would be only 2 (RAA, where R = A or G). Similarly, if one wished to determine the importance of amino acid volume or size at a particular residue, twofold degenerate oligonucleotides encoding glycine (GGG) or tryptophan (TGG) could be used in mutagenesis.

The probability (P) of obtaining a particular mutant can be described by the equation $P = (1 - [(D-R)/D]^S) \times 100\%$. D is the degeneracy of the primers (32 for the NNS series) and R is the redundancy of the mutant sought. For example, for methionine, R = 1 because only one of the 32 possibilities in the NNS series (ATG) codes for methionine. In contrast, other amino acids, such as Thr and Ser, are encoded by two or three nucleotide combinations, making their R values 2 or 3. S is the number of

Schematic of the degenerate oligonucleotide mutagenesis protocol

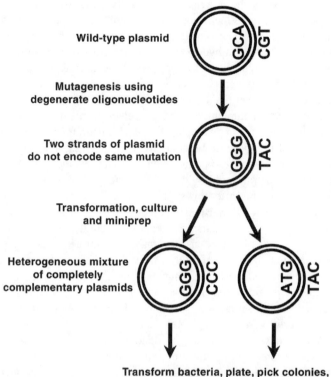

Wild-type plasmid

Mutagenesis using degenerate oligonucleotides

Two strands of plasmid do not encode same mutation

Transformation, culture and miniprep

Heterogeneous mixture of completely complementary plasmids

Transform bacteria, plate, pick colonies, culture, miniprep plasmid and identify mutant by sequencing

FIGURE 3.7 Schematic of a complete mutagenesis procedure using degenerate oligonucleotides. In this example, the wild-type plasmid possesses completely complementary DNA strands. After mutagenesis with degenerate oligonucleotides, complementarity is lost at the mutation site, but is regained after transformation of cells. Plasmid minipreps are then again used to transform bacteria, and colonies picked from an agar plate would be expected to yield a variety of mutants.

miniprep samples sequenced. The probabilities of obtaining a particular mutant in the NNS series are 63.8% if R = 1, 87.3% if R = 2 and 95.7% if R = 3, when 32 miniprep sequences are obtained. Given that R = 1 for 12 amino acids and one stop codon, R = 2 for five amino acids and R = 3 for three amino acids in the NNS series, one would predict that the average probability of obtaining any particular mutant after sequencing 32 minipreps is 74%. In agreement with this, we successfully obtained 15 of 20 mutants at amino acid 288 of the $\alpha1$ glycine receptor subunit using a single pair of mutagenesis primers of NNS degeneracy after sequencing 32 minipreps.

The advantages of an experimental approach involving concurrent multiple mutagenesis at selected amino acids can be easily appreciated. This technique requires only 1 more day to create multiple mutants than it takes to create a single mutation (one extra transformation and miniprep step). Apart from the savings in time and money, another advantage is the increased experimental flexibility afforded by having premade mutants available before they are needed. For example, if an interesting result was observed after mutation of a particular amino acid to serine, one could ask whether the hydroxyl group on the serine residue was the important characteristic mediating that result. This would be quickly tested if the mutations to alanine (lacking the OH group), threonine or tyrosine (both of which have OH groups) had already been premade. The multiple mutagenesis approach thus saves considerable time when mutants do not have to be created anew each time an interesting result is observed.

3.4 APPLICATIONS TO ALCOHOL RESEARCH

3.4.1 WHAT HAVE WE LEARNED?

Mechanistic studies of alcohol actions have a long history, but only recently have these studies been conducted at the molecular level. The recent cloning of a wide variety of enzymes and ion channels has proven a boon, and we are now perhaps entering a golden age for alcohol research. There are many effects of ethanol and potentially many molecular sites at which it can act. An important unanswered question is: what molecular sites underlie specific behavioral effects of ethanol? More specifically, what molecular sites are responsible for the development of tolerance and dependence, what sites underlie ethanol sensitivity and the drive to consume alcohol, and how does ethanol exert its effects on particular proteins? Obtaining answers to these and other questions is greatly facilitated by the use of molecular techniques, such as chimeragenesis and mutagenesis. Some noteworthy findings have already been made. Amino acids of alcohol dehydrogenase participating in the metabolism of alcohols have been identified. For example, the mutation of serine-139 to either alanine or cysteine in the *Drosophila* alcohol dehydrogenase resulted in the formation of enzymes without catalytic activity (Cols et al., 1997). Other studies demonstrated that the mutagenesis of critical amino acids of alcohol dehydrogenase altered the substrate specificities of the mutant enzyme such that it could metabolize larger alcohols than the wild-type enzyme (Creaser et al., 1990; Weinhold and Benner, 1995).

In collaboration with Adron Harris and Neil Harrison, we identified first a 45 amino acid domain and then specific amino acids in transmembrane domains 2 (TM2) and 3 (TM3) of GABA$_A$ and glycine receptor subunits responsible for alcohols and volatile anesthetic enhancement of receptor function (Mihic et al., 1997). The mutation of serine-267 to isoleucine (S267I) in the glycine receptor $\alpha 1$ subunit produced receptors completely insensitive to the enhancing effects of ethanol. Further mutagenesis indicated that the amino acid volume in the residue occupying position 267 in TM2 was a critical factor in demonstrating the potentiation of receptor function by alcohols (Ye et al., 1998), and that changes in alcohol cutoff observed after mutagenesis suggested that this region might constitute the binding site of alcohols on these receptors (Wick et al., 1998).

A chimera composed of the nACh α7 subunit and the 5-HT$_{3A}$ subunit was studied for its sensitivity to ethanol. Yu et al. (1996) found that this chimera, consisting of the N-terminal domain of the nACh subunit and the transmembrane and carboxyl tail domains of the 5-HT$_{3A}$ subunit, displayed ethanol inhibition of receptor function. This inhibition was similar to that produced by ethanol on wild-type nACh α7 receptors, thus suggesting that alcohol inhibits nACh receptor function by actions on its N-terminal domain.

Attempts have also been made in identifying alcohol binding sites on N-methyl-D-aspartate (NMDA)-sensitive glutamate receptors. Anders et al. (2000) demonstrated normal ethanol inhibition of NMDA receptor function when the intracellular C terminal tail of the NR1 subunit was excised (a stop codon was inserted at position 863). However, decreased ethanol effects were observed when a stop codon was instead inserted at position 858, or if amino acids 839-863 were excised, or if a triple mutation (R859A, K860A and N861A) was made in the NR1 subunit (Anders et al., 2000). The F639A mutation in TM3 of the NR1 subunit also markedly decreased ethanol inhibition of NMDA receptor function, when the NR1 subunit was co-expressed with the NR2A, NR2B or NR2C subunits (Ronald et al., 2001).

Ethanol enhances the function of G-protein-coupled inwardly rectifying potassium channels (GIRKs). However, other related inwardly rectifying potassium channels (such as IRK1) are resistant to the effects of ethanol. Chimeras made of the GIRK2/IRK1 channels demonstrated that a region 43 amino acids in length in the carboxyl terminus of these proteins was responsible for determining ethanol sensitivity (Lewohl et al., 1999).

Ethanol potentiates the functioning of the dopamine transporter, but inhibits norepinephrine transporter function. Chimeragenesis of these related parental cDNAs identified the intracellular loop between transmembrane domains 2 and 3 as the critical region for determining the nature of the effects of ethanol. Subsequent mutagenesis studies demonstrated that the mutations G130T, I137F and L138F could each abolish the enhancing effects of ethanol on DAT function (Maiya et al., 2001).

3.4.2 ANALYSIS AND INTERPRETATION OF DATA

The identification of amino acids underlying an effect of ethanol does not necessarily mean that these amino acids constitute part of the ethanol binding site on that protein. These residues may instead be located on a portion of the protein that is important in transducing a signal (e.g., a conformational change) from the site(s) with which ethanol is physically interacting, to those that produce an effect. Differentiating putative ethanol binding sites from transduction sites can be difficult. One approach that has been used to probe ligand-receptor interactions is the substituted-cysteine accessibility method (Foucand et al., 2001). Mascia et al. (2000) used this cysteine substitution method to show that anesthetic thiol reagents, such as propanethiol or propyl methanethiosulfonate, could be irreversibly linked to the glycine receptor α1 subunit at specific amino acids mutated to cysteine. The covalent binding of thiols to the S267C α1 glycine receptor subunit resulted in glycine receptors that displayed irreversible enhancement of receptor function and, more importantly, were resistant to the enhancing effects of enflurane, isoflurane and octanol. This suggests that the irreversible occupation of the S267 site in these receptors prevents other compounds, such as octanol, from interacting with the receptor at that site, further suggesting that alcohol and volatile anesthetic enhancement of glycine receptor function occurs after their binding to that one site.

3.5 FUTURE DIRECTIONS

Perhaps the most significant impact that mutagenesis studies will have on future research into the actions of alcohol will be in guiding the production of knockin mice. This is already starting to occur. Harrison and Homanics (2001) produced knockin mice bearing *in vivo* a mutated GABA$_A$ receptor α1 subunit shown *in vitro* to be resistant to the enhancing effects of ethanol. Unfortunately, this S270H mutation appears to result in an abnormal phenotype even in the absence of ethanol. Harris and colleagues recently generated a line of transgenic mice bearing the S267Q mutation that, *in vitro*, prevents ethanol enhancement of glycine α1 receptor function (Findlay et al., 2001). These animals displayed a slightly decreased sensitivity to the ataxic effects of ethanol, consistent with the hypothesis that at least some of the behavioral effects of ethanol are mediated by the glycine receptor. The creation and future testing of knockin mice will allow us to attribute specific behavioral effects of ethanol to specific receptors, ion channels or enzymes. Strong evidence that this will be a successful approach comes from two elegant studies recently published by Möhler's group (Rudolph et al., 1999; Low et al., 2000). In these two studies, knockin mice were created bearing either α_1 or α_2 GABA$_A$ receptor subunits that were mutated in such a way as to render the resulting GABA$_A$ receptors insensitive to the enhancing effects of benzodiazepines. Mice expressing benzodiazepine-resistant α_1 subunits displayed the anxiolytic and myorelaxant properties of diazepam, but not the sedative or amnestic effects (Rudolph et al., 1999). In contrast, mice expressing benzodiazepine-insensitive α_2 subunits were insensitive to the anxiolytic effects of diazepam (Low et al., 2000). The authors were thus able to attribute specific behavioral effects of benzodiazepines observed *in vivo*, to their enhancing effects on receptors containing specific GABA$_A$ receptor subunits (for knockin technology, also see Chapter 2).

3.6 CONCLUSIONS

Chimeragenesis and mutagenesis are two techniques used to generate information on the molecular mechanisms of action of proteins and to define regions of these proteins with which ligands, such as ethanol, can interact. When sites of ethanol action are identified, one must be cognizant that those amino acids may represent the site with which ethanol physically interacts, or they may represent a site that is important in transducing a signal from an ethanol binding site elsewhere. We anticipate that great strides in alcohol research will be made in the next decade when information obtained from *in vitro* mutagenesis experiments is used to generate knockin mice expressing proteins with altered sensitivities to ethanol.

ACKNOWLEDGMENTS

This chapter was supported by a grant from the Texas Commission on Alcohol and Drug Abuse and NIH grants from National Institute on Alcohol Abuse and Alcoholism (AA11525); National Institute of General Medical Science (GM47818) to SJM and (GM038931); and National Institute on Drug Abuse (DA006634) to KEV.

REFERENCES

Adkins EM, Barker EL and Blakely RD (2001) Interactions of tryptamine derivatives with serotonin transporter species variants implicate transmembrane domain I in substrate recognition. *Mol Pharmacol* 59:514-523.

Anders D et al. (2000) Reduced ethanol inhibition of N-methyl-D-aspartate receptors by deletion of the NR1 C0 domain or overexpression of alpha-actinin-2 proteins. *J Biol Chem* 275:15019-15024.

Boileau AJ and Czajkowski C (1999) Identification of transduction elements for benzodiazepine modulation of the GABA(A) receptor: three residues are required for allosteric coupling. *J Neurosci* 19:10213-10220.

Brock BJ and Waterman MR (2000) The use of random chimeragenesis to study structure/function properties of rat and human P450c17. *Arch Biochem Biophys* 373:401-408.

Chen Z and White MM. (2000) Forskolin modulates acetylcholine receptor gating by interacting with the small extracellular loop between the M2 and M3 transmembrane domains. *Cell Mol Neurobiol* 20:569-577.

Cols N et al. (1997) Drosophila alcohol dehydrogenase: evaluation of Ser139 site-directed mutants. *FEBS Lett* 413:191-193.

Creaser EH, Murali C and Britt KA (1990) Protein engineering of alcohol dehydrogenases: effects of amino acid changes at positions 93 and 48 of yeast ADH1. *Protein Eng* 3:523-526.

Eisele JL et al. (1993) Chimaeric nicotinic-serotonergic receptor combines distinct ligand binding and channel specificities. *Nature* 366:479-483.

Fei YJ et al. (2000) A novel H(+)-coupled oligopeptide transporter (OPT3) from Caenorhabditis elegans with a predominant function as a H(+) channel and an exclusive expression in neurons. *J Biol Chem* 275:9563-9571.

Findlay GS et al. (2001) Acute responses to ethanol in transgenic glycine receptor a1 S267Q mutant mice. *Alcohol Clin Exp Res Suppl* 25: 13A.

Foucaud B et al. (2001) Cysteine mutants as chemical sensors for ligand-receptor interactions. *Trends Pharmacol Sci* 22:170-173.

Harrison NL and Homanics GE (2001) Production and characterization of knock in mice containing a serine to histidine mutation at position 270 of the α1 subunit of the GABAA receptor. *Alcohol Clin Exp Res Suppl* 25: 13A.

Jisaka M et al. (2000) Identification of amino acid determinants of the positional specificity of mouse 8S-lipoxygenase and human 15S-lipoxygenase-2. *J Biol Chem* 275:1287-1293.

Kittler JT et al. (2000) Analysis of GABA$_A$ receptor assembly in mammalian cell lines and hippocampal neurons using γ subunit green fluorescent protein chimeras. *Mol Cell Neurosci* 16:440-452.

Kim JY and Devreotes PN (1994) Random chimeragenesis of G-protein-coupled receptors. Mapping the affinity of the cAMP chemoattractant receptors in Dictyostelium. *J Biol Chem* 269:28724-28731.

Lewohl JM et al. (1999) G-protein-coupled inwardly rectifying potassium channels are targets of alcohol action. *Nat Neurosci* 2:1084-1090.

Low K et al. (2000) Molecular and neuronal substrate for the selective attenuation of anxiety. *Science* 290:131-134.

Martinez-Torres A, Demuro A and Miledi R (2000) GABA ρ1/GABA$_A$ α1 receptor chimeras to study receptor desensitization. *Proc Natl Acad Sci USA* 97:3562-3566.

Martinez-Torres A and Miledi R (2001) Expression of gamma-aminobutyric acid ρ1 and ρ1 Delta 450 as gene fusions with the green fluorescent protein. *Proc Natl Acad Sci USA* 98:1947-1951.

Maiya R et al. (2001) Ethanol-sensitive sites on the human dopamine transporter. *Alcohol Clin Exp Res Suppl* 25:56A.

Mascia MP, Trudell JR and Harris RA (2000) Specific binding sites for alcohols and anesthetics on ligand-gated ion channels. *Proc Natl Acad Sci USA* 97:9305-9310.

Mihic SJ et al. (1997) Sites of alcohol and volatile aneasthetic action on GABA-A and glycine receptors. *Nature* 389:385-389.

Moore KR and Blakely RD (1994) Restriction site-independent formation of chimeras from homologous neurotransmitter-transporter cDNAs. *Biotechniques* 17:130-136.

Nowak MW et al. (1998) *In vivo* incorporation of unnatural amino acids into ion channels in Xenopus oocyte expression system. *Methods Enzymol* 293:504-529.

Parikh A and Guengerich FP (1998) Random mutagenesis by whole-plasmid PCR amplification. *Biotechniques* 24:428-431.

Pikuleva IA, Bjorkhem I and Waterman MR (1996) Studies of distant members of the P450 superfamily (P450scc and P450c27) by random chimeragenesis. *Arch Biochem Biophys* 334:183-192.

Ronald KM, Blevins T and Woodward JJ (2001) Effect of transmembrane domain three mutations on the ethanol sensitivity of NMDA receptors. *Alcohol Clin Exp Res Suppl* 25:9A.

Rudolph U et al. (1999) Benzodiazepine actions mediated by specific gamma-aminobutyric acid$_A$ receptor subtypes. *Nature* 401:796-800.

Satoh T et al. (1999) A chimeric inorganic pyrophosphatase derived from *Escherichia coli* and *Thermus thermophilus* has an increased thermostability. *Biochemistry* 38:1531-1536.

Wang Z et al. (1999) CCR5 HIV-1 coreceptor activity. Role of cooperativity between residues in N-terminal extracellular and intracellular domains. *J Biol Chem* 274:28413-28419.

Weinhold EG and Benner SA (1995) Engineering yeast alcohol dehydrogenase. Replacing Trp54 by Leu broadens substrate specificity. *Protein Eng* 8:457-461.

Wick MJ et al. (1998) Mutations of GABA and glycine receptors change alcohol cutoff: evidence for an alcohol receptor? *Proc Natl Acad Sci USA* 95:6504-6509.

Wyatt MK et al. (1998) Identification of amino acid residues associated with modulation of flavin-containing monooxygenase (FMO) activity by imipramine: structure/function studies with FMO1 from pig and rabbit. *Biochemistry* 37:5930-5938.

Ye Q et al. (1998) Enhancement of glycine receptor function by ethanol is inversely correlated with molecular volume at position 267. *J Biol Chem* 273: 3314-3319.

Yu D et al. (1996) Ethanol inhibition of nicotinic acetylcholine type alpha 7 receptors involves the amino-terminal domain of the receptor. *Mol Pharmacol* 50:1010-1016.

4 Combining Patch-Clamp Recording and Gene Profiling in Single Neurons

Hermes H. Yeh, Stavros Therianos and Shao-Ming Lu

CONTENTS

4.1 INTRODUCTION

It has been almost ten years since the initial wave of studies emerged in the literature reporting success in employing a combination of patch-clamp recording and detection of candidate genes in the same neuron (Eberwine et al., 1992; Lambolez et al., 1992; Surmeier et al., 1992; Mackler and Eberwine, 1993; Jonas and Spruston, 1994). Although these studies rely on different approaches to profile the expression of candidate genes, each with merits and pitfalls, they have in common the theme of correlating function with gene expression in individual identified neurons. Since the first reports appeared, much has been devoted to the refinement of the techniques and to their application to different experimental preparations. It is fair to state that coupling patch-clamp recording to gene profiling has now become popularized because of its power of resolution, and terms such as "single-cell mRNA amplification" and "gene expression profiling" that can trace their origin to these studies have become household terms in the vocabulary of neuroscience researchers.

The need for a functional and molecular biological correlation stems from the realization that there is great diversity not only among the types of neurons found in the nervous system, but also among classes of proteins that mediate functions vital to neuronal function, including the maintenance and regulation of excitability and neurotransmission. In the field of alcohol research, diversity has been a pressing and prevailing issue, as ethanol has been found to exert differential effects on a variety of neurotransmitter receptors that appear specific not only to brain regions and even subpopulations of neurons within a given brain region, but also to be dependent on the subtype of receptors involved (Deitrich et al., 1989; Harris, 1999). Overall, the combinatorial approach of patch-clamp recording and detection of candidate genes encoding receptor subunits has aided in extending ethanol-receptor interactions gleaned from studies employing recombinant receptors to the native receptor. Such efforts yielded important revelations, some of which uncovered differences between recombinant and native receptors (Sapp and Yeh, 1998), while others underscored the need to be mindful of the heterogeneous expression of receptor subtypes within a given neuron in interpreting electrophysiological data (Criswell et al., 1997; Sapp and Yeh, 2000).

Profiling gene expression in single cells is conceptually straightforward yet technically demanding, and most investigators would at least agree that it is tedious and labor-intensive. Why, then, go to the trouble? Clearly, Western blot analysis, immunohistochemistry and *in situ* hybridization have contributed and continue to contribute significantly to our understanding of the distribution of specific genes and proteins in the nervous system. These techniques, however, harbor notable inherent limitations. For example, cellular resolution is necessarily compromised in assessing protein levels by Western blot analysis. Immunohistochemistry and *in situ* hybridization offer cellular resolution in revealing the localization of proteins and messages, respectively, but there is a limit to the extent to which antibodies or molecular probes can be multiplexed to colocalize proteins and mRNA transcripts. Furthermore, in light of the heterogeneity mentioned above, many proteins and transcripts may elude analysis if they are expressed at levels of abundance that are below the threshold of detection in a given type neuron using any one of the three techniques. Indeed, circumventing these potential drawbacks is what motivated the development of the approach to combine patch-clamp recording and detection of gene expression in the same single cell (Eberwine et al., 1992). This was made possible by virtue of the following attributes of the technique:

1. Functional data derived from live individual neurons provide a first-order approximation and reference to guide the selection of a set of candidate genes for subsequent expression profiling.
2. Following electrophysiological recording, the same neuron is retrieved and the mRNA harvested from it is processed for amplification using aRNA- or RT-PCR-based protocols.
3. Both aRNA- and RT-PCR-based procedures (see Appendices A, B and C) produce sufficient amounts of amplified material representing the cell's original mRNA species to generate simultaneously a profile of expression of multiple genes.

In the following sections, we describe and discuss some of the prototypic protocols used in our laboratories for patch-clamp recording and subsequent expression profiling of candidate genes in the same single neuron, including cautionary considerations for the recording and harvesting of neurons, as well as for gene expression profiling. We conclude by looking into the future, and propose a view of how single-cell gene profiling can be enhanced by incorporating some of the technical newcomers that can be applied to the field of alcohol research.

4.2 PATCH-CLAMP RECORDING AND HARVESTING OF SINGLE NEURONS

Figure 4.1A illustrates a set of electrophysiological data representing whole-cell current responses to GABA before, during, and after exposure to carbachol obtained from a pyramidal neuron in a hippocampal slice. Following recording, the neuron was harvested. The series of photomicrographs in Figure 4.1B illustrate the typical sequential observations during the harvesting procedure.

A major tenet contributing to successful patch-clamp recording with subsequent gene profiling in the same single neuron is "RNase-free." This applies to anything that comes in direct contact with the cell or its contents. The RNA content of a single cell is estimated to be in the order of fractions of a picogram, which is readily destroyed by seemingly ubiquitous enzymes that degrade RNA. Special precautionary measures to achieve RNase-free conditions should thus be taken, such as autoclaving and the wearing of gloves. Most investigators have combined gene profiling with whole-cell patch-clamp recording (e.g., Figure 4.5), although other patch-clamp recording configurations appear to be just as feasible.

4.2.1 CONSIDERATIONS IN PATCH-CLAMP RECORDING

It is by no means an exaggeration to state that the steps involved in patch-clamp recording dictate the quality of the outcome of gene profiling. The amount of

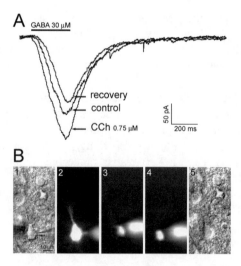

FIGURE 4.1 Demonstration of whole-cell patch-clamp recording and subsequent harvesting of the recorded cell in a hippocampal slice preparation. (A) A pyramidal cell was identified and whole-cell current responses to exogenously applied GABA were monitored before (control), during and after (recovery) application of the muscarinic receptor agonist, carbachol (CCh 0.75 μM). (B) After recording, the cell was harvested by applying negative pressure to the pipette and by visual inspection (B2). Lucifer yellow was included in the recording solution to aid in the visualization of the process. The recorded pyramidal neuron can be seen to enter the pipette progressively (B3 and B4). The vacated spot in the slice after the pyramidal neuron was harvested is seen in B5.

RNA harvested from a single cell is so small that its quality cannot be assessed until the latter has been subjected to amplification. When negative data are obtained, it is often difficult for the experimenter to backtrack and pinpoint the problem. However, it would not be surprising if unfavorable recording conditions were to blame.

4.2.1.1 Pipette Fabrication

As a routine, we heat the glass capillary tubing to be used for fabricating patch-clamp recording pipettes directly over fire using a Bunsen burner. Care is taken to prevent the tubing from bending due to overheating. While some laboratories autoclave the tubing, we prefer heating for two reasons. First, it is faster and more convenient. Second, autoclaving has a tendency to leave residual condensation or other types of accumulations on the inner wall of the capillary tubing, increasing chances of RNase contamination. The pieces of tubing are heated, allowed to air cool and used immediately for pipette fabrication. Forceps are used to mount the capillary tubing onto the pipette puller and to remove the pipettes for storage in a dust-free holder. The type of glass is not critical, and its use is determined based on conventional wisdom associated with patch-clamp recording (Penner, 1995).

4.2.1.2 Solutions

We use sterile disposable plastic ware and, whenever feasible, autoclave all glassware and solutions. Solutions that are not amenable to autoclaving should be filtered through disposable 0.2 μm membrane filters. For example, in most cases, our recording solution contains ATP and GTP, which are included to prevent run-down of currents but are readily inactivated by autoclaving. In this instance, the stock solutions containing ATP and GTP are prepared, filtered, stored frozen in aliquots and added to the recording solution just prior to the experiment. An example of a recording solution that we use routinely to record glutamate-induced whole-cell current is (in mM): KCl 20; K-gluconate 120; EGTA 0.5; HEPES 10; NaCl, 4; MgATP 2.5; NaGTP 0.25; the sodium channel blocker QX 222 ((2-[(2,6-dimethylphemyl)amino]-N,N,N-trimethyl-2-0x0ethanaminium chloride) 5. A standard bath solution that we use to record acutely dissociated or cultured neurons is (in mM): NaCl 137; KCl 5.4; CaCl$_2$ 1.8; MgCl$_2$ 1; HEPES 5 (pH 7.4). When working with thin slices, our artificial cerebral spinal fluid is composed of (in mM): NaCl 125; KCl 2.5; MgCl$_2$ 1; NaHPO$_4$ 1.25; CaCl$_2$ 2; NaHCO$_3$ 25; dextrose 25 (pH 7.4).

The bath and recording solutions, as well as all drug solutions, are prepared in autoclaved ultrapure water. We avoid the use of DEPC-treated water; it is incompatible with RT-PCR-based amplification.

4.2.1.3 Patch-Clamp Recording

The recording pipettes are back-filled with approximately 3 μl of recording solution. To maintain RNase-free conditions, the recording solution is loaded into a sterile disposable plastic syringe that has been heated over fire and pulled to produce a thin tapered tip. The volume of the recording solution can be varied. Ours is set to 3 μl based on our protocol for subsequent aRNA amplification or RT-PCR (see following section). In theory, a minimal volume is preferred to avoid excessive dilution of the cellular content. Accordingly, the length of the silver wire connecting the recording solution with the headstage of the patch-clamp amplifier should be adjusted to accommodate this small volume. The silver wire is rechlorided between recordings. This is conveniently accomplished by dipping the silver wire into Clorox (available in any supermarket) for approximately 30 sec and then rinsing it with sterile distilled water and allowing it to air dry.

Although it can be expected that prolonged whole-cell patch-clamp recording could have adverse effects on the integrity of the RNA harvested and, thus, the outcome of subsequent expression profiling, we are aware of no study that has systematically examined the relationship between these parameters. In previous studies, we have recorded from neurons for up to 15 min and have obtained PCR-amplified products from the same neurons. Inclusion of RNase inhibitor in the recording solution might help, but our experience has been that this hinders patch-clamp recording, presumably due to the glycerol that is added to the RNase inhibitor solution obtained from commercial sources. Similarly, the quality of patch-clamp recording tends to be compromised by attempts to begin reverse transcription *in situ* during recording. Overall, the most critical consideration is the formation of

Recording

Harvest

Reverse Transcription

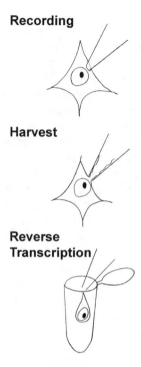

FIGURE 4.2 Schematic representation of the patch-clamp recording and harvesting procedure.

a tight, stable seal that is maintained throughout the recording session until the cell is harvested.

4.2.2 CONSIDERATIONS IN HARVESTING SINGLE NEURONS

Following electrophysiological recording, the neuron under study is "harvested" by applying negative pressure through the recording pipette (Figure 4.2). We routinely apply negative pressure by mouth, but others have successfully used syringes or more automated devices (Penner, 1995). We have harvested individual neurons following acute dissociation, those that have been maintained under long-term culture conditions, and those found in brain slices. Regardless of the preparation and the method used to apply negative pressure, on-line visual inspection is absolutely imperative to ensure that only the cell of interest is harvested. If, at any time during the harvesting procedure, a neighboring cell or surrounding debris is seen entering or adhering to the pipette, the sample should be precluded from further processing.

We typically harvest solitary acutely dissociated or cultured neurons in their entirety. During the aspiration process, it should be possible to observe entry of the cellular contents into the pipette used for harvesting. By applying additional negative pressure, the entire cell can be seen to enter the pipette. The application of negative pressure should be stopped immediately after the cell has entered the pipette to minimize unnecessary contamination. Frequently, the cell to be harvested is too large to be easily picked up (e.g., cerebellar Purkinje cells). In such instances, with slight constant positive pressure applied, the tip of the pipette used for harvesting can be gently broken to approximately 3–5 μm and then used to retrieve the cell. The ability to harvest the entire cell mitigates concern expressed by some investigators that certain mRNA species microcompartmentalized in extrasomatic regions may be misrepresented if only the soma of a given neuron were to be harvested. On the other hand, it is difficult to harvest an entire neuron in brain slice preparations. Monyer and Jonas (1995) have provided an excellent account of this process. Overall, as mentioned above, visual inspection throughout the harvesting process is critical, as individual neurons in slice preparations tend to be densely surrounded by neighboring neurons, glia and other components composing the neuropil.

Following harvesting, the contents of the pipette are transferred into an Eppendorf tube (Figure 4.2). The pipette is inserted deep into the tube, the tip is gently snapped at an angle in the inner wall, and the contents are expelled by blowing. It is important to prevent the solution from clinging onto the outside of the pipette as it is being expelled. The expelled contents should appear as a drop of solution on the inner wall of the Eppendorf tube. Brief centrifugation will bring the drop to the bottom of the tube, and reagents for reverse transcription are then added (see below). The process, from harvesting a cell to starting reverse transcription, should not take longer than 3 min.

Due consideration should be given to negative controls in preparation for expression profiling. We routinely include the following as negative controls of the harvesting process:

1. 3 μl of the recording solution.
2. 3 μl of the bath or perfusion solution.
3. For acutely dissociated or cultured cells, harvesting a cell that does not express the candidate gene of interest. For example, a cerebellar Purkinje cell serves a negative control for the expression of the $GABA_A$ receptor α6 subunit in cerebellar granule cells.
4. For neurons in brain slice preparations, inserting a pipette into the slice without aspirating and harvesting and making certain that no debris adheres to the pipette tip.

4.3 EXPRESSION PROFILING OF CANDIDATE GENES IN SINGLE CELLS

We describe three approaches to profile the expression of candidate genes from single cells, namely, antisense RNA (aRNA) amplification, reverse transcription-polymerase chain reaction (RT-PCR) and multiplex PCR. It should be noted that, although investigators tend to express individual preferences, these approaches to profile gene expression are entirely complementary. In addition, each has its advantages and shortcomings. For example, the aRNA amplification approach augments a cell's endogenous mRNA signal in a relatively linear fashion. Thus, the level of expression of a given mRNA species relative to other mRNAs is preserved, facilitating a direct semi-quantitative assessment of relative abundance of mRNA populations within a cell (Eberwine et al., 1992). Amplification based on RT-PCR, on the other hand, is logarithmic and this diminishes its quantitative power in estimating relative abundance because of difficulties in predicting inherent differences in the efficiency of amplification of different cDNAs. The primary disadvantage of the aRNA amplification approach is the potential cross-hybridization that, if unchecked, may occur among cDNAs with homologous sequences. Although this disadvantage can be avoided if anticipated, it is less of a concern with approaches based on RT-PCR. Another consideration is that aRNA amplification involves many steps and is labor-intensive, while RT-PCR is fast and technically less demanding.

A basic protocol is provided for each of the three approaches (see Appendices 4.1 to 4.3). The reader is advised that continual efforts to refine and improve these protocols continue, and that many variations exist.

4.3.1 SINGLE-CELL aRNA AMPLIFICATION (APPENDIX 4.1)

The major steps involved in aRNA amplification are schematized in Figure 4.3. Avian myeloblastosis virus (AMV) reverse transcriptase (Superscript), 4 dNTPs and an oligodeoxynucleotide containing an oligo-dT region and the T7 RNA polymerase promoter is added after harvesting the electrophysiologically recorded cell. The oligo-dT region primes the cellular mRNAs for cDNA synthesis and the T7 RNA polymerase promoter facilitates subsequent aRNA amplification.

The reverse transcription process produces a single-stranded cDNA that primes itself to form a hairpin loop at the 3' end, facilitating extension of the second strand. Treatment with S1 nuclease subsequently produces a noncovalently linked double-stranded cDNA. The ends of this double-stranded cDNA are then filled in to ensure that the T7 RNA polymerase promoter is complete. This double-stranded cDNA is then ready for the first round of amplification using T7 RNA polymerase. The RNA so synthesized from the cDNA template is antisense to the poly A$^+$RNA (hence the term "aRNA amplification") and can represent up to 2000-fold amplification of the original mRNA population (Eberwine et al., 1992). To prepare for a second round of amplification, the aRNA products will be converted to cDNA using random primers and AMV reverse transcriptase. These single-stranded cDNAs will then be primed with the same oligo-dT-T7 RNA polymerase promoter oligonucleotide used previously and another pool of double-stranded cDNAs is ready to be used as template for a second round of amplification (or for PCR-based amplification, see below).

In the second round of amplification, radiolabeled CTP (cytidine 5' triphosphate) will be incorporated and the resultant radiolabeled aRNA probe will be used to hybridize with slot blots containing different cloned cDNAs that have been linearized by cleavage using the appropriate restriction enzymes (reverse Northern blot). As a routine, irrelevant (e.g., GFAP, a marker of glia cells) and "housekeeping" cDNAs (e.g., β-actin, tubulin) are included in the slot blot as controls. The housekeeping cDNAs also serve as internal standards for semi-quantitative estimates of relative abundance of the hybridized signals. To this end, the aRNA/cDNA hybrid samples are analyzed directly from the reverse Northern blots and the hybridized, radiolabeled

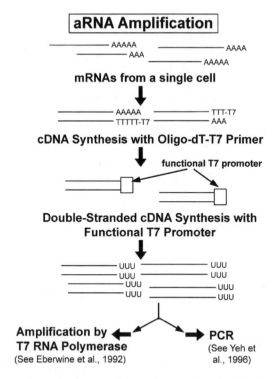

FIGURE 4.3 Schematic representation of the major steps involved in aRNA amplification.

aRNA is quantified by phosphor-imaging. The cpm value for each cDNA is corrected for by subtracting background level from control slots containing Bluescript or other plasmid vectors.

4.3.2 Single-Cell RT-PCR (Appendix 4.2)

Compared with aRNA amplification, single-cell RT-PCR is decidedly more straight-forward and less time-consuming. The compromise is in the ability to make statements about the relative abundance of mRNA species that are reflective of the cell's original levels.

After patch-clamp recording and harvesting the cell, each sample containing a single cell is first treated with RNase-free DNase prior to RT-PCR to eliminate genomic DNA, as it has a tendency to be amplified along with the cDNA during PCR. DNase treatment is carried out in the same reaction mixture as the RT reaction. The solution containing the harvested cell is then added to a reverse transcriptase master mixture without reverse transcriptase. The mixture contains RT buffer, random hexamer primers, dNTPs and RNase inhibitor. To this mixture is added RNase-free DNase and genomic DNA is removed by incubation at 37° C. The DNase is inactivated by heating at 95° C. The sample is then cooled to room temperature and reverse transcriptase is added; synthesis of single-stranded cDNA is carried out for 1 h at 42° C. Alternatively, the cDNA produced after the first round of aRNA amplification (discussed above) can be used as a template for PCR. "No-RT" controls are routinely run with water added in lieu of either the reverse transcriptase or the sample. Other samples are also processed by adding RNase as control. PCR (usually 35–40 cycles) is carried out with 50 pmol of each of primers, 1 or 2 µl of the reverse transcribed cDNA and Taq DNA polymerase. A 10 µl aliquot of the reaction product is then electrophoresed parallel to a molecular weight ladder through 2.5% NuSieve 3:1 agarose and visualized with ethidium bromide. For Southern blot analysis, oligonucleotide probes are 5'-end labeled with ^{32}P α-dCTP using T4 polynucleotide kinase and transferred onto nylon membrane. Unincorporated oligonucleotides are removed by centrifugation with a Bio-gel column. Prehybridization and hybridization are performed in a rotating oven and the blots are incubated in cylinders with 15 ml hybridization mix, containing 0.5 M phosphate buffer, 7% SDS, 10X Denhardt solution and 100 µg/ml salmon sperm DNA. The probes are added after 1 h of prehybridization and the blots hybridized overnight at 52° C.

4.3.3 Single-Cell Multiplex RT-PCR (Appendix 4.3)

The application of single cell multiplex RT-PCR to neurobiological research has been described some time ago (Lambolez et al., 1992). It consists of a reverse transcription step followed by two rounds of PCR amplification. Multiplex PCR allows the simultaneous amplification of multiple genes of interest (routinely 20–25) from a single cell. As schematically represented in Figure 4.4, during the first round of PCR amplification, all primers of interest are mixed in a same tube. In the second round, using the product from the first round PCR as template, specific primers amplifying each gene of interest are included in separate PCR tubes. In our view, this procedure is conceptually very complementary to the aRNA amplification approach. The choice between the two approaches depends on the number of candidates to be explored and the speed by which the "final product" is to be obtained. Most of all, it relies to a large extent on "experimental style."

The considerations mentioned above concerning patch-clamp recording, harvesting single neurons and negative controls apply to mPCR. It should be noted that the first critical consideration in mPCR is to design appropriate and compatible primer sets. As all primers are mixed together during the first round of PCR, it is critical to avoid a design strategy in which complementary oligonucleotide sequences are present simultaneously. The annealing temperature of all primers should be the same. Generally, we design specific "20mers" which contain 50% of Gs or Cs that share an annealing temperature of 60° C. It is desirable to design forward and reverse primers that yield amplicons less than a Kb in size. The second consideration that, to us, is most critical throughout this procedure, is to test and optimize the primer design. We perform this test using a first strand cDNA derived from the tissue of interest (e.g., the cerebellum) at a concentration of 50–100 ng. Using this tissue, mPCR is performed under the same conditions as is used for single cells (see below). If any one of the primer sets does not yield an amplicon, or if any of the resultant amplicons does not have the expected molecular weight, it is more efficient to redesign the primer set rather than spend time changing the annealing temperature or the Mg^{++} concentration. This is a

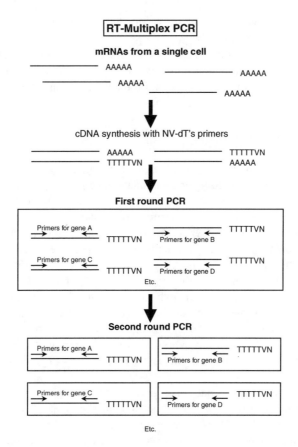

FIGURE 4.4 Schematic representation of the major steps involved in multiplex PCR.

critical step that should not be avoided even if the best of primer design software is employed.

After harvesting single neurons, the tip of the microelectrode is broken into a 500 µl PCR tube containing Sensiscript Reverse Transcriptase (Qiagen), RNase inhibitor, 4 dNTPs, a 5' $(T)_{24}$ VN 3' DNA primer, and the appropriate buffer in a total volume of 20µl. The first strand DNA can be obtained at 42° C for 2 h. We routinely leave the reverse transcription overnight at 37° C without noticeable DNA degradation. The first round of PCR is performed in a total volume of 100 µl in the same tube that contains the reverse transcribed cDNA. The PCR and primer mixtures are added using a 15-min hot start protocol. It is important to use a hot start protocol to insure that degradation of the reverse transcriptase enzyme is complete. An incomplete degradation of this enzyme tends to induce smears during the second round of PCR. A 2µl aliquot of this round of amplification is then used as template in the second round of PCR. In a total volume of 50 µl, Taq polymerase, 4 dNTPs, specific forward and reverse primers and the appropriate buffer are added on ice to the template. Keeping the samples on ice between the two rounds of PCR also avoids smears. The second round is performed using the same PCR conditions (see protocol). At the end, the amplicons are visualized by 1% agarose gel electrophoresis.

Once the optimal conditions have been worked out, multiplex PCR can be used quite routinely for profiling genes in single cells. At present, its main limitation is in its quantitative power. However, such protocols are being developed and can be expected to be available in the near future.

4.4   EXPERIMENTAL RESULTS

In this section, we provide an example of an experiment that employed the combination of patch-clamp recording and candidate gene profiling in a single neuron.

The GABA$_A$ receptor α6 subunit is unique in at least two respects. First, from a functional standpoint, studies in expression systems indicate that α6 subunit-containing recombinant GABA$_A$ receptors are insensitive to modulation by diazepam, a benzodiazepine that is normally a potent positive modulator of GABA$_A$ receptor function (Luddens et al., 1990; Gunnersen et al., 1996). Second, in the cerebellum, the α6 subunit is uniquely expressed in granule cells (Gutierrez et al., 1996). Studies employing *in situ* hybridization and Western blot analysis have demonstrated that, in the rodent, neither the α6 subunit transcript nor protein can be detected until well after the first postnatal week in cerebellar development (Zheng et al., 1993). Thus, the expression of the α6 subunit is developmentally regulated. Prior to its expression, the α6 subunit appears to predominate in immature granule cells.

In light of the migratory pattern of granule cells during cerebellar development, the late onset of expression of the α6 subunit implies that it should be absent in immature granule cells that are in the external granule cell layer and that it should emerge once the granule cells have migrated to reach their destinations in the internal granule cell layer. To test this idea, GABA$_A$ receptor-mediated responses were monitored in granule cells from thin slices of postnatal cerebellum. As illustrated in Figure 4.5A, premigratory granule cells could be readily identified in the external granule layer for patch-clamp recording. The cell under study was labeled by Lucifer yellow, which was included in the recording solution and readily diffused into the cell during patch-clamp recording in the whole-cell mode.

In monitoring whole-cell current, the goal was to monitor GABA-activated current responses, assess the response profile in light of pharmacological agents that could provide clues as to whether the GABA$_A$ receptor α6 subunit is present early in the maturational process of the granule cells. The continuous penwriter record (Figure 4.5b) was obtained from the immature granule cell situated in the external granule layer (Figure 4.5a). It can be readily seen that the GABA responses recorded from this immature granule cell were suppressed in amplitude during exposure to furosemide and potentiated during exposure to diazepam. Because furosemide selectively inhibits α6 subunit-containing GABA$_A$ receptors, these observations suggest that immature cerebellar granule cells in fact display α6-like pharmacology quite early in development. As well, because diazepam modulated the GABA responses of this cell, it would be expected that there is a component in the whole-cell GABA response that was sensitive to modulation by benzodiazepines.

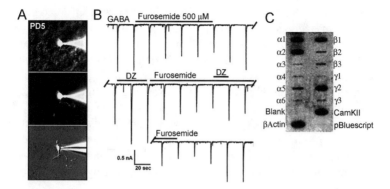

FIGURE 4.5 The cerebellar granule cell (A) was tested for sensitivity to modulation of its GABA response to furosemide and diazepam (DZ; panel B). Intracellular filling with Lucifer yellow during recording (top and middle photomicrographs in panel (A) illustrate an immature granule cell situated in the external granule cell layer of a slice derived from postnatal day 5 (PD5) rat cerebellum. The cell was subsequently lifted from the slice preparation and harvested (bottom photomicrograph in panel A). Panel C illustrates the reverse Northern blot obtained from the cell illustrating the expression of $GABA_A$ receptor subunit transcripts after aRNA amplification.

Following recording and pharmacological characterization of the GABA response, the granule cell was isolated from the slice and harvested (bottom panel, Figure 4.5A). Profiling of $GABA_A$ receptor subunits using aRNA amplification revealed the expression of a great number of subunit mRNAs. Indeed, *in situ* hybridization has revealed numerous $GABA_A$ receptor subunit transcripts in cerebellar granule cells (Laurie et al., 1992). Noteworthy is that the reverse Northern blot (Figure 4.5C) indicates the simultaneous presence of the $\alpha1$ and $\alpha6$ transcripts in this granule cell, reflecting the electrophysiological finding of sensitivity to modulation by diazepam and to inhibition by furosemide. Overall, these results point to the possibility that granule cells either express at least two $GABA_A$ receptor isoforms or that they can express during development an isoform that displays mixed pharmacological properties.

4.5 CONCLUSIONS AND FUTURE DIRECTIONS

4.5.1 WHERE ARE WE NOW?

The combination of patch-clamp recording and single-cell mRNA profiling melded techniques from two different disciplines to yield an approach that transformed from concept to practice the correlation of function with gene expression. The approach was born out of the need to ascribe neuronal function to the expression of and regulation by specific genes in identified neurons. Given the bewildering diversity of neuronal populations, genes and gene products, as well as their heterogeneous expression and distribution in the nervous system, the efforts devoted to developing strategies to achieve a functional and molecular correlation at the level of a single identified neuron was, in retrospect, worthwhile and valuable.

Not only was such a correlation possible, these strategies afforded the simultaneous analysis of many genes from a single cell, and allowed questions to be asked about the relationship between neuronal function and the "coordinate" expression of genes. However, despite its attributes, the approach may not be ideal in all cases. As a specific example, investigators who must rely on a "high-throughput" system may find the combination of patch-clamp recording and single-cell gene profiling falling short of this requirement. In addition, as already mentioned, the techniques can be labor-intensive. With patch-clamp recording being the common denominator, PCR-based gene profiling offers speed while aRNA amplification offers the prospect of assessing relative abundance of cellular messages. Ultimately, the value and adoption of any one of the approaches depends on the biological question at hand.

4.5.2 Where Do We Go Next?

Recent technological advances offer exciting opportunities to refine and extend the established protocols for amplification and profiling of genes in single cells. Particularly noteworthy among these are the developments in real-time quantitation of PCR-amplified products, cDNA microarrays and in profiling proteins.

4.5.2.1 Real-Time Quantitative PCR

Real-time quantitative PCR has considerable promise in accelerating the steps involved in single-cell gene profiling and, at the same time, facilitates quantitative analyses of the PCR-amplified products. These factors combine to enhance data throughput beyond functional analyses by electrophysiology. A limitation at present appears to be that the minuscule starting material derived from a single cell is below the level that can be processed and detected by currently available real-time quantitative PCR systems. A major challenge in incorporating this approach to single-cell PCR, therefore, lies in the ability to develop strategies to amplify the cell's original mRNA species while preserving its quality prior to processing by real-time quantitative PCR.

4.5.2.2 cDNA Microarrays

In essence, the gene profiling approaches described in the previous sections are a form of "macroarrays" (see Figure 4.5C) that can be regarded as having foreshadowed the cDNA microarrays. Indeed, the cDNA microarray technology has been largely responsible for leading the state of molecular biological research into a full-blown genomic era. The power of this technology is truly impressive, but its current potential drawbacks lie in the cost and quality control associated with setting up and running microarray experiments (Knight, 2001), as well as in the availability of expert bioinformatics that is optimized toward accurate interpretation of the biological significance of gene expression profiles. Nonetheless, it would not be surprising if the cDNA microarray technology will be applied successfully to, and gain popularity for, profiling genes expressed in single cells. As with real-time quantitative PCR, an important consideration will be to obtain sufficient amounts of high-quality starting material.

4.5.2.3 Profiling Proteins

There may be little correlation between mRNA levels and protein levels, and changes in gene expression may not necessarily reflect parallel changes in the post-translational levels of the corresponding gene products. In this light, a critical companion to genomic analysis is the burgeoning technology of proteomics. Indeed, we view the generation of technologies to assess protein expression, function and interaction as the next major conquest in the "postgenomic" era. As protein arraying gains sophistication and becomes routine, parallel analysis of hundreds or even thousands of proteins can be achieved.

ACKNOWLEDGMENTS

This work was supported by the National Institute of Neurological Disorders and Stroke (grant NS24830) and National Institute on Alcohol Abuse and Alcoholism (grant AA03510). The authors thank the collaborative efforts of Drs. Jim Eberwine, and Corinne Spencer. The authors also acknowledge the technical assistance of Pamela Yeh.

REFERENCES

Criswell HE et al. (1997) Action of zolpidem on responses to GABA in relation to mRNAs for GABA(A) receptor alpha subunits within single cells: evidence for multiple functional GABA(A) isoreceptors on individual neurons. *Neuropharmacology* 36:1641-52.

Deitrich RA et al. (1989) Mechanism of action of ethanol: initial central nervous system actions. *Pharmacol Rev* 41:489-537.

Eberwine JH et al. (1992) Analysis of gene expression in single live neurons. *Proc Natl Acad Sci USA* 89:3010-4.

Gunnersen D, Kaufman CM and Skolnick P (1996) Pharmacological properties of recombinant "diazepam-insensitive" GABA$_A$ receptors. *Neuropharmacology* 35:1307-14.

Gutierrez A, Khan ZU and De Blas AL (1996) Immunocytochemical localization of the alpha 6 subunit of the gamma-aminobutyric acid A receptor in the rat nervous system. *J Comp Neuro* 365:504-10.

Harris RA (1999) Ethanol actions on multiple ion channels: which are important? *Alcohol Clin Exp Res* 23:1563-70.

Jonas, P, Spruston N (1994) Mechanisms shaping glutamate-mediated excitatory postsynaptic currents in the CNS. *Curr Opin Neurobiol* 4:366-72.

Knight J (2001) When the chips are down. *Nature* 410:860-861.

Lambolez B et al. (1992) AMPA receptor subunits expressed by single Purkinje cells. *Neuron* 9:247-58.

Laurie DJ, Seeburg PH and Wisden W (1992) The distribution of 13 GABA$_A$ receptor subunit mRNAs in the rat brain. II. Olfactory bulb and cerebellum. *J Neurosci* 12:1063-76.

Luddens H et al. (1990) Cerebellar GABA$_A$ receptor selective for a behavioural alcohol antagonist [see comments]. *Nature* 346:648-51.

Mackler SA and Eberwine JH (1993) Diversity of glutamate receptor subunit mRNA expression within live hippocampal CA1 neurons. *Mol Pharm* 44:308-15.

Monyer H and Jonas P (1995) Polymerase chain reaction analysis of ion channel expression in single neurons of brain slices. In: *Single-Channel Recording* (2nd ed., Sakmann B and Neher E, Eds.) pp 357-374 New York:Plenum Press.

Penner R (1995) A practical guide to patch-clamping. In: *Single-Channel Recording* (2nd ed., Sakmann B and Neher E, Eds) pp 3-30 New York:Plenum Press.

Sapp DW and Yeh HH (1998) Ethanol-GABA(A) Receptor interactions - a comparison between cell lines and cerebellar Purkinje cells. *J. Pharm. Exp Therap* 284:768-776.

Sapp DW and Yeh, HH (2000) Heterogeneity of GABA(A) receptor-mediated responses in the human IMR- 32 neuroblastoma cell line. *J Neurosci Res* 60:504-10.

Surmeier DJ et al. (1992) Dopamine receptor subtypes colocalize in rat striatonigral neurons. *Proc Nat Acad Sci USA* 89:10178-82.

Zheng T et al. (1993) Developmental expression of the alpha 6 $GABA_A$ receptor subunit mRNA occurs only after cerebellar granule cell migration. *Brain Res Dev Brain Res* 75:91-103.

Appendix 4.1
Protocol for Single-Cell mRNA Amplification/Reverse Northern Analysis

A-1 FIRST ROUND AMPLIFICATION

A. FIRST STRAND CDNA SYNTHESIS

1. Combine:
   - Cell contents plus recording solution   3 µl
   - 1 × RT buffer   17.5 µl
   - dGTP (10 mM)   1 µl
   - dATP (10 mM)   1 µl
   - dTTP (10 mM)   1 µl
   - dCTP (10 mM)   1 µl
   - Oligo-dT-T7 (100 ng/µl)   1 µl
   - DTT (100 mM)   3 µl
   - RNasin   0.5 µl
   - AMV-RT (25U/µl)   1 µl
   - Total volume   30 µl
2. Mix gently; incubate at 42° C for 90 min
3. Phenol/chloroform extraction cDNA:
   - DD-WATER   105 µl
   - 3 M Na acetate   15 µl
   - Chloroform   75 µl
   - Buffer-saturated phenol   75 µl
   - Final volume   300 µl
4. Vortex for 10 sec. Centrifuge for 3 min. Carefully remove ~145 µl aqueous (top) phase to a new tube.
5. Precipitate with ethanol. Add 300 µl 100% ice-cold ethanol, 1 µl tRNA (5 µg).
6. Leave on dry ice for least 20–30 min. Centrifuge at 18,000 rpm at 4° C for 30 min.

B.   SECOND STRAND cDNA SYNTHESIS

7. Dry and resuspend pellet in 20 µl DD-WATER.
8. Heat at 90–95°C for 3 min to denature RNA:DNA hybrid.
9. Cool quickly on ice; centrifuge briefly to bring down condensation.
10. Add to sample: 20 µl)
    - 10 × 2nd strand buffer   5 µl
    - 100mMDTT   2 µl
    - 4 dNTPs (2.5mM each)   5 µl
    - Random hexamer (100ng/ µl)   1 µl
    - T4 DNA polymerase (5U/ µl)   0.2 µl   (1U final)
    - Klenow (5U/ µl)   0.5 µl l (2U final)
    - DD-water   16.3 µl
    - Final volume   50 µl
11. Mix gently; incubate at 14°C for a minimum of 5 h, or overnight.

C.   BLUNT-END TREATMENT

12. S1 nuclease cut or blunt-end repair (for reverse Northern): Dilute 1 µl of S1 nuclease in 400 µl of 10 × S1 buffer.
13. Add to 2nd strand:
    - DD-water   400 µl
    - 10 × S1 buffer   50 µl
    - Diluted S1 nuclease (1U)   1 µl
    - tRNA   1 µl
14. Incubate at 37°C for 5 min.
15. Extract with phenol/chloroform (0.5 ×):
    - Add to blunt-end treated 2nd strand sample   452 µl
    - Chloroform   226 µl
    - Buffer-saturated phenol   226 µl
16. Vortex 10 sec centrifuge for 3min extract top aqueous layer into a new tube.
17. Precipitate with 1 ml 100% ethanol on dry ice for 20–30 min.
18. Centrifuge 18,000 rpm at 4°C for 30 min, dry pellet.
19. Resuspend pellet in 20 µl of DD-WATER. Add:
    - DD-WATER   21 µl
    - 10 × KFI buffer   3 µl
    - 100mM DTT*   3 µl
    - 4 dNTPs   3 µl
    - T4 DNA polymerase (5U/ µl)   0.2 µl
    - Final volume   50 µl
20. Incubate at 37°C for 15–30 minutes.
21. Phenol-chloroform extract ds-cDNA. Add:
    - DD-WATER   85 µl

---

* Replace with DD-WATER if already included in the 10 × KFI buffer.

- 3M sodium acetate   15 µl
- Chloroform   75 µl
- Buffer-saturated phenol   75 µl
- Final volume   300 µl

22. Vortex for 3 sec. Centrifuge for 3 min. Carefully transfer ~145 µl top aqueous layer to a new tube.
23. Add 300 µl 100% ice-cold ethanol on dry ice for 20–30 min. Centrifuge at 18,000 rpm at 4°C for 30 min.
24. Re-suspend pellet in 20 µl of DD-WATER. Drop dialyze 10–20 µl sample against 50ml DD-WATER for at least 4 h.

D. FIRST ROUND aRNA AMPLIFICATION (COLD REACTION)

25. Use 1/10 of total sample and add:
    - Dialyzed sample   2.0 µl
    - 10 × TSC buffer (RNA amplification buffer)   2.5 µl
    - 20mM spermidine*   2.0 µl
    - (with UTP)(2.5mM each)   2.5 µl
    - 100mM DTT   1 µl
    - RNasin   0.5 µl
    - T7 RNA polymerase (1000 U/ µl)   1 µl (1000 U final)
    - DD-WATER   8.5 µl
    - Final volume   20 µl
    - Replace with DD-WATER if already included in the 10 × RNA amplification buffer.
26. Mix gently; incubate at 37°C for 4 h.
27. Phenol/chloroform extract aRNA. Add to tube:
    - DD-WATER   95 µl
    - 3M ammonium acetate*   15 µl
    - Chloroform   75 µl
    - Buffer-saturated phenol   75 µl
    - Final volume   300 µl
28. Vortex for 10 sec. Centrifuge for 3 min. Carefully remove 145 µl aqueous (top) layer to a new tube.
29. Add:
    - tRNA** (1 µg/ µl)   1 µl
    - 100% ice-cold ethanol   450 µl
30. Precipitate on dry ice for 20–30 min. Centrifuge 18,000 rpm at 4°C for 30 min. Resuspend the pellet in 19.5 µl of DD-WATER.

---

* Ammonium acetate is more efficient than sodium acetate at removing free NTPs. Adding salt at this step (instead of subsequent ethanol step) improves separation of the aqueous and organic phases.
** Use of a carrier is highly recommended to aid precipitation of minuscule amounts of RNA (e.g., from single cells.)

A-2 CONVERSION OF aRNA TO DOUBLE-STRANDED cDNA

A. First Strand cDNA Synthesis
   31. Denature 19.5 μl aRNA sample at 90–95°C for 3 min.
   32. Quick cool, quick spin to bring down condensation.
   33. Add to denatured aRNA sample: (19.5 μl)
      • 5 × first strand buffer 3.0 μl
      • 4 dNTPs (2.5mM each) 3.0 μl
      • Random hexamer (100ng/ μl) 1.0 μl
      • 100mM DTT 2.0 μl
      • RNasin 0.5 μl
      • Superscript RT 1.0 μl
      • Final volume 30 μl
   34. Mix gently; incubate at 42°C for 90 min.
   35. Phenol-chloroform extract ss-cDNA. Add:
      • DD-WATER 105 μl
      • 3M sodium acetate 15 μl
      • Chloroform 75 μl
      • Buffer-saturated phenol 75 μl
      • Final volume: 300 μl
   36. Vortex for 10 sec. Centrifuge for 3 min. Carefully transfer 145 μl aqueous (top) layer to a new tube.
   37. Add 300 μl 100% ice-cold ethanol to precipitate on dry ice for 20–30 min.
   38. Centrifuge at 18,000 rpm at 4°C for 30 min. Re-suspend the pellet in 12.3 μl of DD-WATER.

B. Second strand cDNA synthesis
   39. Heat at 90–95°C for 3 min to denature aRNA:DNA hybrid.
   41. Quick cool, quick spin to bring down condensation.
   42. Add to sample:
      • 10 × KFI buffer 2 μl 1
      • 100 mM DTT* 2 μl
      • 4 dNTPs (2.5 mM each) 2 μl
      • T7-oligo (dT) 24 primer (100 ng/ μl) 1 μl
      • T4 DNA polymerase (5U/ μl) 0.2 μl (1U final)
      • Klenow (5U/ μl) 0.5 μl (2 U final)
      • Final volume: 20 μl
   43. Mix gently; incubate at 14°C overnight.
   44. Centrifuge briefly to bring down condensation.

C. Blunt-End Reaction
   45. S1 nuclease cut or blunt-end repair (if go into reverse Northern):

---

* Replace with DD-WATER if already included in 10 × KFI buffer.

Dilute 1 μl of S1 nuclease in 400 μl of 10 × S1 buffer.

46. Add to 2nd strand:
    - DD-WATER   400 μl
    - 10 × S1 buffer   50 μl
    - Diluted S1 nuclease (1U)   1 μl
    - tRNA   1 μl
47. Incubate at 37°C for 5 min
48. Extract with phenol/chloroform (0.5 ×):
    - Add to blunt-end treated 2nd strand sample 452 μl
    - Chloroform   226 μl
    - Buffer-saturated phenol   226 μl
49. Vortex 10 seconds. Centrifuge for 3 min. Extract top aqueous layer into a new tube.
50. Precipitate with 1ml 100% ethanol on dry ice for 20–30 min.
51. Centrifuge 18,000 rpm at 4°C for 30 min, dry pellet.
52. Resuspend pellet in 20 μl of DD-WATER.

    If planning to do reverse Northern, blunt-end to eliminate possible "wrap-around" RNA synthesis, which could reduce (via RNA-RNA self-hybridization) the amount of aRNA probe available for hybridization to cDNA on the blot.

53. Add to 2nd strand reaction: (20 μl)
    - DD-WATER   21 μl
    - 10 × KFI buffer   3 μl
    - 100mM DTT*   3 μl
    - 4 dNTPs (2.5mM each)   3 μl
    - T4 DNA polymerase (5U/μl)   0.2 μl
    - Final volume   50 μl
54. Incubate at 37° C for 15–30 min.
55. Phenol-chloroform extract ds-cDNA. Add:
    - DD-WATER   85 μl
    - 3M sodium acetate   15 μl
    - Chloroform   75 μl
    - Buffer-saturated phenol   75 μl
    - Final volume: 300 μl
56. Vortex for 10 seconds. Centrifuge for 3min. Carefully transfer 145 μl (top) aqueous phase to a new tube.
57. Add 300 μl 1 100% ice-cold ethanol to precipitate on dry ice for 20–30 min.
58. Centrifuge at 18,000 rpm at 4° C for 30 min. Resuspend the pellet in 20 μl of DD-WATER. At this point, the sample can be used for PCR analysis in a 1:100 final dilution.

---

* Replace with DD-WATER if already included in the 10 × KFI buffer.

A-3 SECOND ROUND ᴀRNA AMPLIFICATION (HOT REACTION)

51. Drop-dialyze 10–20 µl sample against 50ml DD-WATER for at least 4 h.
52. Combine:
    - 1/10 dialyzed sample plus DD-WATER   7.7 µl
    - 10 × RNA amplification buffer   2.0 µl
    - 20mM spermidine*   2.0 µl
    - 100mM DTT   1.0 µl
    - 3 rNTPs (ATP,GTP,UTP) (2.5mM each)   2.0 µl
    - 100µM CTP   0.8 µl
    - RNasin   0.5 µl
    - alpha-[^{32}P]-CTP(3000 Ci/mmol;1 mci/100 µl) 3.0
        (4mM final = 80pmol)
    - Remove 0.5 µl and spot onto 1 mm Whitman paper (= TCA before)
    - Add T7 RNA polymerase (1000U/µl)   1.0 µl
    - Final volume   20 µl
53. Mix gently; incubate at 37° C for 4-8 h. (6 h recommended.) Prehybridize blots (go to step 57).
54. Centrifuge briefly to bring down condensation.
55. Remove 0.5 µl and spot onto 1 mm Whatman paper (= TCA after)
56. TCA precipitation
(a) Wash TCA-before and TCA-after samples in 10% TCA for 5 min on a rotating platform. Allow a sufficient volume of TCA for the samples to flow freely. (b) Replace with fresh 10% TCA and wash for 5 min. Wash once more for 20 min. (c) Air-dry before measuring radioactivity levels, either by counting in a scintillation counter or by listening with a Geiger counter.

There should be a significant increase in the amount of radioactivity measured in the TCA-after samples relative to the TCA-before samples, signifying successful a RNA synthesis as measured by incorporation of ^{32}P-CTP. An audible difference using a Geiger counter should be detected.

A-4 REVERSE NORTHERN BLOTTING

A. Prehybridization
57. Prehybridization solution: final concentration
    - Ultrapure formamide   50ml   50%
    - 20 × SSC   20ml   4 ×
    - 50% dextran sulfate   20ml   10%
    - 50 × Denhardt's   10ml   5 ×

---

* Replace with DD-WATER if already included in the 10 × RNA amplification buffer.

- Salmon sperm DNA   1ml   100 μg/ml
- Total Volume   101ml

58. Place blot(s) into a 15-ml or 50-ml conical tube (cDNA side toward the lumen of the tube). Allow sufficient room to avoid overlapping membrane (unless using nylon mesh), while minimizing volume of prehybridization solution. Using a minimal volume will result in a higher concentration of probe and better hybridization.
59. Add prehybridization solution to blots. 4 ml per 15-ml tube or 8 ml per 50-ml tube is sufficient. Thoroughly wet the membranes and remove large bubbles between the membrane(s) and the side of the tube.
60. Prehybridize at least 3 h at 42°C in the hybridization oven. When using 50-ml conicals, it is recommended that the tubes be sealed with parafilm to avoid leakage.

B. Hybridization
61. Heat aRNA probe at 90–95°C for 5 min.
62. Cool quickly on ice; centrifuge briefly to bring down condensation.
63. Keep all samples on ice to minimize renaturation.
64. Add the probe to prehybridizatiion solution in the tube containing the blot(s). Do NOT let the probe come in direct contact with the blot.
65. Recap the tube (and reparafilm); mix well before returning to 42°C oven.
66. Hybridize for at least 16 h when using dextran sulfate (otherwise 2 d)

C. Washing
67. After hybridization, remove blots directly into a large plastic container with 500–800ml of $2 \times$ SSC, 0.1% SDS. Place on a rotation platform to wash for 30 min at room temperature.
68. Remove wash. Do a second wash in $2 \times$ SSC, 0.1% SDS for 30 min.
69. Remove second wash. Add $0.2 \times$ SSC, 0.1% SDS and wash for at least 1 h.

D. Autoradiography
70. Remove blots from third wash solution and place directly into plastic sealable bags. Keep blots moist in case more washing is required, or if you desire to strip blots and reuse.
71. Seal plastic bag. Expose blots to x-ray film or phosphorimaging screen.

Notes:

A. Amplification Buffers (for 50 ml in DEPC-H_2O):
10 × RT
- 500 mM Tris-base, pH 8.3   3.03 g (pH with HCl)
- 1.2 M KCl   4.47 g
- 100 mM $MgCl_2$ 1.02 g
- Mix. Filter through 0.2μm filter. Store in 1 ml aliquots at −20°C.

10 × 2nd Strand
- 1 M Tris-base, pH 7.4   6.06 g (pH with HCl)
- 200 mM KCl   0.746 g
- 100 mM $MgCl_2$ 1.02 g
- 400 mM $(NH_4)_2SO_4$ 2.64 g
- Mix. Filter through 0.2μm filter. Store in 1 ml aliquots at –20°C.
- 50 mM DTT   add later*

10 × RNA Amplification
- 400 mM Tris-base, pH 7.5   2.42 g (pH with HCl)
- 70 mM $MgCl_2$ 0.712 g
- 100 mM NaCl   0.292 g
- Mix. Filter through 0.2 μm filter. Store in 1 ml aliquots at –20°C.
- 20mM spermidine   add later*

10 x KFI
- 200 mM Tris-base, pH 7.5   1.21 g (pH with HCl)
- 100 mM $MgCl_2$ 1.02 g
- 50 mM NaCl   0.146 g
- Mix. Filter through 0.2 μm filter. Store in 1-ml aliquots at –20°C.
- 50 mM DTT   add later*

B. Preparation of cDNA Blot

1. Linearize cDNAs of interest using appropriate restriction enzyme. Check for complete digestion before proceeding. There is no need to ethanol-precipitate cDNAs. Include linearized plasmid vector as background control.

2. DEPC-treated the top slotted portion of the blotting apparatus (Milliblot-S System).

3. Cut three pieces of 3MM Whatman paper and one piece of nitrocellulose or nylon membrane to fit within rubber gaskets of apparatus.

4. Using clean gloves or DEPC-treated forceps, wet Whatman papers and membrane in 10 × SSC.

5. Assemble apparatus from the bottom. Place three sheets Whatman paper on top of middle slotted piece, then the membrane. Make sure paper and membrane stay within gaskets to ensure tight seal. Roll out any air bubbles that may have formed between layers before placing top slotted piece onto apparatus and completing assembly. Add 10 × SSC to wells to moisten. Set aside.

If using vacuum application, check at this time to ensure that suction is gentle enough. (i.e., at least 5 min for 100 μl volume) before loading samples.

6. Prepare 0.5 μg of linearized cDNA in 100μl 10 x SSC per well.

7. (a) Heat denature samples at 85–90°C for 5 min. (b) Cool quickly on ice. (c) Centrifuge briefly to bring down condensation. (d) Keep samples on ice while loading.

8. Remove excess 10 × SSC from wells of blotting apparatus (Shake vigorously into sink). Add 100μl sample to each well.

It is a good idea to carefully write out the layout of cDNAs before loading and note any deviations immediately after loading.

9. Apply a gentle vacuum to draw samples through slots. This should take at least 5 min or you risk pulling samples through membrane. Alternatively, use gravity to draw samples through (this will take several h).

10. Once samples have been drawn through and wells are completely dry, disassemble apparatus. Take note of orientation of blot (perhaps cut a corner) before removing membrane and placing on Whatman paper to dry.

11. UV-crosslink cDNA to membrane using Stratalinker.

Appendix 4.2
Protocol for Single-Cell
RT-PCR

B-1 RT REACTION

Materials needed:
- Enzyme buffer 5 × First Strand Buffer (GibcoBRL)
- Superscript Enzyme Reverse Transcriptase
- dNTPs (Neucleotide phosphate)
- Random Hexamers Primer
- RNase Inhibitor
- 0.1 M DTT (Dithiothreitol)

Procedure:

	Volume (Final)/tube	
RNA 500ng	10 μl	
5 × First Strand Buffer (10 × for AMV)	6 μl	1 ×
Superscript RT 1000U or AMV	1 μl	1000 U
DNTPs 2.5 mM	6 μl	0.5 mM
Random Hexamer Primer 300 ng or T7	3 μl	30 ng
RNase inhibitor	0.5 μl	20 U
DTT 100 mM	3 μl	10 mM
Hybridization solution	0.5 μl	
Total volume	30 μl	

Incubate solution at 42° C for 90 min.
Store at −20° C if cannot continue on to PCR amplification.

B-2 PCR AMPLIFICATION

Materials needed:
- $MgCl_2$
- Thermal Buffer
- DNTPs
- Taq enzyme
- Specific Primers

Procedure:
1. Add to cDNA the following:

	Volume	Final
Thermal buffer 10 ×	5 μl	1 ×
MgCl2 50 mM	5 μl	2.5 mM

109

- dNTPs 2.5 mM   5 µl   0.25 mM
- Taq enzyme   0.25 µl
- Primers   1 + 1 µl
- cDNA   1 µl
- Hybridization solution   31.75 µl
- Total volume   50 µl
- For PCR control, add 1 µl of hybridization solution.

2. Insert PCR tubes into PCR machine and run appropriate program.
3. Agarose gel analysis of PCR product.
   - 1.5% agarose gel in 1 × TAE running buffer

Appendix 4.3
Protocol for Single-Cell
Multiplex RT-PCR

C-1 RT REACTION:

Materials needed:
- 10 × Buffer RT (Qiagen)
- Sensiscript Reverse Transcriptase
- dNTP Mix
- Oligo-dT primer (we recommend 5' $(T)_{24}$ VN 3')
- RNasin
- RNase free water

Procedure:	Volume	(Final)/tube
• 10 × Buffer RT	2 µl	1×
• Sensiscript Reverse Transcriptase	1 µl	
• dNTP Mix (5 mM each dNTP)	2 µl	0.5 mM each
• Oligo-dT primer	1 µl	1 µM
• RNasin	1µl	10 U
• RNase free water	10 µl	
• Template RNA	3 µl	
• Total Volume	20 µl	

- Carry out reverse transcription by incubating for 120 min at 42° C or overnight at 37°C.
- First strand cDNA can be stored at –20° C.

C-2 FIRST ROUND OF PCR AMPLIFICATION:

Materials needed:
- 10 × PCR Buffer containing 15 mM $MgCl_2$
- Forward and reverse primers
- HotStarTaq DNA polymerase (Qiagen)
- Distilled water
- Note: There is no need to add dNTPs at this step due to their presence in the RT mix.

Procedure:
1. Add on ice to the 20 µl first strand cDNA the following:

	Volume	(Final)
• 10 × PCR buffer	10 µl	1×
• HotStarTaq	0.5µl	2.5 U
• First strand cDNA	20 µl	
• Distilled water	49.5 µl	

- Total volume   80 µl
- For negative PCR control, add 20 µl of distilled water

2. Insert PCR tubes into the thermocycler and begin the hot start protocol. After 10 min at 95°C. Add:
    - Forward and reverse primers   10 µl   0.5 µM
    - Distilled water   10 µl
    - Total volume   20 µl
    - The total volume of this round of PCR is 100 µl
    - First round of PCR cycling program:
    - Initial activation step   15 min   95° C
    - 3-step cycling:
    - Denaturation   45 sec
    - Annealing   45 sec   60° C
    - Extension   45 sec   72° C
    - Number of cycles   25
    - Final extension   10 min   72° C
    - Pause   4° C

C-3 SECOND ROUND OF PCR AMPLIFICATION:

Materials needed:
- $10 \times$ PCR Buffer containing 15 mM $MgCl_2$
- dNTP Mix
- Specific forward primer
- Specific reverse primer
- Taq DNA polymerase
- Distilled water

Prepare on ice a mix for each gene of interest as following:

	Volume	Final
10 × PCR Buffer	5 µl	1 ×
dNTP Mix (10 mM of each)	1 µl	200 µM each
Specific forward primer	1 µl	0.5 µM
Specific reverse primer	1 µl	0.5 µM
Taq DNA polymerase	0.5 µl	2.5 U
Distilled water	39.5µl	

Add on ice:
- First round template   2µl
- The total volume of this round of PCR is 50 µl

Second round of PCR cycling program:
- Initial denaturation   2 min   94°C
- 3-step cycling:
- Denaturation   45 sec   94°C
- Annealing   45 sec   60°C
- Extension   45 sec   72°C
- Number of cycles   35
- Final extension   10 min   72° C
- Pause   4° C

5   Quantitative Analysis of Neurotransmitter Receptor Expression with Antibody-Based Techniques

C. Fernando Valenzuela and Daniel D. Savage

CONTENTS

5.1 INTRODUCTION

An important question in alcohol-related neuroscience research is whether ethanol exposure results in maladaptive changes in the expression of neurotransmitter receptors.
Alterations in neurotransmitter receptor number, receptor subtypes or receptor subunit
composition, appear to have important roles in the development of alcohol tolerance,
alcohol dependency and prenatal alcohol-associated changes in neural function. Thus,
alcohol researchers are often faced with the need to quantify neurotransmitter receptor
expression levels in cultured cells exposed to ethanol, neural tissue from ethanol-treated
subjects and also from genetically selected or engineered rodent lines. Quantification of
receptor expression in the central nervous system (CNS) is not a trivial task and it usually
requires the use of a combination of techniques to ensure the accuracy of results (for
instance, see Wu et al., 1995; Petrie et al., 2001). Over the past 20 years, a myriad of
laboratory techniques have been used for this purpose, including radioligand binding
assays, mRNA quantification methods and antibody-based techniques.

Early studies used radioligand binding to membrane homogenates to assess the
effects of ethanol on neurotransmitter receptor levels (Deitrich et al., 1989). Subsequent
studies have used receptor autoradiography to localize and quantify, at the histological
level, the effects of chronic ethanol exposure on neurotransmitter receptor expression in
mature and developing animals (Savage et al., 1991; Negro et al., 1995; Woods et al.,
1995; Rothberg et al., 1996; Cowen and Lawrence, 1999; Daubert et al., 1999). To date,
receptor autoradiography techniques are still widely used in alcohol-related neuroscience
research because of their relative technical ease and the ability to produce quantitatively
reliable and reproducible results. In general, the main steps involved in receptor autoradiography are incubation of tissue sections with the radioligand of choice under optimal
conditions, removal of unbound ligand by washing, and autoradiographic detection of
the bound ligand using highly sensitive films or phosphorimager screens (Chabot et al.,
1996; Charon et al., 1998). However, as knowledge of metabotropic receptor subtypes
and ionotropic receptor subunit heterogeneity increased over the past 10 years, the
development of subtype- or subunit-selective radioligands for more detailed analysis of
neurotransmitter receptors has not kept pace, thus limiting the utility of this approach
for studies of receptor subtypes and subunits.

In the absence of subunit- and subtype-selective radioligand markers for measuring receptor protein, investigators have relied on measures of messenger RNA
(mRNA) to study the effects of ethanol on receptor subtype or subunit expression.
This approach has provided some important insights on how ethanol exposure can
alter receptor subtype or subunit expression. Currently, the most widely used techniques to quantify mRNA levels are Northern blot analysis, ribonuclease protection
assay, and quantitative reverse transcriptase-coupled polymerase chain reaction (RT-

PCR) (Reue, 1998). Northern blot is the RNA counterpart of the Southern blot technique developed for DNA analysis (Southern, 1975) and it involves the electrophoretic separation of RNA species on the basis of size and their transfer onto a blotting membrane. RNA sequences of interest are detected by hybridization to a radiolabeled probe followed by autoradiography or phosphorimaging. Intensities of the bands of interest can be quantified by densitometry or scintillation counting of excised bands (Reue, 1998). Northern blots have been used to quantify $GABA_A$ receptor subunit mRNA levels in rats chronically exposed to ethanol (Keir and Morrow, 1994) and in ethanol-naive withdrawal-seizure-resistant vs. withdrawal-seizure-prone mouse brains (Montpied et al., 1991). However, it should be emphasized that this technique is generally regarded as semiquantitative because it is not usually possible to completely control RNA transfer and probe hybridization efficiencies (Reue, 1998). Moreover, this technique is not suitable to detect low abundance mRNA species. Low abundance mRNAs can be detected with either ribonuclease protection assays or RT-PCR.

Ribonuclease protection assays involve hybridization in solution of total RNA with a labeled probe followed by ribonuclease digestion of single stranded probe and RNA. Digestion-protected hybrids can then be analyzed by denaturing polyacrylamide gel electrophoresis. Titration reactions with unlabeled RNA sense transcripts allow absolute mRNA quantification in this type of assay (Reue, 1998). For examples on the use of ribonuclease protection assays to measure the effects of long-term ethanol exposure on neurotransmitter receptor expression, see the papers by Ticku and collaborators (Follesa and Ticku, 1995; Hu et al., 1996; Kalluri et al., 1998) and Wilce and collaborators (Hardy et al., 1999).

To quantify smaller amounts of mRNA, the most sensitive technique currently available is quantitative RT-PCR (see Chapter 5 for details on single-cell RT-PCR). This technique involves the production of a single-strand complementary DNA copy of the RNA of interest by using reverse transcriptase followed by amplification of this DNA by PCR. Absolute quantification of PCR products can be achieved by co-amplification of a series of dilutions of standard RNA (competitive RT-PCR) or by monitoring the kinetics of the PCR reaction in real time by using fluorescence-based detection devices (Freeman et al., 1999). RT-PCR techniques have been used to study the effects of chronic ethanol consumption on $GABA_A$ receptor subunit mRNA expression in rat cerebral cortex (Devaud et al., 1995)

It should be emphasized that none of the techniques described above provides histological information on mRNA expression levels. However, quantification of the effects of ethanol on mRNA levels in specific brain regions and neuronal populations can be accomplished by using *in situ* hybridization. This technique involves the hybridization of tagged DNA or RNA probes to the mRNA of interest in tissue sections (Chabot et al., 1996). For example, this technique has been used to quantify, in discrete regions of the rodent brain, the long-term effects of ethanol on $GABA_A$ (Mahmoudi et al., 1997; Petrie et al., 2001) and NMDA receptor subunit expression (Darstein et al., 2000), brain-derived neurotrophic factor mRNA expression (Tapia-Arancibia et al., 2001), calmodulin gene expression (Vizi et al., 2000) and metabotropic glutamate receptor expression (Simonyi et al., 1996). Moreover, this technique has been used to quantify differences in NMDA receptor subunit mRNA expression in ethanol-withdrawal seizure-prone and seizure-resistant mice (Mason et al., 2001) and also differences in opioid propeptide mRNA expression in the brains of ALKO alcohol and non-alcohol (AA/ANA) rats (Marinelli et al., 2000).

Although mRNA-based techniques can provide important information on the effects of ethanol on neurotransmitter receptor expression, it must be kept in mind that results obtained with these techniques must be interpreted carefully, because changes in mRNA expression are not always predictive of corresponding changes in receptor protein expression or function. Therefore, it is of paramount importance to also measure protein expression levels when performing this type of study. Fortunately, substantial progress in the generation and commercialization of many antibodies for receptor proteins, coupled with major technological improvements in computer-assisted microscopy, densitometry and image analysis techniques has made it increasingly possible to assess ethanol-induced effects on neurotransmitter receptor protein levels using immunoblotting and immunohistochemical techniques. For example, immunoblotting assays have been used to determine the effects of long-term ethanol exposure on the expression of $GABA_A$ receptor subunits (Mehta and Ticku, 1999), ionotropic glutamate receptor subunits (Trevisan et al., 1994; Follesa and Ticku, 1996; Kumari and Ticku, 1998; Chandler et al., 1999) and the effects of fetal ethanol exposure on ionotropic receptor subunits (Costa et al., 2000). Immunofluorescence techniques have been used to assess the effect of long-term ethanol exposure on glycine receptor expression in cultured spinal neurons (van Zundert et al., 2000). However, studies on the effects of ethanol on the expression levels of other neurotransmitter receptors remain to be performed.

In the next sections, we provide a brief description of methodologies that can be used to chronically expose culture cells or animals to ethanol. We then provide a general discussion on the generation and use of different types of antibodies and the application of immunoblotting and immunohistochemical techniques to quantify the effects of long-term ethanol exposure on neurotransmitter receptor protein levels.

5.2 LONG-TERM EXPOSURE TO ETHANOL OF CULTURED CELLS AND ANIMALS

Cultured cells can be useful to study the effects of long-term ethanol exposure on neurotransmitter receptor levels. For example, fibroblast-like cells stably expressing $GABA_A$ receptors subunits (Harris et al., 1998), clonal neural cells (Messing et al., 1986), neuroblastoma-glioma cells (Charness et al., 1986) and primary neuronal cultures from different CNS regions have all been used for this purpose (Iorio et al., 1992; Heaton et al., 1994; Hoffman et al., 1995; van Zundert et al., 2000). In principle, exposure of these cells to ethanol can be easily accomplished by adding ethanol to the culture media. However, caution must be exercised to prevent or compensate for evaporative loss of ethanol, which can be significant at 37°C (Eysseric et al., 1997). For example, this can be achieved by allowing the culture media to pre-equilibrate to the incubator atmosphere, adding the desired ethanol concentration, and then placing the culture dishes or plates in sealed plastic bags containing a beaker with water plus the same ethanol concentration that was added to the media (Harris et al., 1998). To further compensate for evaporative loss, 50% of the ethanol concentration added the first day of treatment can be added to the culture media every subsequent day (Iorio et al., 1992). For a discussion of alternative methods to chronically expose cultured cells to ethanol, see Eysseric et al., 1997.

The long-term consequences of ethanol exposure can be also studied in animals such as rats, mice, or guinea pigs. A number of methods can be used to chronically administer ethanol to these animals. The most commonly used methods to achieve appreciable blood alcohol levels are the liquid diet technique, the vapor chamber technique and the intragastric intubation technique. In the liquid diet technique, rats are given nothing to eat or drink but an ethanol-containing formula that is introduced progressively (Lieber et al., 1989). This diet can produce blood alcohol levels of 0.08–0.15 g/dl (Lieber et al., 1989; Allan et al., 1998) or even higher (0.2–0.3 g/dl; Trevisan et al., 1994). Controls are pair-fed with an isocaloric formula without ethanol. To determine if the liquid diet has an effect on its own, a control group receiving solid food ad libitum should also be included in the experimental design. The liquid diet technique has been used extensively to study the long-term effects of ethanol on neurotransmitter receptor expression in adult and fetal animals (for example, see Frye et al., 1981 and Costa et al., 2000).

An alternative method to deliver ethanol to animals is the vapor chamber technique (Rogers et al., 1979). In this method, animals are placed in a sealed vapor inhalation chamber and exposed to ethanol vapor in combination with air. Control rats are maintained in the same type of chambers but are exposed to air only. This method can produce blood alcohol levels as high as 0.2–0.3 g/dl (Gruol et al., 1998; Nelson et al., 1999), and it has been widely used to study the chronic effects of ethanol exposure on neurotransmitter receptors in developing and adult animals (Gruol et al., 1998; Jarvis and Becker, 1998; Nelson et al., 1999).

Ethanol can also be administered to animals by gastric intubation, but it must be kept in mind that blood alcohol levels cannot be sustained throughout the day with this method unless the animals are intubated multiple times (Lieber et al., 1989). Relatively short-term treatments (three daily intubations for 6 d) have been successfully used by Ticku and collaborators to study the effects of long-term ethanol exposure on NMDA receptor subunit expression in the rat brain (for example see Kalluri et al., 1998). It is noteworthy, however, that longer treatments are also feasible. Bailey et al. (2001) administered ethanol to pregnant guinea pigs throughout gestation (two daily intubations between gestational days 2 and 67) to assess its effects on $GABA_A$ receptor subunit expression in the cerebral cortex. In this study, pregnant guinea pigs received 4 g/kg of ethanol per day divided into two equal doses given 2 h apart. This treatment resulted in peak blood alcohol levels of ~0.3 g/dl (Bailey et al., 2001). Control animals were gavaged with an isocaloric solution without ethanol.

For a more detailed discussion of the advantages, disadvantages and applications of each of the ethanol administration methods discussed above, the reader is referred to Rogers et al. (1979); Lieber and DeCarli (1982); Lieber et al. (1989).

5.3 GENERATION AND USE OF DIFFERENT TYPES OF ANTIBODIES

The first step in the generation of antibodies is the preparation of the antigen for immunization of the appropriate animal. Antigens of the highest possible purity should be used to obtain antibodies with high specificity. If the antigen DNA sequence is known, then synthetic peptides or bacterially expressed proteins can be used to immunize animals. Synthetic peptides coupled to a carrier protein usually elicit a good immune response. However, antibodies raised against these peptides do not always recognize the native protein. Antibodies raised against bacterially expressed proteins may yield better results if this problem is encountered.

Once the antigen is ready, the appropriate animal for immunization must be chosen. The choice of animal depends on many factors including how much serum is needed and what type of antibody is going to be produced (Harlow and Lane, 1988). For instance, if a large amount of polyclonal antibodies is required, then rabbits or goats should be used. On the other hand, mice are the species of choice for the generation of monoclonal antibodies. Immunization of the animals usually requires the use of an adjuvant to nonspecifically stimulate the immune response. A commonly used adjuvant is the Freund's adjuvant that is a water-in-oil emulsion containing killed *Mycobacterium Tuberculosis* (complete Freund's adjuvant). The first injection of antigen is usually done with complete Freund's adjuvant. Subsequent injections (boosts) are usually performed with incomplete Freund's adjuvant (i.e., without *M. Tuberculosis*). The production of antibodies is checked by obtaining serum samples 7–14 d after a boost injection. Antibody titers are usually checked by radioimmunoassay (RIA), enzyme-linked immunosorbent assay (ELISA), or Western immunoblotting. If polyclonal antibodies are being generated, animals are repeatedly boosted and bled at appropriate intervals to obtain sera for antibody purification once a good titer of antibodies has been developed (Harlow and Lane, 1988).

If monoclonal antibodies are being generated, spleen cells from the immunized mice are fused to myeloma cells to produce hybridomas, which are screened for antibody production and those that are positive are expanded and single-cell cloned. Hybridomas can be maintained in a wide range of culture media and readily stored in liquid nitrogen. Monoclonal antibodies can be obtained from tissue culture supernatants or as ascitic fluid. If required, monoclonal antibodies can be further purified as described elsewhere (Harlow and Lane, 1988). Purified antibodies should be stored at relatively high concentrations (>1 mg/ml) in aliquots at −20°C in an isotonic solution such as phosphate buffered saline (PBS) at neutral pH. More diluted antibody solutions should be stored in PBS plus 1% bovine serum albumin (Harlow and Lane, 1988).

It is noteworthy that polyclonal and monoclonal antibodies each have their advantages and disadvantages that should be kept in mind before using them to quantify neurotransmitter receptor expression by Western immunoblotting or immunohistochemical techniques (Harlow and Lane, 1988). In general, most polyclonal antibodies usually yield a stronger signal than monoclonal antibodies in this type of assays. Moreover, most polyclonal antibodies recognize denatured antigens, whereas many monoclonal antibodies do not. However, monoclonal antibodies usually yield lower background signals and have greater specificity than most polyclonal antibodies. Importantly, a major advantage of monoclonal antibodies is their virtually unlimited supply. In contrast, polyclonal antibody production is limited to the lifetime of the source animal.

For a more detailed discussion on the production, purification, storage, uses and the advantages and disadvantages of monoclonal and polyclonal antibodies, the reader is referred elsewhere (Harlow and Lane, 1988).

5.4 QUANTITATIVE IMMUNOBLOTTING

5.4.1 BACKGROUND

In 1975, Southern developed a blotting technique (Southern blot) to immobilize DNA onto nitrocellulose membranes for further analysis by *in situ* hybridization (Southern, 1975). Others then applied this technique to the study of RNA by *in situ* hybridization (Northern blot) and proteins (Western blot or immunoblot). Immunoblotting is a versatile technique that permits the analysis of a protein based on both its size and binding to specific antibodies (Towbin et al., 1979). The fact that this technique allows the determination of the molecular weight of the immunoreactive species makes it superior to other types of immunoassays, such as ELISA, RIA or dot-blot assay. Although immunoblotting was originally developed as a qualitative technique, it has since been refined and it is now frequently used for quantitative protein analyses. However, several important factors must be taken into consideration when performing quantitative immunoblotting analyses to obtain interpretable data.

A factor that must always be determined prior to performing a quantitative immunoblotting assay is its dynamic working range. This range corresponds to the set of conditions where immunolabeling increases linearly with respect to increasing concentrations of the protein of interest. An immunoassay performed with protein concentrations that fall below or above this range will not provide quantitative information on the expression levels of this protein. The dynamic working range for a given receptor antigen–antibody interaction is specific and must be determined empirically as described in Section 5.4.2. and Appendix 5.1.

Another important point is that a standard curve must be generated and used to calculate relative units of protein expression. Standard curves are widely used in other types of protein determination assays but are often overlooked in immunoblotting studies. The standard curve must be run in the same gel and transferred under the same set of conditions as the samples of interest, i.e., every gel must include a standard curve. If the number of samples requires that multiple gels be run, the standard curve in each gel should be generated from a single sample kept in aliquots at –70°C. This procedure is necessary to allow comparisons among different gels. In all cases, the immunolabeling of the standard curve bands must increase linearly with increasing sample concentrations. This standard curve can be subsequently analyzed by linear regression analysis, from which relative expression levels of the protein of interest can be straightforwardly determined. Importantly, immunolabeling of the samples must fall within the dynamic working range of the standard curve.

Finally, quantitative immunoblots must include internal controls to ensure adequate sample loading and transfer and also for normalization purposes. Determination of the levels of housekeeping genes is a common practice in Southern and Northern blotting assays and it should also be performed in immunoblotting assays. A commonly used control is the determination of the expression levels of a housekeeping protein that is unaffected by the treatment in question (Liao et al., 2000). Among the housekeeping proteins that are frequently used as internal controls are β-actin and tubulin. Ideally, levels of a housekeeping protein should be measured in the same membrane (i.e., by reprobing). If this is not possible, duplicate membranes can be tested. It should be kept in mind, however, that the use of duplicate membranes does not completely rule out pipetting and transferring errors. Alternatively, normalization can be performed with respect to average intensity values obtained from several prominent bands visualized by Coomassie staining of the blotting membranes.

5.4.2 METHODOLOGY

5.4.2.1 Preparation of the Antigen Sample

An important strength of immunoblotting is that virtually any protein sample can be analyzed by this technique. However, the protein sample must be in a buffer that is compatible with the electrophoresis system (i.e., it should have a nearly neutral pH and salt concentrations below 200 mM) and its protein concentration must not exceed the loading capacity of the gel. Samples for immunoblotting do not require elaborate preparation steps in most cases. For instance, samples can be homogenized by sonication in PBS with protease inhibitors. In some instances, however, it may be necessary to prepare cell membranes or to fractionate the samples by using centrifugation or chromatographic methods. Purification by immunoprecipitation techniques can also be used if an antigen is not particularly abundant. In all cases, samples for the generation of the standard curve and for the determination of the levels of the protein of interest should be prepared under identical conditions. One of the most common methods to prepare samples for immunoblotting is to directly lyse cells or tissues (see Appendix 5.1).

5.4.2.2 Electrophoresis

Samples can be separated by different types of electrophoretic techniques, including discontinuous sodium dodecylsulfate polyacrylamide gel electrophoresis (SDS-PAGE) (Laemmli, 1970), non-denaturing electrophoresis and two-dimensional electrophoresis. We will discuss here only the details of discontinuous SDS-PAGE, as this is the most commonly used type of electrophoresis. Before separating the samples by discontinuous SDS-PAGE, two important issues should be considered. First, the amount of protein to be loaded per lane must be determined empirically for each particular antibody and should never exceed 150 μg of total protein per lane. A recommended starting point would be to dilute an aliquot of the standard curve sample to a concentration of 1 mg/ml of total protein and to load 5, 10, 15, 20, 25, 30, 35 and 40 μl of this sample on a gel (i.e., 5-40 μg/lane). This pilot gel should then be blotted and analyzed as described below to determine if the dynamic working range of the assay falls within this range of total protein concentrations. Second, some consideration should be given to the type of gel to be used. In general, the lower the percentage of bis-acrylamide in the gel and the thinner the gel (commonly between 1 and 1.5 mm), the easier the transfer will be (Harlow and Lane, 1988). Moreover, high molecular weight proteins are difficult to transfer unless they run at least one third of the distance from the top to the bottom of the gel, which can be achieved with lower percentage bis-acrylamide gels (Harlow and Lane, 1988). For discontinuous SDS-PAGE, precasted gels can be purchased from several commercial suppliers such as BioRad (Hercules, CA), Fisher Scientific (Pittsburgh, PA) and others. Alternatively, gels can be prepared as described in Appendix 5.1.

5.4.2.3 Blotting

Before initiating Western immunoblotting experiments, the appropriate blotting membrane must be chosen. It is recommended that an initial comparison be made of the performance of different types of membranes for a particular antigen–antibody pair. The two types of membranes most commonly used today are nitrocellulose and polyvinylidene difluoride (PDVF). Nitrocellulose has high binding protein capacity ($80 \mu g/cm^2$) and nonspecific protein binding to this type of membrane can be easily blocked. It does not require pre-activation. It is commercially available in 0.45 and $0.22 \mu m$ pore sizes. The smaller pore size nitrocellulose is useful for the transferring of low molecular weight proteins (i.e., <20 kDa). The transfer buffer used with nitrocellulose membranes must contain methanol, as it increases the binding of denatured proteins to this type of membrane. Two disadvantages encountered when using nitrocellulose membranes are the poor transfer of high molecular weight proteins onto these membranes and their lack of mechanical strength. In contrast, PVDF membranes have higher mechanical strength and binding protein capacity ($160 \mu g/cm^2$) than nitrocellulose. However, nonspecific binding can be more difficult to block in PVDF membranes than nitrocellulose membranes. Importantly, PVDF membranes must be wetted in 100% methanol to become activated but transfer buffer must not contain alcohol.

Another important issue to consider is the choice of transfer buffer. A widely used transfer buffer contains 25 mM Tris Base (pH 8.3), 190 mM glycine, and 20% reagent grade methanol. This buffer is appropriate for the transfer of most proteins. In our experience, this buffer is usually adequate for proteins <130 kDa. For the transfer of higher molecular weight proteins, 0.1 % SDS w/v can be added to the transfer buffer. SDS will improve the elution of high molecular weight proteins but it may reduce binding efficiency to nitrocellulose. The use of PVDF membranes with transfer buffer containing SDS but not methanol may result in improved transfer efficiency of high molecular weight proteins. It should be kept in mind, however, that SDS may precipitate below 10°C and that it may increase the conductivity of the transfer buffer, resulting in increased current and heating. Poorly controlled buffer temperature can affect transfer and can also be a safety hazard. The manufacturer's instructions must be consulted to determine the conditions necessary to appropriately control temperature in a particular immunoblotting apparatus. Transfer of proteins to membranes is usually achieved by electrophoretic elution. Two types of apparatuses can be used for this purpose: semi-dry or wet (also called tank transfer). Wet immunoblotting apparatuses are currently widely used around the world. A detailed protocol for electrotransferring with this type of set up is provided in the Appendix 5.1. For immunolabeling and detection, these steps can be followed:

1. Non-specific binding to the blotting membranes must be prevented prior to probing them with primary antibodies. A solution commonly used for this purpose is known as Blotto/Tween, which contains 5% w/v nonfat dry milk, 0.2 % w/v Tween-20 and 0.02% w/v sodium azide in Tris-buffered saline (TBS), which is composed of 50 mM Tris (pH 7.5) and 150 mM NaCl. If necessary, the concentrations of nonfat dry milk and Tween-20 can be increased to 10% w/v and 0.4% w/v, respectively. Incubation in Blotto/Tween solution for 1–2 h at room temperature is usually sufficient to prevent nonspecific binding. However, membranes can be incubated overnight at 37°C if shorter incubations at room temperature yield unsatisfactory results. If Blotto/Tween-20 does not work under any of the conditions described above, non-specific binding could be blocked with 3% w/v bovine serum albumin in TBS plus 0.02% w/v sodium azide; 0.2% w/v Tween-20 in TBS plus 0.02% w/v sodium azide; or 10% v/v horse serum in TBS plus 0.02% w/v sodium azide. It is noteworthy that milk and albumin have phosphotyrosine, which may cause high background when probing membranes with anti-phosphotyrosine antibodies (Harlow and Lane, 1988).

2. Remove the membrane from the blocking solution and wash it 2 × 5 minutes with TBS.

3. Incubate with primary antibody (i.e., specific for the protein of interest) in 3% bovine serum albumin in TBS. Polyclonal antibodies that have been purified by antigen affinity chromatography or monoclonal antibodies that recognize denatured epitopes are the preparations of choice for immunoblotting assays. It is noteworthy, however, that polyclonal antibodies are frequently supplied as crude serum or purified immuno-globulin fractions. These preparations could contain virtually all the repertoire of antibodies from the source animal, which can lead to background problems. Thus, an important control that must be performed when using these preparations of polyclonal antibodies is to preabsorb them with the antigen, which should prevent immunolabeling of the protein of interest. Final antibody concentrations usually range between 1 and 50 μg/ml, but optimal dilution levels should be determined empirically in a set of pilot experiments. Moreover, the duration of the incubation must be determined in the same manner. Most antibodies work well when incubated for 12–18 h at 4°C with constant shaking. Other antibodies give optimal results with incubations as short as 1 h at room temperature with constant shaking.

4. Wash 2 × 10 minutes with TBS plus 0.1% w/v Tween-20 (TBST) followed by 2 × 10 minutes washes with 3% w/v bovine serum albumin in TBS.

5. Incubate with the appropriate secondary (i.e., labeled) reagent. The optimal concentration of the secondary antibody and the duration of incubation must be determined empirically in preliminary pilot experiments. Detection can be accomplished with a variety of reagents, including ^{125}I-labeled protein A or protein G or ^{125}I-labeled species-specific anti-immunoglobulin antibodies. Another group of secondary reagents includes antibodies coupled to enzymes that produce either chemiluminescence (Thorpe et al., 1989; Huang and Amero, 1997) or color reaction products (Harlow and Lane, 1988). Detection of antibodies by chemiluminescence is perhaps the most widely used method of detection today. Kits for chemiluminescence detection are available from several commercial suppliers (for instance, Roche Molecular Biochemicals, Indianapolis, IN or Pierce, Rockford, IL). Detection of fluorophore-coupled secondary antibodies by phosphorimager analysis has been recently put forward as a reliable method to detect antibodies in immunoblotting assays (Fradelizi et al., 1999). In all cases, a control experiment in the absence of primary antibody must be performed with any of these secondary reagents to determine their background signal.

6. Wash the membrane 4×15 minutes with TBST with constant shaking.

7. Radiolabeled secondary reagents and antibodies coupled to enzymes that produce chemiluminescence can be detected with x-ray film or the appropriate phosphorimager screens. Fluorophore-coupled secondary antibodies can also be detected with a phosphorimager. Immunolabeling with enzyme-labeled secondary antibodies can be detected by the addition of chromogenic substrates. For more details, consult the manufacturer's instructions for each specific secondary reagent.

8. Densitometry of the bands of interest is then performed. X-ray films and membranes exposed to chromogens can be scanned and analyzed by a number of commercially available programs such as Image-Pro® Plus (Media Cybernetics, L.P., Silver Spring, MD) and others. Phosphorimager acquired images are usually analyzed with software provided by the manufacturers of this type of equipment.

9. Reprobe the blot with an antihousekeeping protein antibody. This step could be accomplished in different manners. It the membrane shows only a single band corresponding to the protein of interest, it could be rewetted in TBS and directly reprobed with the antihousekeeping protein antibody. However, the molecular weight of this protein should be significantly different from the molecular weight of the protein of interest. Under some circumstances, it may be possible to probe the membrane with a mixture of antibodies against the protein of interest and the housekeeping protein. If chemiluminescence detection methods are used, horseradish peroxidase antibodies bound to the protein of interest can be inactivated by treatment with hydrogen peroxide prior to reprobing (Liao et al., 2000). Alternatively, membranes can be stripped by incubation at 60°C in TBS plus 2% SDS and 0.1 M 2–mercaptoethanol for 30–60 min. Immunoblotting stripping kits are commercially available from, for example, Chemicon

(Temecula, CA) and Pierce (Rockford, IL). Recipes for other stripping solutions can be found elsewhere (Renart et al., 1996). After membrane stripping, wash extensively with TBS, reblock, reprobe and detect as described above. Another possibility would be to stain the membranes, for example with Coomassie blue, and to normalize values of interest with respect to major bands in each lane.

10. Perform densitometric and statistical analyses.

5.4.3   Potential Results, Analysis and Interpretation

Following the steps described above, it is rather straightforward to determine the relative concentration of a protein by immunoblotting analysis. Results of this type of assay may appear the same as those shown in Figure 5.1.

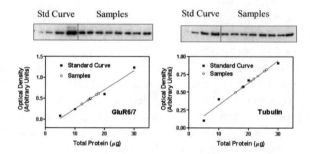

FIGURE 5.1 A quantitative immunoblotting assay for GluR6/7 subunits, which belong to the kainate family of glutamate receptors. It was performed with homogenates prepared by sonication of eight different hippocampi from adult mice. Samples (5–30 µg of total protein per standard curve lane and 15 µg of the unknowns) were separated by discontinuous SDS-PAGE on 7.5 % minigels and transferred onto 0.45 µm pore size nitrocellulose using transfer buffer containing 25 mM Tris Base (pH 8.3), 190 mM glycine, and 20% methanol. Nonspecific binding was blocked with Blotto/Tween-20 for 2 h at room temperature. Polyclonal antibodies against GluR6/7 subunits were purified by antigen affinity chromatography (Upstate Biotechnology, Lake Placid, NY) and were used at a final concentration of ~1 µg/ml. Membranes were stripped with a Re-blot Western Blotting Recycling Kit from Chemicon (Temecula, CA) and then reprobed with anti-tubulin monoclonal antibodies from Sigma (St. Louis, MO) at a 1:5,000 dilution. Detection was performed with a chemiluminescence kit from Roche Molecular Biochemicals (Indianapolis, IN). X-ray films were scanned with a Hewlett-Packard ScanJet II scanner, then analyzed with Image-Pro® Plus image analysis system (Media Cybernetics, L.P., Silver Spring, MD). Linear regression analysis was performed with GraphPad software (San Diego, CA).

Note that immunolabeling of standard curve samples increases linearly, and immunolabeling of samples of unknown concentrations fall within the range of this standard curve. From this standard curve, relative levels of protein expression can be determined. It should be emphasized that this assay cannot determine the actual protein concentration of the subunit of interest because the standard curve values that are entered in the abscissa correspond to the total protein concentration of the hippocampal homogenates that were loaded per lane. Therefore, once linear regression analysis and interpolation are performed to determine unknown values, total protein concentrations should be denoted as relative units of protein (GluR6/7 in the example). To determine the actual concentration of a protein of interest, it would be necessary to use a 100% pure preparation of the protein of interest to construct the standard curve.

Also shown in Figure 5.1 is the immunoblotting analysis of a housekeeping protein in the same membranes that had been previously probed with anti-GluR6/7 antibodies. The protein shown in the example is β-tubulin. As can be observed, there is some variability in the immunolabeling of this protein, which suggests that there were small differences in sample preparation or gel loading. In order to correct for these intersample differences, relative units of GluR6/7 protein levels could be divided by the relative units of β-tubulin or a correction factor generated from these tubulin levels.

In the example shown in Figure 5.1, 15-well combs were used, which permit the loading of four samples to construct the standard curve plus eight samples. If one were examining samples from a control and an ethanol-exposed group, this would allow studying just four samples from each group. A sample number of four would be insufficient to reach statistical significance if a small ethanol-induced change in subunit levels is expected. A power calculation performed with StatMate software (GraphPad Inc, San Diego, CA) showed that to obtain a $p < 0.05$ by unpaired t-test with a standard deviation of 20%, the difference in the mean expression levels between samples would have to be ~60% with an n = 4. Therefore, it is usually necessary to study larger number of samples, which require analysis by multiple immunoblots. In this case, standard curves should be constructed with a single sample kept in aliquots in a deep freezer to allow comparisons among samples in different membranes.

5.5 QUANTITATIVE IMMUNOHISTOCHEMISTRY

Immunohistochemistry is a widely used technique in neuroscience research. This technique has been mainly used in qualitative studies to detect the presence and localization of neurotransmitter receptors and many other proteins in the nervous system. In the past, immunohistochemistry was not commonly used for quantitative studies. This type of study was generally regarded as difficult because of confounding variables such as variability in sample and antibody preparations, inconsistency in the performance of enzymatic detection systems and microscopes, and lack of appropriate imaging analysis tools (Appel, 1997). In recent years, however, significant progress has been achieved in the generation of reliable secondary immunoreagents and substantial improvements have taken place in the areas of fluorescence and confocal microscopy as well as in the technology for image acquisition and analysis. These technical advances have made it increasingly possible to perform quantitative immunohistochemical studies.

A major advantage of immunohistochemistry over other types of immunoassay is that this technique is performed on tissue with relatively well preserved cytoarchitecture, which permits the evaluation of discrete cell groups and, in some cases, of specific subcellular compartments. However, a disadvantage of this technique is that the molecular weight of the protein of interest cannot be determined in immunohistochemical experiments. Thus, in contrast to immunoblotting, size cannot be used to identify the specificity of immunolabeling when a particular antibody cross-reacts with more than one protein. Consequently, the use of high quality antibodies is essential for the success of this type of experiment. Moreover, appropriate control experiments, such as the pre-absorption of primary antibodies, must be performed to ensure that immunolabeling is specific for the antigen of interest.

In the next sections, we have elected to review the technical details of quantitative confocal microscope and radioimmunohistochemistry. However, it is important to emphasize that at least one other method has been described that can provide quantitative immunohistochemical information. This method involves analysis of background and stained structures by using the peroxidase-antiperoxidase (PAP) method (Sternberger and Sternberger, 1986).

5.5.1 QUANTITATIVE CONFOCAL MICROSCOPY

5.5.1.1 Background

Because of its high resolution, confocal microscopy has emerged as an important technique to quantify receptor expression in immunolabeled tissues. Optimal illumination, along with the ability to scan samples in all axes, make the laser scanning confocal microscope (LSCM) an ideal tool for measuring receptor levels in immunohistochemical studies. Moreover, imaging illumination, scanning and acquisition parameters are computer controlled, which makes it relatively straightforward to standardize such parameters for the analysis of multiple samples. In addition, images acquired with an LSCM can be subsequently processed with computer software to accurately determine immunofluorescence intensity levels for specific neurotransmitter receptor subunits. Good et al. (1992) were the first to apply confocal microscopy to the quantification of receptor levels in single cells. These investigators demonstrated that epidermal growth factor receptor density could be determined as accurately with confocal microscopy techniques as with ^{125}I-EGF binding techniques. LSCM was subsequently used by Dodge et al. to quantify intracellular levels of a Clara cell secretory 10 kD protein in rat bronchi (Dodge et al., 1994). More recently, Gazzaley et al. used LSCM to quantify the regulation of NMDAR1 subunit protein expression by estradiol in the rat hippocampus (Gazzaley et al., 1996b; Gazzaley et al., 1996a). This technique is currently being used to quantify the long-term effects of ethanol on ionotropic glutamate receptor expression in hippocampal neurons (Valenzuela and Savage, unpublished observations).

5.5.1.2 Methodology

5.5.1.2.1 Tissue Preparation

Sectioning freshly frozen tissues and sectioning tissues prepared from perfused animals are two commonly used methods to prepare samples for quantitative confocal experiments. These two methods will be considered in detail below. It is recommended that a pilot experiment be performed comparing the staining obtained with tissue prepared by using these two methods to determine which one works best for the antigen–antibody pair of interest. Another method that can be used to prepare samples for quantitative confocal experiments is the sectioning of paraffin-embedded tissues. Although this method has been successfully used in confocal microscopy studies (Nichol et al., 1999), it involves harsh fixation and embedding procedures that could affect the antigen of interest. Therefore, the methods for the preparation of sections from paraffin embedded tissues will not be considered further in this chapter. Details on this technique can be found elsewhere (Harlow and Lane, 1988).

1. For the preparation of frozen tissue sections, begin by rapidly euthanizing animals following the guidelines of the American Veterinary Medical Association (Euthanasia APO, 2001). Remove brains or other tissues of interest rapidly and carefully to prevent damage. Wash by submerging into ice-cold PBS. Wipe off excess PBS with filter paper, gently place it at the end of a spatula bent at a 90-degree angle and submerge it slowly (to prevent cracking) into isopentane that has been chilled in a dry ice–methanol bath. After ~4–5 min (for an adult rat brain), remove brain, wrap loosely in aluminum foil and place into a prelabeled airtight plastic bag. Store immediately at –70°C. Keep on dry ice until transferring to the deep freezer. Frozen samples can be kept in airtight bags at –70°C for several months. Cooling isopentane for freezing whole brains can be achieved by submerging a beaker containing it into a larger container with dry ice-cooled methanol. When submerging the beaker with isopentane, make sure it is covered to prevent drops of methanol from falling into the isopentane. To cool with dry ice, start slowly adding small pieces of it into a suitable container (for instance a Dewar) with methanol that had been pre-cooled overnight at –20°C. The methanol will be ready when adding dry ice pieces causes minimal bubbling. Plan in advance, as cooling the methanol could take about 1–2 hours. Smaller samples than a whole brain could also be directly frozen in liquid nitrogen (Harlow and Lane, 1988).

2. For the preparation of tissue from perfused animals, please first consult with the attending veterinarian at your Institutional Animal Care Facility for guidelines on this type of procedure. Animals should be deeply anesthetized with a suitable agent and then transcardially perfused with cold 1% paraformaldehyde in 0.1 M PBS followed by cold 4% paraformaldehyde in PBS. Brains are then carefully removed and postfixed for 2–3 h with 4% paraformaldehyde in PBS at 4°C. Brains can be stored for several months at 4°C in 5% sucrose in PBS in closed containers to prevent evaporation. To prepare 4% paraformaldehyde, dissolve 8 grams in 100 ml of water, add a few drops of 1 N NaOH, heat at 60°C and stir it in a fume hood until clear. Let this solution cool down to room temperature and mix it with 100 ml of 2X PBS. Paraformaldehyde should be prepared fresh daily.

3. Frozen samples can be sectioned (5–10 μm thickness) with a cryostat following the manufacturer's instructions. Cryostat-prepared sections can be collected on pretreated glass slides such as Superfrost® Plus selected precleaned micro slides (VWR Scientific, West Chester, PA). Alternatively,

clean glass slides can be coated with 1% gelatin (Harlow and Lane, 1988). Sections can be stored in suitable containers kept in airtight bags at –70°C for a few weeks.

4. Once preserved, tissue from perfused animals can be sectioned by freezing as described above or with a vibratome (40 μm sections). Vibratome prepared sections can be stored in PBS plus 0.1% sodium azide at 4°C.

5.5.1.2.2 Immunolabeling

1. For cryostat prepared sections, thaw them for 3–5 min. Draw a ring around each section with a hydrophobic slide marker specifically designed for staining procedures (also known as a PAP pen; Research Biochemicals, Natick, MA). The purpose of this ring is to hold antibody solutions in place. Incubate sections in fresh 4% paraformaldehyde-PBS (prepared as described above) for 20 min at room temperature. Multiple slides can be incubated simultaneously by using Coplin jars or other commercially available staining dishes (for instance, see Fisher Scientific or VWR Scientific catalogs). Rinse with PBS 3 × 10 min. Incubation of vibratome-cut sections can be performed in multi-well plates under constant shaking.

2. If the antibody of interest recognizes an intracellular epitope, permeabilize the sections with PBS plus 0.2-0.3% Triton X-100 at room temperature for 20 min. Rinse with PBS 3 × 10 minutes and dip in water three times.

3. To prevent nonspecific antibody binding, sections are blocked by preincubation (1 h) with PBS (with 0.2–0.3% Triton X-100 if it was used to permeabilize before) plus 2–4% bovine serum albumin or 1–2% purified antibodies from the same species as the detection reagent.

4. Incubate with primary antibody diluted in PBS plus 2–4% BSA (with 0.2–0.3% Triton X-100 if it was used to permeabilize before). Put enough antibody solution to cover the section but make sure it does not overflow the PAP pen ring. Optimal antibody dilutions and incubation times should be determined empirically in a set of preliminary experiments. Whenever possible, polyclonal antibodies purified by antigen chromatography or monoclonal antibodies (purified or in hybridoma supernatants) should be used for this type of experiment. Monoclonal antibodies in ascites fluid, unpurified sera and polyclonal antibodies purified by methods other than antigen chromatography should be used with caution, as these preparations contain nonspecific antibodies from the source animal. Virtually any of these preparations will give nonspecific immunolabeling if used at high enough concentrations. Therefore, these preparations should be used as diluted as possible. Moreover, the specificity of the immunolabeling should be confirmed by incubating dilutions that give optimal results with an excess of the immunizing antigen (i.e., peptide, fusion protein or purified protein). This procedure should result in elimination of the immunolabeling signal coming from the antigen of interest. Another

important control should be to test pre-immune serum. Incubation with the primary antibodies can take as long as 24–48 h and it should preferably be performed in a humidified chamber at 4°C.

5. On another set of sections from the same animal, probe with an antibody that recognizes a housekeeping protein. This procedure will be useful to establish the specificity of changes in the levels of the protein of interest.

6. Gently shake primary antibody solution off the slides. Wash 3 × 10 minutes with PBS in a Coplin or any other type of jar. Re-apply PAP pen if necessary.

7. Incubate in the dark with a fluorescently labeled secondary immunoreagent, which could be fluorescently labeled protein A or G, species-specific biotinylated antibodies used in conjunction with fluorescently conjugated avidin, or the appropriate species-specific fluorescently labeled secondary antibody. Species-specific antibodies recognize the Fc region of immunoglobulins from a particular species. These immunoreagents are available from several commercial sources, such as Molecular Probes (Eugene, OR), Jackson Immunoresearch Laboratories (West Grove, PA), and others. Incubation times and optimal dilutions for these secondary immunoreagents should be determined empirically in a set of preliminary experiments. Start by incubating them for 1 h at room temperature in PBS plus 2–4% BSA (with 0.2–0.3% Triton X-100 if it was used to permeabilize before). Importantly, these reagents should be tested in the absence of primary antibodies to determine their background signal contribution. Manufacturers of these reagents often provide recommendations for dilution ranges that can be used as a starting point in histochemical experiments.

8. Gently shake secondary antibody solution off the slides. Wash 3 × 10 minutes with PBS in Coplin or any other type of jar. Dip three times in water.

9. Add a small drop of mounting media to the specimen. Use of an anti-fading mounting media is recommended (for instance, Fluormount-G from Southern Biotechnology, Birmingham, AL; Slowfade from Molecular Probes, Eugene, OR or AF1 from Citifluor, London, England). Gently, place a coverslip on top avoiding bubbles. Remove excess of mounting medium with a tissue paper. Wait 2 h and then add a small layer of nail polish at the edges of the coverslip. Let the mounting medium set overnight.

5.5.1.2.3 *Image Acquisition with Laser Scanning Confocal Microscope (LSCM)*

1. Specimens from control and treated animals to be analyzed by LSCM should be stained as described above on the same day and under identical experimental conditions. The operator of the LSCM should be blinded to

which sections correspond to which experimental group. At least three non-adjacent sections must be analyzed from each animal.

2. Optimal parameters for confocal imaging acquisition should be established at the beginning of the day; i.e., images from control and treated sample pairs should be acquired under identical conditions. Ideally, these settings should remain constant throughout the project. In some cases, settings must be changed to analyze fields in different regions. However, the same region must be always analyzed under the same settings in all samples. Among the parameters that must be standardized are the objectives and filters to be used, laser output, scan speed, pixel depth, scan direction, scan averaging, pinhole, zoom factor, amplitude offset and detector gain settings. Images must be optimized for contrast, brightness and background levels. Z-axis distance from the surface of the section must be standardized. Software that controls the LSCM is designed to facilitate these standardization procedures (i.e., parameters can be stored in the computer that controls the LSCM). Consult the manufacturer's instructions of each particular LSCM for details.

3. Fields should be selected randomly but within a standardized region. For instance, in a recent study on estrogen regulation of hippocampal NMDA receptor subunit expression, fields analyzed in the CA1 region were always located in the middle of this region (Gazzaley et al., 1996a). Special attention should be given to the prevention of fluorescencebleaching. Fields of interest should be located rapidly using either bright field or standard fluorescence illumination, making sure to set the lamp intensity to the lowest possible level that permits localization of the field. Make every attempt to expose all specimens to bright field or standard fluorescence illumination for the same length of time.

5.5.1.3 Potential Results, Analysis and Interpretation

Fluorescence intensity analysis of images obtained with an LCSM can be performed with software packages designed for each specific confocal microscope. Alternatively, images can be saved in a number of file formats that can be imported into many analysis programs such as Image-Pro® Plus (Media Cybernetics, L.P., Silver Spring, MD) or Bioquant® (R & M Biometrics, Nashville, TN). With the appropriate experimental and statistical analysis design, this type of experiment can detect relatively small differences in receptor levels. For instance, in the study of Gazzaley et al. (1996a), the authors obtained mean intensity values from six fields per region in three nonconsecutive hippocampal sections per animal and determined the percent change with respect to control. Data was subsequently analyzed by two-way ANOVA at a <0.05 level of significance and a Sheffe's *post hoc* test. Using this experimental design, it was possible to detect estrogen-induced increases of 10–20% in NMDAR1 subunit immunolabeling in the somatic and dendritic fields of the CA1 and dentate gyrus. Importantly, it was found in this study that estrogen did not affect expression levels of a housekeeping protein (i.e., MAP2), indicating that the effect of this hormone was specific for NMDAR1 subunits (Gazzaley et al., 1996a).

5.5.2 RADIOIMMUNOHISTOCHEMISTRY

5.5.2.1 Background

Radioimmunohistochemistry, also known as immunoautoradiography, is another method that can be used to perform quantitative analysis of receptor expression in tissue sections from ethanol-exposed animals. Studies have shown that this technique can provide accurate quantitative information on brain receptor protein expression levels. For example, Correa et al. (1991) demonstrated that quantification of angiotensin-converting enzyme in rat brain by enzymatic, ligand binding and radioimmunohistochemistry yielded comparable results. Using this technique, Eastwood and Harrison (1995) demonstrated a decrease in synaptophysin expression levels in samples from the medial temporal lobe of schizophrenics. More recently, Duelli et al.(1998) detected a decrease in the density of the Glut1 and Glut3 glucose transporters in visual structures of the rat brain during chronic deprivation, and Herman and Spencer (1998) showed adrenalectomy-induced effects on glucocorticoid receptor immunoreactivity in the rat hippocampus. This technique is also being used to assess the effects of prenatal ethanol exposure on metabotropic and ionotropic glutamate receptor expression in the hippocampus (Savage, unpublished observations). Figure 5.2 illustrates the distrubution hippocampal AMPA receptors visualized by radioligand binding with ^{3}H-fluorowillardiine compared with the distribution of GluR subunits 1, 2/3 and 4 visualized by radioimmunohistochemical techniques. Whereas the distribution of antibody binding to subunits is similar in some regions, there are subtle differences in subunit distribution in other areas, as illustrated by examining GluR1 and GluR4 binding in the apical dendritic field of the hippocampal CA3 region. It is important to note that, while quantitative comparisons can be made in multiple brain regions and sections processed by the same radioimmunohistochemical procedure, quantitative comparisons of different antibody binding reactions are not possible using these methods.

5.5.2.2 Methodology

1. Tissue preparation, sectioning, fixing and incubation with primary antibodies can be performed as described in section 5.5.1.2.
2. Incubate sections with secondary immunoreagents. Radiolabeled species-specific secondary antibodies are frequently used for this type of procedure. Anti-mouse, -rabbit and -rat antibodies labeled with ^{125}I, or, more commonly, ^{35}S, can be purchased, for example, from Amersham (Piscataway, NJ) or NEN (Boston, MA). The molarity of a particular batch of antibody should be calculated using the specific activity and concentration information provided by the manufacturer. For instance, if an antibody batch has a specific activity of 600 Ci/mmol and a concentration of 0.1 µCi/µl, then the molarity of this batch would be 166 nM. Make sure to correct for isotope decay as necessary. Optimal dilutions and incubation times for radiolabeled secondary antibodies should be determined empirically in a set of preliminary experiments. Start by testing 0.1–1 nM concentrations for 1 h at room temperature

GluR$_1$ GluR$_{2/3}$

GluR$_4$ [^{3}H]-Fluorowillardiine

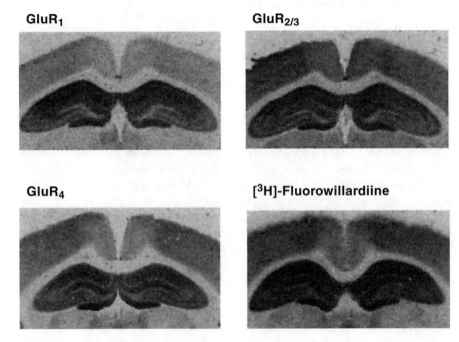

FIGURE 5.2 Radioimmunohistochemisty detection of immunolabeling with anti-AMPA receptor subunit antibodies, performed with coronal frozen sections from adult rats. Immunoaffinity purified polyclonal anti-GluR1, GluR2/3 and GluR4 antibodies were obtained from Upstate Biotechnology (Lake Placid, NY). Donkey anti-rabbit ^{35}S-labeled secondary antibody was from Amersham-Pharmacia Biotech (Piscataway, NJ). For comparison, a ligand binding autoradiogram with the AMPA receptor ligand, ^{3}H-fluorowillardiine (Tocris, Ballwin, MO), is also shown.

in PBS plus 2–4% BSA (with 0.2–0.3% Triton X-100 if it was used to permeabilize sections before). Importantly, radiolabeled secondary antibodies should be tested in the absence of primary antibody to determine their background signal contribution.

3. Remove secondary antibody solution and dispose of radioactive waste according to the guidelines of your institutional radiation safety department. Wash sections 3 × 10 minutes with PBS at room temperature in Coplin or other type of jar. Dip three times in water and then air-dry sections.

4. Load sections into film cassettes along with standards containing known amounts of radioactivity. Standards can be created by blotting varying amounts of labeled secondary antibody onto nitrocellulose. Load film (for example, Kodak Biomax MR or MS film), close cassettes and store at –20°C during exposure.

5. For each new radioimmunohistochemical procedure, the film exposure period will need to be determined empirically and will vary as a function of receptor density and the specific activity and incubation concentration of secondary antibody.

6. Follow the manufacturer's instruction for developing the film. Dry the film overnight. For ease of handling and protection of the film emulsion, autoradiograms can be cut into strips and mounted onto glass microscope slides using Permount® (Fisher Scientific, Pittsburgh, PA). Images from mounted film sections can be acquired with a light microscope equipped with a digital camera.
7. Alternatively, radiolabeled sections can be exposed to screens that can be further analyzed by a phosphorimager or similar device (Charon et al., 1998).

5.5.2.3   Potential Results, Analysis and Interpretation

Densitometric analysis of radioimmunohistochemistry autoradiograms can be performed with programs such as the Image-Pro® Plus (Media Cybernetics, L.P., Silver Spring, MD). First, a standard curve of optical density for each standard of known radioactivity per unit area is generated. Then, antibody binding per unit area is determined by measuring the optical density and converting the data to moles per unit area based on the standard curve. For many brain regions, optical density measurements of ^{35}S-labeled antibody binding on the right and left side of four sections can be averaged to obtain a single measure of total antibody binding over each area of interest. The right and left side of two sections incubated in the absence of primary antibody can be used to obtain a single measure of nonspecific antibody binding. Specific ^{35}S-labeled antibody binding can be defined as the difference between total and nonspecific binding.

5.6   CONCLUSIONS AND FUTURE DIRECTIONS

In this chapter, we have described three methods that can be used to quantify the effects of ethanol on the protein expression levels of neurotransmitter receptor subunits. Using one or more of these methods in conjunction with ligand binding and mRNA-based techniques should provide important information on the short- and long-term effects of ethanol on the expression of neurotransmitter receptor subunits. A recent Pubmed search of the literature revealed that quantitative immunoblotting and immunohistochemical assays have been applied to study the effects of ethanol only on a very small subset of neurotransmitter receptor subunits, particularly to those belonging to the GABA$_A$ and ionotropic glutamate family of receptors. Therefore, the effect of ethanol on the expression of subunits belonging to other families of neurotransmitter receptors remains an open question. We hope that this chapter kindles the interest of many investigators to pursue studies on this important area of alcohol-related neuroscience research.

ACKNOWLEDGMENTS

We thank Vania Ferreira and Shanti Frausto for assistance with immunoblotting assays and Linda Beyer-Smith for assistance with the radioimmunohistochemical procedures. This work has been supported by U.S. Army grant DAMD17-00-1-0579

(C.F.V.), a grant from the Alcoholic Beverage Medical Foundation (C.F.V) and NIH grants from the National Institute on Alcohol Abuse and Alcoholism — AA12684 (C.F.V.), AA06548 (D.D.S.) and AA12400 (D.D.S.).

REFERENCES

Allan AM et al. (1998) Prenatal ethanol exposure alters the modulation of the gamma-aminobutyric acidA1 receptor-gated chloride ion channel in adult rat offspring. *J Pharmacol Exp Ther* 284:250-257.

Appel NM (1997) Classical and contemporary histochemical approaches for evaluating central nervous system microanatomy. *Ann N Y Acad Sci* 820:14-28.

Bailey CD, Brien JF and Reynolds JN (2001) Chronic prenatal ethanol exposure increases GABAa receptor subunit protein expression in the adult guinea pig cerebral cortex. *J Neurosci* 21:4381-4389.

Chabot JG, Kar S and Quirion R (1996) Autoradiographical and immunohistochemical analysis of receptor localization in the central nervous system. *Histochem J* 28:729-745.

Chandler LJ, Norwood D and Sutton G (1999) Chronic ethanol upregulates NMDA and AMPA, but not kainate receptor subunit proteins in rat primary cortical cultures. *Alcohol Clin Exp Res* 23:363-370.

Charness ME, Querimit LA and Diamond I (1986) Ethanol increases the expression of functional delta-opioid receptors in neuroblastoma ×glioma NG108-15 hybrid cells. *J Biol Chem* 261:3164-3169.

Charon Y, Laniece P and Tricoire H (1998) Radio-imaging for quantitative autoradiography in biology. *Nucl Med Biol* 25:699-704.

Correa FM, Guilhaume SS and Saavedra JM (1991) Comparative quantification of rat brain and pituitary angiotensin-converting enzyme with autoradiographic and enzymatic methods. *Brain Res* 545:215-222.

Costa ET, Savage DD and Valenzuela CF (2000) A review of the effects of prenatal or early postnatal ethanol exposure on brain ligand-gated ion channels. *Alcohol Clin Exp Res* 24:706-715.

Cowen MS and Lawrence AJ (1999) The role of opioid-dopamine interactions in the induction and maintenance of ethanol consumption. *Prog Neuropsychopharm Biol Psych* 23:1171-1212.

Darstein MB, Landwehrmeyer GB and Feuerstein TJ (2000) Changes in NMDA receptor subunit gene expression in the rat brain following withdrawal from forced long-term ethanol intake. *Naunyn Schmiedebergs Arch Pharmacol* 361:206-213.

Daubert DL et al. (1999) Changes in angiotensin II receptors in dopamine-rich regions of the mouse brain with age and ethanol consumption. *Brain Res* 816:8-16.

Deitrich RA et al. (1989) Mechanism of action of ethanol: initial central nervous system actions. *Pharmacol Rev* 41:489-537.

Devaud LL et al. (1995) Chronic ethanol consumption differentially alters the expression of gamma-aminobutyric acidA receptor subunit mRNAs in rat cerebral cortex: competitive, quantitative reverse transcriptase-polymerase chain reaction analysis. *Mol Pharmacol* 48:861-868.

Dodge DE, Plopper CG and Rucker RB (1994) Regulation of Clara cell 10 kD protein secretion by pilocarpine: quantitative comparison of nonciliated cells in rat bronchi and bronchioles based on laser scanning confocal microscopy. *Am J Respir Cell Mol Biol* 10:259-270.

Duelli R, Maurer MH and Kuschinsky W (1998) Decreased glucose transporter densities, rate constants and glucose utilization in visual structures of rat brain during chronic visual deprivation. *Neurosci Lett* 250:49-52.

Eastwood SL and Harrison PJ (1995) Decreased synaptophysin in the medial temporal lobe in schizophrenia demonstrated using immunoautoradiography. *Neuroscience* 69:339-343.

Euthanasia APo (2001) 2000 Report of the AVMA Panel on Euthanasia. *J Am Vet Med Assoc* 218:669-696.

Eysseric H et al. (1997) There is no simple method to maintain a constant ethanol concentration in long-term cell culture: keys to a solution applied to the survey of astrocytic ethanol absorption. *Alcohol* 14:111-115.

Follesa P and Ticku MK (1995) Chronic ethanol treatment differentially regulates NMDA receptor subunit mRNA expression in rat brain. *Brain Res Mol Brain Res* 29:99-106.

Follesa P and Ticku MK (1996) Chronic ethanol-mediated up-regulation of the N-methyl-D-aspartate receptor polypeptide subunits in mouse cortical neurons in culture. *J Biol Chem* 271:13297-13299.

Fradelizi J et al. (1999) Quantitative measurement of proteins by western blotting with Cy5-coupled secondary antibodies. *Biotechniques* 26:484-486, 488, 490 passim.

Freeman WM, Walker SJ and Vrana KE (1999) Quantitative RT-PCR: pitfalls and potential. *Biotechniques* 26:112-122, 124-115.

Frye GD et al. (1981) Effects of acute and chronic 1,3-butanediol treatment on central nervous system function: a comparison with ethanol. *J Pharmacol Exp Ther* 216:306-314.

Gazzaley AH et al. (1996a) Differential regulation of NMDAR1 mRNA and protein by estradiol in the rat hippocampus. *J Neurosci* 16:6830-6838.

Gazzaley AH et al. (1996b) Circuit-specific alterations of N-methyl-D-aspartate receptor subunit 1 in the dentate gyrus of aged monkeys. *Proc Natl Acad Sci USA* 93:3121-3125.

Good MJ et al. (1992) Localization and quantification of epidermal growth factor receptors on single cells by confocal laser scanning microscopy. *J Histochem Cytochem* 40:1353-1361.

Gruol DL et al. (1998) Neonatal alcohol exposure reduces NMDA induced Ca2+ signaling in developing cerebellar granule neurons [in process citation]. *Brain Res* 793:12-20.

Hardy PA, Chen W and Wilce PA (1999) Chronic ethanol exposure and withdrawal influence NMDA receptor subunit and splice variant mRNA expression in the rat cerebral cortex. *Brain Res* 819:33-39.

Harlow E and Lane D (1988) *Antibodies: A Laboratory Manual.* Cold Spring Harbor, NY: Cold Spring Harbor Laboratory Publications.

Harris RA et al. (1998) Adaptation of gamma-aminobutyric acid type A receptors to alcohol exposure: studies with stably transfected cells. *J Pharmacol Exp Ther* 284:180-188.

Heaton MB et al. (1994) Ethanol neurotoxicity *in vitro*: effects of GM1 ganglioside and protein synthesis inhibition. *Brain Res* 654:336-342.

Herman JP and Spencer R (1998) Regulation of hippocampal glucocorticoid receptor gene transcription and protein expression *in vivo*. *J Neurosci* 18:7462-7473.

Hoffman PL et al. (1995) Attenuation of glutamate-induced neurotoxicity in chronically ethanol- exposed cerebellar granule cells by NMDA receptor antagonists and ganglioside GM1. *Alcohol Clin Exp Res* 19:721-726.

Hu XJ, Follesa P and Ticku MK (1996) Chronic ethanol treatment produces a selective upregulation of the NMDA receptor subunit gene expression in mammalian cultured cortical neurons. *Brain Res Mol Brain Res* 36:211-218.

Huang D and Amero SA (1997) Measurement of antigen by enhanced chemiluminescent western blot. *Biotechniques* 22:454-456, 458.

Iorio KR et al. (1992) Chronic exposure of cerebellar granule cells to ethanol results in increased N-methyl-D-aspartate receptor function. *Mol Pharmacol* 41:1142-1148.

Jarvis MF and Becker HC (1998) Single and repeated episodes of ethanol withdrawal increase adenosine A1, but not A2A, receptor density in mouse brain. *Brain Res* 786:80-88.

Kalluri HS, Mehta AK and Ticku MK (1998) Up-regulation of NMDA receptor subunits in rat brain following chronic ethanol treatment. *Brain Res Mol Brain Res* 58:221-224.

Keir WJ and Morrow AL (1994) Differential expression of GABAA receptor subunit mRNAs in ethanol-naive withdrawal seizure resistant (WSR) vs. withdrawal seizure prone (WSP) mouse brain. *Brain Res Mol Brain Res* 25:200-208.

Kumari M and Ticku MK (1998) Ethanol and regulation of the NMDA receptor subunits in fetal cortical neurons. *J Neurochem* 70:1467-1473.

Laemmli UK (1970) Cleavage of structural proteins during the assembly of the head of bacteriophage T4. *Nature* 227:680-685.

Liao J, Xu X and Wargovich MJ (2000) Direct reprobing with anti-beta-actin antibody as an internal control for Western blotting analysis. *Biotechniques* 28:216-218.

Lieber CS and DeCarli LM (1982) The feeding of alcohol in liquid diets: two decades of applications and 1982 update. *Alcohol Clin Exp Res* 6:523-531.

Lieber CS, DeCarli LM and Sorrell MF (1989) Experimental methods of ethanol administration. *Hepatology* 10:501-510.

Mahmoudi M et al. (1997) Chronic intermittent ethanol treatment in rats increases GABA(A) receptor alpha4-subunit expression: possible relevance to alcohol dependence. *J Neurochem* 68:2485-2492.

Marinelli PW, Kiianmaa K and Gianoulakis C (2000) Opioid propeptide mRNA content and receptor density in the brains of AA and ANA rats. *Life Sci* 66:1915-1927.

Mason JN et al. (2001) NMDA receptor subunit mRNA and protein expression in ethanol-withdrawal seizure-prone and -resistant mice. *Alcohol Clin Exp Res* 25:651-660.

Mehta AK and Ticku MK (1999) An update on GABAA receptors. *Brain Res Brain Res Rev* 29:196-217.

Messing RO et al. (1986) Ethanol regulates calcium channels in clonal neural cells. *Proc Natl Acad Sci USA* 83:6213-6215.

Montpied P et al. (1991) Prolonged ethanol inhalation decreases gamma-aminobutyric acidA receptor alpha subunit mRNAs in the rat cerebral cortex. *Mol Pharmacol* 39:157-163.

Negro M, Fernandez-Lopez A and Calvo P (1995) Autoradiographical study of types 1 and 2 of benzodiazepine receptors in rat brain after chronic ethanol treatment and its withdrawal. *Neuropharmacology* 34:1177-1182.

Nelson TE, Ur CL and Gruol DL (1999) Chronic intermittent ethanol exposure alters CA1 synaptic transmission in rat hippocampal slices. *Neuroscience* 94:431-442.

Nichol KA, Depczynski BB and Cunningham AM (1999) Characterization of hypothalamic neurons expressing a neuropeptide receptor, GALR2, using combined *in situ* hybridization- immunohistochemistry. *Methods* 18:481-486.

Petrie J et al. (2001) Altered gabaa receptor subunit and splice variant expression in rats treated with chronic intermittent ethanol. *Alcohol Clin Exp Res* 25:819-828.

Renart J et al. (1996) Immunoblotting techniques. In: *Immunoassay* (Diamandis E and Christopoulos TK, Eds), pp 537-554. San Diego, CA: Academic Press.

Reue K (1998) mRNA quantitation techniques: considerations for experimental design and application. *J Nutr* 128:2038-2044.

Rogers J, Wiener SG and Bloom FE (1979) Long-term ethanol administration methods for rats: advantages of inhalation over intubation or liquid diets. *Behav Neural Biol* 27:466-486.

Rothberg BS et al. (1996) Long-term effects of chronic ethanol on muscarinic receptor binding in rat brain. *Alcohol Clin Exp Res* 20:1613-1617.

Savage DD et al. (1991) Prenatal ethanol exposure decreases hippocampal NMDA-sensitive [3H]- glutamate binding site density in 45-day-old rats. *Alcohol* 8:193-201.

Simonyi A et al. (1996) Chronic ethanol on mRNA levels of IP3R1, IP3 3-kinase and mGluR1 in mouse Purkinje neurons. *Neuroreport* 7:2115-2118.

Southern EM (1975) Detection of specific sequences among DNA fragments separated by gel electrophoresis. *J Mol Biol* 98:503-517.

Sternberger LA and Sternberger NH (1986) The unlabeled antibody method: comparison of peroxidase-antiperoxidase with avidin-biotin complex by a new method of quantification. *J Histochem Cytochem* 34:599-605.

Tapia-Arancibia L et al. (2001) Effects of alcohol on brain-derived neurotrophic factor mRNA expression in discrete regions of the rat hippocampus and hypothalamus. *J Neurosci Res* 63:200-208.

Thorpe GH et al. (1989) Chemiluminescent enzyme immunoassay of alpha-fetoprotein based on an adamantyl dioxetane phenyl phosphate substrate. *Clin Chem* 35:2319-2321.

Towbin H, Staehelin T and Gordon J (1979) Electrophoretic transfer of proteins from poly-acrylamide gels to nitrocellulose sheets: procedure and some applications. *Proc Natl Acad Sci USA* 76:4350-4354.

Trevisan L et al. (1994) Chronic ingestion of ethanol up-regulates NMDAR1 receptor subunit immunoreactivity in rat hippocampus. *J Neurochem* 62:1635-1638.

van Zundert B, Albarran FA and Aguayo LG (2000) Effects of chronic ethanol treatment on gamma-aminobutyric acid(A) and glycine receptors in mouse glycinergic spinal neurons. *J Pharmacol Exp Ther* 295:423-429.

Vizi S, Palfi A and Gulya K (2000) Multiple calmodulin genes exhibit systematically differential responses to chronic ethanol treatment and withdrawal in several regions of the rat brain. *Brain Res Mol Brain Res* 83:63-71.

Woods JM, Ricken JD and Druse MJ (1995) Effects of chronic alcohol consumption and aging on dopamine D1 receptors in Fischer 344 rats. *Alcohol Clin Exp Res* 19:1331-1337.

Wu CH et al. (1995) Differential expression of GABAA/benzodiazepine receptor subunit mRNAs and ligand binding sites in mouse cerebellar neurons following in vivo ethanol administration: an autoradiographic analysis. *J Neurochem* 65:1229-1239.

Appendix 5.1
Detailed Protocols
for Western Immunoblotting

ANTIGEN SAMPLE PREPARATION

1. Measure the weight or volume of the sample to be lysed.
2. For each 1 ml or 1 g of sample add 5–10 volumes of phosphate-buffered saline (PBS) plus a protease inhibitor cocktail (for instance, protease inhibitor cocktail P-8340 from Sigma Chemical Co, St. Louis, MO).
3. Sonicate with an immersible tip sonicator for 15–30 sec to shear chromosomal DNA. Alternatively, pass the sample repeatedly through a 20-gauge needle and then through a 26-gauge needle.
4. Save an aliquot of the sonicated sample to determine total protein concentration by using an assay such as the Lowry or Bradford assays.
5. Mix the rest of the sonicated sample 1:1 v/v with 2X SDS-PAGE sample buffer (125 mM Tris-Cl pH 6.8, 4% w/v SDS, 20% w/v glycerol, 10% v/v 2-mercaptoethanol, and 0.05–0.1% w/v bromophenol blue). Vortex vigorously.
6. Most samples should be boiled for 3 min using screw cap tubes to prevent sample loss. Vortex again.
7. Spin the sample at 10,000 g for 10 min to pellet insoluble material. Store the supernatant in aliquots at –20°C (or at –70°C for long-term storage). The samples are now ready for electrophoresis.

SDS-PAGE GEL CASTING

The following recipe applies to a BioRad minigel system only. Other systems may require larger volumes of solutions. The reader should consult the manufacturer instructions.

1. Assemble the glass plate sandwich following the manufacturer instructions.
2. To prepare a 10% stacking gel, mix 2.41 ml of deionized water, 2 ml of 30 acrylamide/2.7 % w/v bis solution, 1.5 ml of 4X separating (lower) gel buffer (1.5 M Tris-base, pH 8.8), and 60 µl of 10% SDS. It is recommended that this mixture be degassed under a vacuum for 15 min to speed up polymerization. However, this step is not absolutely required. Unpolymerized acrylamide is toxic and it should be handled

with the appropriate safety precautions. Under constant stirring, add 30 μl of 10% w/v ammonium persulfate (prepared fresh daily in deionized water) and 2 μl of N,N,N',N'-tetramethylethylenediamine (TEMED). Allow this mixture to stir for 30 sec and then pour it smoothly into the glass plate sandwich to a level that should be about 1 cm below the teeth of the comb. Immediately overlay the solution with water saturated n-butanol (50 ml of n-butanol plus 5 ml of deionized water). Saturated n-butanol should be applied gently. Allow the gel to polymerize for 45 min to 1 hour. After the gel has polymerized, rinse the n-butanol thoroughly with deionized water and dry the area above the separating gel with filter paper.

3. The stacking gel should be prepared as the separating gel by mixing 2.44 ml of deionized water, 532 μl of 30 acrylamide/2.7 % w/v bis solution, 1.0 ml of 4X stacking (upper) gel buffer (0.5 M Tris-Cl, pH 6.8), and 40 μl of 10% w/v SDS. Under constant stirring, add 20 μl of fresh 10% w/v ammonium persulfate and 3 μl of TEMED. Allow the mixture to stir for 30 sec and then pour it on top of the polymerized separating gel. Insert the well-forming combs avoiding air bubbles from being trapped under the teeth. Allow the gel to polymerize for 30–45 min and then pull the comb straight up slowly and gently. Rinse the wells with running buffer (see below for composition).

4. Following the instructions of the manufacturer, attach the gel sandwich to the remaining components of the electrophoresis unit, fill the chambers appropriately with running buffer, and load the samples. As mentioned above, a gel should be loaded with 5, 10, 15, 20, 25, 30, 35 and 40 μl of the standard curve sample (1 mg/ml) to establish the dynamic working range of the assay. As an example, assume that the dynamic working range for the detection of a protein of interest by a particular antibody is between 10–40 μg of total protein. Then, subsequent gels could be loaded as follows:

TABLE 5.1

EXAMPLE OF GEL LOADING FOR A QUANTITATIVE IMMUNOBLOT EXPERIMENT

Molecular Weight Markers	Standard Curve				Samples			
	Std	Std	Std	Std	Control #1	Control #2	Treatment #1	Treatment #2
10-15 μl*	10 μg	20 μg	30 μg	40 μg	20 μ g	20 μ g	20 μ g	20 μ g

* This amount is for BioRad kaleidoscope prestained markers. Their use is recommended because they can be useful to determine blotting transfer efficiency. Molecular weight of the protein of interest should fall within the range of the markers.

Using 15-well combs, it would be possible to load marker and standard curve lanes as shown in Table 5.1 plus four control and four treatment samples (Figure 5.1). Wells 1 and 15 (i.e., at each end of the gel) should not be used because these lanes usually get distorted.

5. Run the minigel following the manufacturer's instructions. For BioRad minigels, running the gels at a constant voltage setting of 150 or 200 V for 1 h or 45 min, respectively, usually gives optimal resolution and minimal distortion. Running buffer should contain 25 mM Tris-base (pH 8.3), 192 mM glycine and 0.1% SDS. Make sure the bromophenol blue dye front is close to the bottom of the gel before stopping the run.

WET ELECTROPHORETIC TRANSFER

The following protocol applies to a BioRad Mini Trans-Blot® electrophoretic transfer gel system only. The reader should consult the manufacturer's instructions for other apparatuses.

1. Prepare the transfer buffer as described above and cool it at 4°C overnight.
2. Cut the membrane of choice and four sheets of Whatman 3-MM filter paper (or equivalent) to the size of the gel.
3. If using a nitrocellulose membrane, equilibrate it in transfer buffer for 5–15 min. If using a PVDF membrane, wet it in methanol for about 1 min, rinse with water and then equilibrate in transfer buffer. Equilibrate gel in transfer buffer for same length of time. Remember that membranes and gels must be handled with gloves to avoid contamination with skin oils and proteins.
4. Prepare the sandwich containing the gel and blotting membrane at its core. Submerge the clamping cassette with the dark side down into a tray containing transfer buffer. Orientation of the cassette is critical. The dark side in the cassette is color coded to match the negative electrode (cathode) in the BioRad Mini Trans-Blot® system. Submerge a filter pad and place it on top of the dark side of the cassette. Submerge two sheets of Whatman 3-MM filter paper and place on top of filter pad. Gently submerge pre-equilibrated gel and place on top of filter paper. Submerge pre-equilibrated nitrocellulose or PVDF membrane and place on top of gel. Gently roll out any air bubbles with a glass tube. Submerge two sheets of Whatman 3-MM filter paper and place on top of membrane. Submerge a filter pad and place on top of filter papers.
5. Close and lock cassette. Insert into transfer apparatus making sure that is is properly oriented. the BioRad Mini Trans-Blot® system requires the use of a cooling pack, which must be inserted into the unit now.
6. Fill tank with cold transfer buffer, add stirring bar to help maintain even transfer temperature, close lid, connect to power supply and run at 100 V for 1 h. It is not recommended that transfer buffer be re-used, as it looses buffering capacity after each use.

7. Turn off power supply and disassemble cassette. Gently remove membrane and place in a shallow tray or box. Rinse 2 × 5 minutes with Tris-Buffered Saline (TBS), which is composed of 50 mM Tris-base (pH 7.5) and 150 mM NaCl.

8. Determine the completeness of transfer by comparing the amount of prestained molecular weight markers present in the gel vs. the membrane. Alternatively, stain membranes following a method that does not distort antigen detection, such as Ponceau S (Harlow, 1988 #5857). (This step is optional.)

6 Artificial Bilayer Techniques in Ion Channel Study

Steven N. Treistman, Robert O'Connell, and John Crowley

CONTENTS

6.1 INTRODUCTION

Many of the early studies of effects of alcohol in the nervous system have been framed and interpreted to determine whether the primary target of alcohol's action is the lipid or the protein components of brain. We suggest that it is necessary to consider functioning membrane proteins and their lipid environment (as well as the various interfaces between them, such as lipid–protein, lipid–protein–water, and protein–lipid–protein, etc.) as a dynamic system, in which the small amphiphilic alcohol molecule will interact simultaneously with a number of targets.

That lipids play an important role in the effects of alcohol on neural function is suggested by the great diversity of lipids in neuronal membranes, the large influence of lipid composition on channel protein function, and apparent compensatory changes in lipid composition that occur as a function of chronic drug exposure. A vast body of work now addresses the influence of chronic ethanol exposure on membrane lipid composition and function. However, results from different laboratories are often at odds. The complexity of natural membranes and the numerous and interlinked lipid metabolism pathways make a reasonable analysis of lipid involvement in alcohol's actions in intact animals and tissue difficult.

Understanding how the interactions among protein subunits, lipids, and water are affected by alcohol is best accomplished in a very simplified system, where the function of an isolated channel protein is studied in a reconstituted planar lipid bilayer. This technique has proven to be a very powerful method to elucidate the role of lipids in channel modulation. If an artificial membrane is used to incorporate cloned channels, then the protein and the lipid environment can be manipulated allowing the control of enough variables to permit a meaningful assessment of the role of lipid environment on drug action on protein targets.

6.2   BUILDING A PLANAR LIPID BILAYER SETUP

Due to the fragile nature of the lipid bilayer, and the need to maintain an adequate signal-to-noise ratio, the recording chamber must be free of electrical interference, vibrations, and mechanical disturbances. As such, it is important to construct the bilayer rig in an area of the laboratory with little foot traffic and background noise. To eliminate floor vibrations, place the recording chamber on a vibration isolation table. The table is constructed of a large metal top that is supported on a bed of compressed nitrogen gas. They are commercially available from several sources, including Technical Manufacturing Company (Peabody, MA) and Kinetic Systems (Boston, MA). Cost-effective alternatives include placing the metal slab on a partially inflated motorcycle innertube (Alvarez, 1986), tennis balls, or pneumatic shock absorbers (Hanke and Schlue, 1993a). To eliminate electrical interference, the chamber is placed in a Faraday cage, along with the probe of the patch-clamp amplifier used to measure the currents. The cages are also commercially available, but adequate homemade versions can be constructed. Cages can be built inexpensively with wood frames and copper or aluminum mesh, or from plate aluminum attached directly to the vibration isolation table. Additionally, the recording chamber is placed in an aluminum box with a closing top to further reduce electrical interference. Large cages are convenient when a microscope is required for viewing the chamber. It is recommended that any piece of equipment that will contact the chamber, notably the perfusion system, also be shielded within the Faraday cage. Finally, the interior of the Faraday cage can be lined with sound-deadening foam, to prevent the bilayer recording setup from picking up acoustic noise from laboratory equipment and personnel.

Bilayer recording chambers come in a variety of styles. In this section, only chambers for use with painted bilayers will be considered. Figure 6.1 shows schematic drawings of two types of bilayer recording chambers. The chambers are commonly milled from a block of Teflon or Delrin plastic. Vertical bilayer chambers consist of two intersecting circular holes, one of which is fitted with a polycarbonate

Horizontal Bilayer Chamber

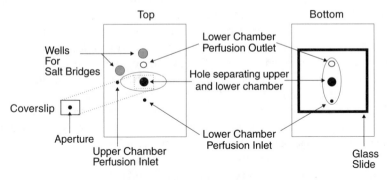

Vertical Bilayer Chamber

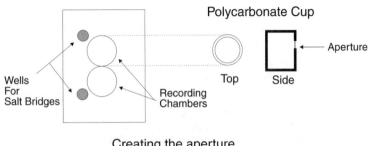

Creating the aperture

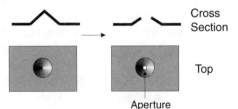

FIGURE 6.1 Schematic of bilayer recording chambers.

cup. The cup contains the aperture across which the bilayer is painted. The other chamber is constructed with a clear window to allow visualization of the bilayer with a microscope. Horizontal bilayer chambers contain two holes milled on either side of the plastic block, separated by a thin layer of plastic with a hole in the center. The bottom chamber is enclosed with a glass microscope slide, affixed to the Teflon/Delrin using a silicone sealant. A plastic microscope coverslip is attached to the bottom of the upper chamber. This coverslip acts as the divider between the two chambers and contains the aperture across which the bilayer will be formed. A 7:3 mixture of wax:Vaseline is used to attach the coverslip to the plastic chamber. Both chamber styles feature wells to insert the Ag/AgCl electrodes, and salt bridges that connect them to the recording chambers. In addition, ports can be added to attach perfusion and vacuum lines. Perfusion of the bilayer should be designed in a way that minimizes the mechanical disturbance of the chamber. When working with horizontal bilayer chambers, a gravity-driven setup combined with a vacuum line is sufficient. The perfusion input and the vacuum line are placed on opposite sides of the chamber, and the rate of perfusion is best maintained below 1 ml/min. It is easier to perfuse the grounded chamber, whether upper or lower, as even a shielded perfusion line can act as an antenna. This method of bilayer perfusion produces noise in the current trace, and is useful mainly for bath exchange of the chamber during breaks in the recording period. If low-noise records are required during the perfusion of the chamber, a more elaborate perfusion system is required. Hanke and Schlue (1993a) present a push/pull motor driven syringe system designed to maintain the volume of the chamber very precisely during perfusion.

The apertures in the polycarbonate cups or plastic coverslips can be formed using a technique developed by Wonderlin, Finkel and French (1990). A small metal cone, or stylus, is attached to a power supply, which heats it. A foot pedal is used to activate the heating element, leaving both hands free for manipulating the plastic. The cup or coverslip is pressed against the heated metal cone until a small depression forms. Once the depression is formed, the foot pedal is released and the plastic is held in place manually until it cools. After cooling, the cone-shaped depression in the cup or coverslip is shaved with a razorblade under a microscope to produce a circular hole. It is important to hold the blade level while shaving across the top of the cone. Use only apertures that are round, level, and smooth, to promote bilayer stability. Hole size is controlled by shaving at higher points on the cone, and working downward to increase the diameter. Larger apertures will produce bilayers of larger capacitance, promoting incorporation, but yielding noisier recordings. Smaller holes allow more stable recordings, but the decrease in bilayer surface area reduces the incorporation efficiency. Holes on the order of 100 μm in diameter produce bilayers that yield a nice balance of stability and incorporation efficiency.

The electrodes in the chamber are attached to the headstage of a patch-clamp amplifier. Bilayer-specific models are commercially available, though any patch-clamp amplifier can be used, provided the feedback resistance is on the order of 10 GΩ. The large capacitance of the bilayer produces considerable background noise, and allows only poor temporal resolution of voltage steps due to large capacitive transients. Bilayer-specific amplifiers circumvent this problem by implementing either integrating headstages or switching-resistive headstages that allow rapid charging of the bilayer membrane. A typical resistive-headstage patch amplifier will perform adequately when recording steady-state activity of a channel at a particular voltage. A review of methods for maximizing bandwith and the resolution of voltage steps, while reducing background noise, can be found in Chapter 5 of the Axon Guide (Sherman-Gold, 1993).

The design and analysis of electrophysiological experiments are commonly performed using a suite of software programs such as pClamp, designed by Axon Instruments. The software communicates directly with the amplifier, allowing manipulation of the holding potential, and the generation of voltage pulses. As a result, a direct memory access (DMA) interface is required to allow communication between the digital signals of the personal computer (PC) and the analog signals of the amplifier. The interface connects to the PC, and usually runs off of the PC power supply. If extremely low-noise recordings are necessary, it may be beneficial to use an interface with an isolated power supply. The output of the amplifier can be stored directly on the computer, or on a tape recorder (Bezanilla, 1985). When the voltage protocol involves voltage pulses with short durations, storage directly onto the PC is acceptable. However, a tape recorder is recommended when obtaining long records of steady-state channel activity at a particular voltage. Data can be stored directly onto a digital audio tape (DAT) recorder, or onto a Betamax tape with the use of a pulse code modulator (PCM) to convert the amplifier analog output to a digital signal. A DAT recorder designed to collect data in this fashion is available from Dagan Corporation (Minneapolis, MN). The advantages of this mode of recording include large storage capacity and ease of data retrieval, as records are stored as separate tracks on the tape. The Betamax/PCM combination works adequately, though Betamax tape availability is somewhat limited. An eight-pole low-pass Bessel filter is useful for resolving the data during an experiment, though it is best to store the data at the bandwith at which it leaves the amplifier. This allows the data to be filtered at a desired frequency when the records are played back from the tape and stored onto the PC. It is useful to view the voltage pulses and corresponding currents on an oscilloscope, both during the experiment and during playback of data stored on tape. Figure 6.2 is a schematic of the electronics and connections in a typical bilayer recording setup.

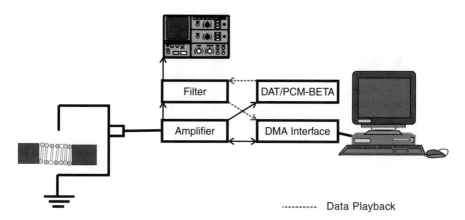

FIGURE 6.2 Planar lipid bilayer recording setup.

6.3 BIOCHEMICAL PREPARATIONS FOR RECONSTITUTION

The planar bilayer technique has a number of advantages. It allows control of both the lipid environment of the channel and the aqueous phases at the intra- and extracellular face of the channel. In addition, reconstitution of channel proteins from biochemical preparations circumvents the issue of accessibility when studying channels in membranes not easily patched with a glass micropipette. Investigators can isolate a native channel from a specific tissue, or transfect a cell line with a cloned channel of interest. The use of cloned channels adds another level of control, because manipulation of the channel protein is possible before insertion into the bilayer. Crude membrane fractions containing the channel of interest are suitable for incorporation into artificial membranes. Below, a procedure is outlined for isolating both native and cloned channels. The channel type discussed in each case is the large conductance Ca^{++}-activated K^+ channel (BK), studied extensively using reconstitution techniques. For the ethanol researcher, this channel represents a functionally relevant and easily accessible target. Data from our laboratory have demonstrated both acute (Dopico et al., 1996; 1998; 1999) and chronic (Knott et al., 2000) channel regulation by ethanol. In addition, acute regulation of the drug persists in the planar lipid bilayer, when addressed using both native (Chu et al., 1998; Walters et al., 2000) and cloned (Crowley et al., 2000) channels. The crude membrane fractions derived from these sources yield a plentiful supply of channel proteins that readily incorporate into painted bilayers. The reader is encouraged to refer to the original citations for experimental detail, as well as an article that covers ion channel reconstitution (Favre et al., 1999).

6.3.1 RAT SKELETAL MUSCLE T-TUBULE PREPARATION

A protocol for isolating T-tubule membranes from rabbit skeletal muscle was first described by Rosemblatt and colleagues (1981). This technique was adapted by Moczydlowski and Latorre (1983a) to isolate rat skeletal muscle T-tubules, from which they described in detail the gating kinetics of a Ca^{++}-activated K^+ channel reconstituted into planar lipid bilayers (Moczydlowski and Latorre, 1983b). This preparation was also used to quantitate differences in BK channel activity in bilayers composed of neutral phosphatidylethanolamine (PE) vs. negatively charged phosphatidylserine (PS) (Moczydlowski et al., 1985). This study demonstrates not only the importance of bilayer surface charge for channel function, but that lipid exchange occurs between the artificial membrane and the incorporated membrane fragments. In our laboratory, rat t-tubule BK channels incorporated into PE/PS (3:1 w/w) bilayers show a dose-dependent increase in activity upon addition of ethanol (Chu et al., 1998).

6.3.2 CRUDE MEMBRANE FRACTIONS FROM CULTURED CELLS

The expression of cloned channels in cultured cells provides a number of experimental advantages for reconstitution experiments. The high protein expression levels and relative ease of culture maintenance provide an abundant supply of channels for incorporation. Experiments undertaken with cloned proteins also have the obvious advantage of manipulation of the protein sequence before expression and reconstitution.

HEK-293 cells are a common line used for expression of cloned BK channels. The cells express an array of endogenous chloride and potassium channels, but no Ca^{++}-dependent currents are detected. In addition, neither BK blocker charybdotoxin nor small-conductance Ca^{++}-activated K^+ channel (SK) blocker apamin can inhibit endogenous outward currents (Yu and Kerchner, 1998; Zhu et al., 1998). Therefore, confirmation that a cloned channel has incorporated into a bilayer is easily accomplished by altering the free Ca^{++} in the recording solution. Chinese hamster ovary (CHO) cells and COS cells are also commonly used as expression systems.

The array of endogenous channels, ease of transfection, and the protein expression levels are all important considerations when choosing a cell line to express a cloned channel for reconstitution. An effective protocol for isolation of membrane fragments from cultured cells is described by Sun, Naini, and Miller (Sun et al., 1994).

6.3.3 LIPIDS USED IN RECONSTITUTION EXPERIMENTS

The lipids used for bilayer reconstitution experiments are commercially available from Avanti Polar Lipids (Alabaster, AL), in both natural and synthetic forms. The lipids are purchased either as a powder or dissolved in chloroform. For a reconstitution experiment, an appropriate phospholipid mixture is aliquotted and mixed under N_2 gas. The selection of the lipid mixture is influenced both by the experimental design and by the need to promote vesicle incorporation. The lipid mixture is vortexed and dried under N_2 gas. The dried lipid is resuspended at the desired concentration in decane. The decane solvent is air-sensitive and should be stored under N_2 gas.

Most reconstitution experiments employ mixtures of PE, PS, and phosphatidylcholine (PC). PE/PS bilayers mixed in a 3:1 (w/w) ratio are stable, and readily incorporate channels from both membrane preparations described above. In general, mixtures should contain PE and acidic lipids such as PS and phosphatidylinositol (PI) to promote vesicle incorporation. In addition, certain channels require specific lipid species to function properly, such as the requirement of cholesterol for the nicotinic acetylcholine receptor (Barrantes, 1989). These are important factors to consider when choosing the bilayer composition. Experimentally, diverse lipids are often used as probes to investigate their influence on ion channel function through alterations in physical properties such as the headgroup size and charge, as well as the degree of acyl chain saturation.

6.4 RECORDING FROM THE PLANAR LIPID BILAYER

To prepare the cup or coverslip for a reconstitution experiment, the aperture is pretreated with the lipid mixture dissolved in decane. A sable-hair paintbrush, size 000 or 0000, is used to paint a drop of the mixture across the hole. The brush is most effective when only two bristles, positioned directly next to one another, are left in the tip. To prepare the brush, use a microscope and a pair of microdissection scissors to snip the bristles. To reduce contamination across experiments, a separate brush should be prepared for use with a particular lipid mixture. Capillary action allows a small amount of the decane and lipid mixture to stick to the bristles, which is deposited across the aperture. The drop should fill the entire opening without flooding the coverslip. Over the course of 1 min, the drop will dry and leave a rim of dried lipid around the edge of the hole. Allow the cup or coverslip to dry for 5 min before recording.

Once the coverslip is dry, the recording chamber can be assembled with the electrodes and salt bridges. Both chambers are filled with the appropriate recording solutions, and the amplifier is switched on. The lipid mixture is again brushed across the aperture. Instead of a paintbrush, a glass pipette with a rounded tip (formed by rotating the pipette over a Bunsen burner) is used. Once cooled, the pipette tip is dipped in the decane/lipid mixture, and brushed across the aperture.

While attempting to form a bilayer, a repetitive triangular voltage waveform is maintained across the aperture to monitor bilayer formation. The amplitude and rate are set to a desired value (20 mV/25 ms, for example). The capacitance of the bilayer can be determined using the equation $I = (C)(dV/dt)$, where I is the capacitive current amplitude, C is the capacitance, and dV/dt corresponds to the change in voltage over time (known from the size and duration of the triangle pulse). Before recording from the bilayer, the triangle pulse is presented using a series of known capacitors. The size of the resulting capacitive current is plotted against the known capacitance to produce a standard curve. This is used to determine bilayer capacitance when the triangle pulse is run across the artificial membrane. The size of the current will be proportional to the size of the aperture, and it should be very square in shape. The capacitive current should retain its size and shape for several min before adding the channel preparation.

The method for incorporation of channel proteins is slightly different in vertical vs. horizontal bilayer chambers. For horizontal chambers, a pipette is used to drop 0.5 µl of channel preparation above the aperture. A long gel-loading pipette tip is recommended, as it allows a slow, controlled release of the membrane preparation. The force of gravity will pull the channel-containing fragments down onto the bilayer. For vertical bilayer setups, a stirring mechanism is required to drive channel incorporation. Regardless of the chamber type used, a number of experimental conditions will promote channel incorporation. The lipid mixture used to form the bilayer should contain PE and some proportion of negatively charged lipids such as PI or PS. The *cis* chamber, to which the membrane preparation is added, should contain some free Ca^{++}, and be hyperosmotic relative to the *trans* chamber (Miller and Racker, 1976; Miller et al., 1976). In cases where vesicles or liposomes are added, they should be prepared such that the interior is hyperosmotic relative to the bath solution. The mechanisms underlying these requirements are not completely understood, and remain rigorously studied because vesicle fusion plays a vital role in many biological processes. All of these conditions can be optimized for a given membrane preparation, with some trial and error. More-extensive explanations of the principles of channel incorporation are available in reviews by Labarca and Latorre (1992), and chapters in Ion Channel Reconstitution (Hanke, 1986) and Planar Lipid Bilayers: Methods and Applications (Hanke and Schlue, 1993b).

Once channel openings appear in the bilayer recordings, further incorporation events can be prevented by neutralizing the osmotic gradient or dropping the $[Ca^{++}]_{Free}$ in the *cis* chamber. This can be accomplished by perfusion, or by addition of salts or Ca^{++}-chelators to the appropriate chamber. In cases where multiple channel openings are visible, the number of channels can be determined by manipulating conditions (such as membrane potential) to maximize the open probability (P_o) of the channel. At high potentials, the likelihood of all channels simultaneously entering the open state is very high, and the number of channels can be determined by computing the number of elemental contributions necessary to reach the cumulative current. It is useful to record the steady-state channel activity at a number of voltage steps before testing the effects of ethanol to ensure that basic channel properties, such as conductance and voltage sensitivity, are normal.

Before adding ethanol, a control record should be recorded for at least 1 min at a particular voltage. It is important that the activity of the channel be stable, as any rundown or increase can skew the effects of the drug. After the addition of ethanol, periodically monitor the capacitive current to determine the stability of the bilayer. Record the channel activity at several voltage steps in the presence of ethanol, to allow comparison of the slope conductance and voltage sensitivity in the presence and absence of the drug.

6.5 ANALYSIS OF SINGLE CHANNEL DATA

Data stored on a tape recorder is reacquired in the gap-free recording mode of a program such as Fetchex, available in the pClamp suite of software from Axon Instruments. It is recommended that data generated during reconstitution experiments be stored unfiltered at the bandwith at which it leaves the amplifier (10 kHz, for example). The records can be passed through an eight-pole low-pass Bessel filter during reacquisition, and filtered at a desired frequency. This allows one record to be used for detailed kinetic analysis, by resolving short events with minimal filtering, or to be low-pass filtered at a lower frequency for display.

Records of steady-state single channel activity yield a wealth of information. Programs such as Fetchan, another pClamp module, are designed to facilitate the analysis of the data. The first step in analyzing a channel record is the construction of an all-points amplitude histogram. The histogram will show a peak, with a Gaussian distribution, for both the closed and open state of the channel. A least-squares function is used to fit the histogram generated from the channel activity record at a given voltage. This will yield both the open probability (P_o) of the channel and the amplitude of a single channel event (i) at that voltage. This exercise is repeated at a number of holding potentials, so that both i and P_o can be plotted against voltage. From this, both single channel conductance (g) and voltage sensitivity can be determined.

Comparison of the channel P_o in the presence and absence of ethanol, under otherwise identical conditions, provides a straightforward means of assessing the overall effect of the drug on channel activity. In single-channel bilayers, the analysis can be carried a step further by constructing and analyzing an events list. An events list is displayed as a histogram with the number of observations plotted vs. event duration. This histogram is fit with a series of exponential functions, using a maximum-likelihood estimator, to uncover the time constants for a particular channel state. An F-statistic table is used to determine the minimum number of exponentials that adequately fit the histogram. In practical terms, an open state histogram that is best fit with two exponentials indicates two kinetically distinct channel openings, a short and a long open state. The fit of the histogram provides not only the duration of these kinetic states, but also the proportion of time the channel spends in them. The analysis

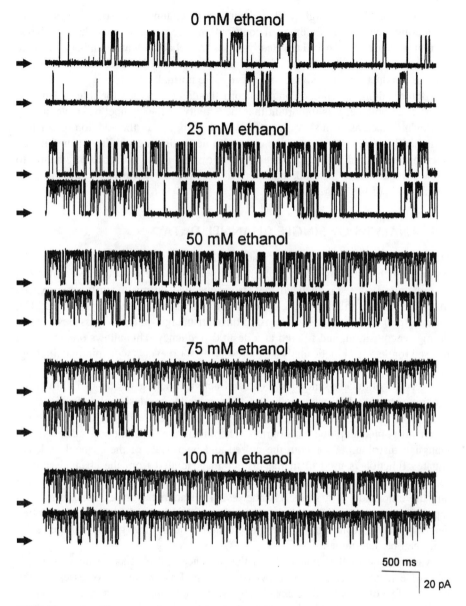

FIGURE 6.3 Ethanol enhances the activity of a single BK channel, from rat skeletal muscle t-tubule, incorporated into a PE/PS bilayer in a concentration-dependent manner (holding potential = 0 mV, ~8.1 µM free $[Ca^{2+}]_{ic}$). Arrows, closed levels.

is performed for both the open and closed events. Events lists can be generated from channel records before and after the addition of ethanol. Comparison of this data demonstrates how the drug modifies the open and closed states of the channel.

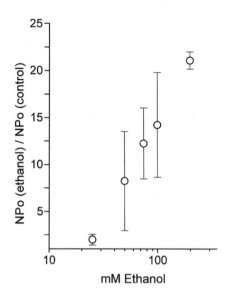

FIGURE 6.4 Cumulative concentration-response relation for ethanol's activation of BK channels. The data are from 11 bilayers with either a single channel, or two channels.

6.6 EXAMPLES OF DATA OBTAINED USING THE PLANAR LIPID BILAYER

Figure 6.3 shows representative traces of a rat t-tubule BK channel incorporated into a PE/PS bilayer, in the presence of increasing concentrations of ethanol. The drug elicits a concentration-dependent increase in channel activity, under the same conditions of $[Ca^{++}]_{Free}$ and voltage. Figure 6.4 is a concentration-response curve that summarizes the extent to which the channel is activated by ethanol. The data are plotted as the ratio NP_0 (ethanol)/NP_0 (control) for increasing concentrations of the drug, where N is the number of channels in the bilayer and P_0 is channel activity. Figure 6.5 provides an example of an events list analysis of the effect of ethanol on a single BK channel. The data demonstrate that the most prominent effect of ethanol is a reduction in the duration and proportion of time spent in the long closed kinetic state. Figure 6.6 shows representative traces of a cloned hSlo BK channel (hbr1, graciously provided by Dr. P.K. Ahring) before and after the addition of 50 mM EtOH. Ethanol enhances the activity of the cloned channel to a similar extent to that seen for the native rat t-tubule channel. These data demonstrate that the BK protein (or protein complex), in the absence of complex lipid architecture or cytoplasmic elements, is capable of responding to ethanol in a manner similar to that of the channel *in situ*, and makes obvious the ability to manipulate the lipid environment and assess the consequences on drug action.

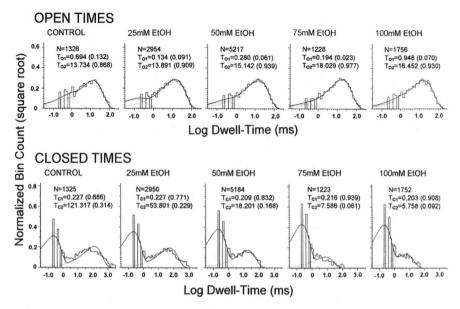

FIGURE 6.5 Open and closed time distributions in the absence or presence of ethanol. Figure shows data from a representative BK channel reconstituted into a PE/PS bilayer, before and after exposure to different concentrations of ethanol. Each panel shows total number of events (N), duration of each particular component (T, in msec), and relative contribution of each particular component to total fit (in parentheses).Number of events were normalized before applying a Sigworth and Sine transformation (Sigworth & Sine, 1987), which allows a better resolution of individual components. Dotted line = individual fitted components; solid line = composite fit.

6.7   ADVANCED APPLICATIONS

6.7.1   ASYMMETRIC BILAYERS

Most of the planar bilayer work described in the literature, and above, utilizes symmetric bilayers, in which each of the leaflets is identical. However, biological membranes are not typically symmetrical, and a more accurate representation of drug action on membranes and associated proteins will involve the generation of asymmetric bilayers. Described simplistically, the formation of asymmetric bilayers is often accomplished by the apposition of two different monolayers in troughs separated by a movable partition containing a small aperture. As the partition is removed, the monolayers are forced together to form an asymmetric bilayer, often referred to as a folded bilayer (Montal and Mueller, 1972; Tancrede et al., 1983; Heywang et al., 1998; Cassia-Moura et al., 2000). It is important to recognize that the final organization of the folded bilayer may not represent total segregation of the monolayer components from each other.

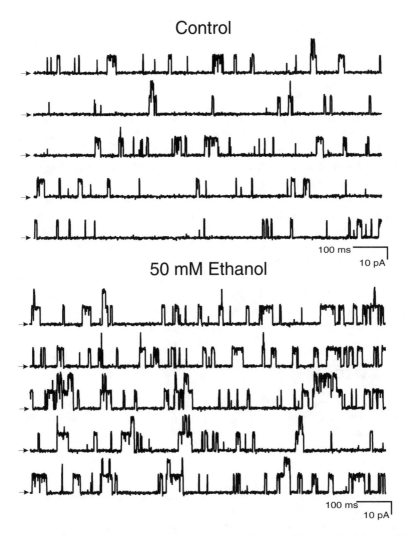

FIGURE 6.6 Representative traces of a cloned hSlo BK channel incorporated into a PE/PS bilayer in the absence and presence of 50 mM EtOH (holding potential = 0 mV, ~25 μM $[Ca^{2+}]_{ic}$). The bilayer contains two channels. Arrows = closed levels.

6.7.2 SUPPORTED BILAYERS

Another "advanced" technique involves the formation of the bilayer on solid or semi-solid platforms, enhancing their stability and allowing conditions and measurements not possible with the previously described painted bilayer. These include ordering the lipids in the bilayer (Heywang et al., 1998), simultaneous electrical and optical measurement of single ion channels (Ide and Yanagida, 1999), and the use of a polymer support platform associated with tethered bilayer-imbedded proteins, allowing lateral movement and monitoring of the protein (Wagner and Tamm, 2000).

6.8   CONCLUSIONS

The techniques and data described in this chapter illustrate artificial planar bilayer techniques, and demonstrate how these techniques can be used to study the influence of the lipid environment on protein function and drug action. We have focused on electrophysiological measurements of neuronal channels incorporated into bilayers. Although the techniques of planar bilayer formation and protein incorporation must be practiced before they become routine, they contain enough unknowns to still be something of an art. However, once incorporation and recordings are reliably obtained, the analysis of the emerging records is not different from those of any single channel experiment utilizing patch clamp technology. Of course, the nature of the artificial bilayer experiment is very reductionist, and as with any reductionist experiment, caution must be exercised in the interpretation of results, because proteins (probably) do not exist in a one- or two-lipid environments. However, the information obtained from these simple experiments can be used to derive hypotheses testable in more complicated lipid environments.

ACKNOWLEDGMENTS

This work is supported by NIH grants from the National Institute on Alcohol Abuse and Alcoholism AA12054 (S.N.T.) and Predoctoral Fellowship AA05548 (J.C.).

REFERENCES

Alvarez, O. (1986) How to set up a bilayer system, in: *Ion Channel Reconstitution* (Miller, C Ed) pp.115-139. New York:Plenum Press.

Barrantes FJ (1989) The lipid environment of the nicotinic acetylcholine receptor in native and reconstituted membranes. *Crit Rev Biochem Mol Biol* 24:437-478.

Bezanilla F (1985) A high capacity data recording device based on a digital audio processor and a video cassette recorder. *Biophys J* 47:437-441.

Cassia-Moura R et al. (2000) The dynamic activation of colicin Ia channels in planar bilayer lipid membrane. *J Theor Biol* 206:235-241.

Chu B et al. SN (1998) Ethanol potentiation of calcium-activated potassium channels reconstituted into planar lipid bilayers. *Mol Pharmacol* 54:397-406.

Crowley, J, Dopico, AM and Treistman, SN Ethanol Potentiation of Cloned BK Channels Incorporated into Planar Lipid Bilayers. *Soc Neurosci Abstr* 26:1402. (2000)

Dopico, AM. Anantharam, V and Treistman, SN (1998) Ethanol increases the activity of Ca^{++}-dependent K^+ (mslo) channels: functional interaction with cytosolic Ca^{++}. *J Pharmacol Exp Ther* 284:258-268.

Dopico, AM Lemos, JR and Treistman, SN (1996) Ethanol increases the activity of large conductance, Ca^{++}-activated K^+ channels in isolated neurohypophysial terminals. *Mol Pharmacol* 49:40-48.

Dopico AM R et al. (1999) Rat supraoptic magnocellular neurones show distinct large conductance, Ca^{++}-activated K+ channel subtypes in cell bodies vs. nerve endings. *J Physiol* 519 Pt 1:101-114.

Favre I Sun YM and Moczydlowski E (1999) Reconstitution of native and cloned channels into planar bilayers. *Methods Enzymol* 294:287-304.

Hanke W (1986) Incorporation of ion channels by fusion, in: *Ion Channel Reconstitution* (Miller C Ed) pp.141-153 New York:Plenum Press.

Hanke W and Schlue WR (1993a) Technical details of bilayer experiments, in: *Planar Lipid Bilayers: Methods and Applications* pp. 24-43 San Diego:Academic Press.

Hanke W and Schlue WR (1993b) Incorporation of proteins into planar lipid bilayers, in: *Planar Lipid Bilayers:Methods and Applications* pp.79-92 San Diego:Academic Press.

Heywang CR et al. (1998) Orientation of anthracyclines in lipid monolayers and planar asymmetrical bilayers: a surface-enhanced resonance Raman scattering study. *Biophys J* 75:2368-2381.

Ide T and Yanagida T (1999) An artificial lipid bilayer formed on an agarose-coated glass for simultaneous electrical and optical measurement of single ion channels. *Biochem Biophys Res Commun* 265:595-599.

Knott TK et al. (2000) Chronic Exposure to Ethanol (EtOH) Reduces Sensitivity of BK Channels to Acute EtOH in Neurohypophysial Terminals. *Soc Neurosci Abstr* 26:1819.

Labarca P and Latorre R (1992) Insertion of ion channels into planar lipid bilayers by vesicle fusion. *Meth Enzymol* 207:447-463.

Miller C et al. (1976) Ca^{++}-induced fusion of proteoliposomes: dependence on transmembrane osmotic gradient. *J Membr Biol* 30:271-282.

Miller C and Racker E (1976) Ca^{++}-induced fusion of fragmented sarcoplasmic reticulum with artificial planar bilayers. *J Membr Biol* 30:283-300.

Moczydlowski, E et al. (1985) Effect of phospholipid surface charge on the conductance and gating of a Ca^{++}-activated K^+ channel in planar lipid bilayers. *J Membr Biol* 83:273-282.

Moczydlowski, E and Latorre, R (1983a) Saxitoxin and oubain binding activity of isolated skeletal muscle membrane as indicators of surface origin and purity. *Biochim Biophys Acta* 732:412-420.

Moczydlowski E and Latorre R (1983b) Gating kinetics of Ca^{++}-activated K^+ channels from rat muscle incorporated into planar lipid bilayers. Evidence for two voltage- dependent Ca2+ binding reactions. *J Gen Physiol* 82:511-542.

Montal M and Mueller P (1972) Formation of bimolecular membranes from lipid monolayers and a study of their electrical properties. *Proc Natl Acad Sci USA* 69:3561-3566.

Rosemblatt M et al. (1981) Immunological and biochemical properties of transverse tubule membranes isolated from rabbit skeletal muscle. *J Biol Chem* 256:8140-8148.

Sherman-Gold, R (1993) Advanced methods in electrophysiology, in: *The Axon Guide* 122-132. Axon Instruments.

Sun T Naini AA and Miller C (1994) High-level expression and functional reconstitution of Shaker K^+ channels. *Biochemistry* 33:9992-9999.

Sigworth FJ and Sine SM (1987) Data transformations for improved display and fitting of single-channel dwell time histograms. *Biophys J* 52:1047-1054.

Tancrede P et al. (1983) Formation of asymmetrical planar lipid bilayer membranes from characterized monolayers. *J Biochem Biophys Meth* 7:299-310.

Wagner ML and Tamm LK (2000) Tethered polymer-supported planar lipid bilayers for reconstitution of integral membrane proteins: silane-polyethyleneglycol-lipid as a cushion and covalent linker. *Biophys J* 79:1400-1414.

Walters FS, Covarrubias M and Ellingson JS (2000) Potent inhibition of the aortic smooth muscle maxi-K channel by clinical doses of ethanol. *Am J Physiol Cell Physiol* 279:C1107-C1115.

Wonderlin WF, Finkel A and French RJ (1990) Optimizing planar lipid bilayer single-channel recordings for high resolution with rapid voltage steps. *Biophys J* 58:289-297.

Yu SP and Kerchner GA (1998) Endogenous voltage-gated potassium channels in human embryonic kidney (HEK293) cells. *J Neurosci Res* 52:612-617.

Zhu G et al. (1998) Identification of endogenous outward currents in the human embryonic kidney (HEK 293) cell line. *J Neurosci Meth* 81:73-83.

7 Rapid Drug Superfusion and Kinetic Analysis

David M. Lovinger, Suzanne Sikes and Qing Zhou

CONTENTS

7.1 INTRODUCTION

Ligand-gated ion channels are a family of proteins specialized for the very rapid generation of transmembrane electrical current in response to binding of a neurotransmitter or other ligand. The ligand-gated ion channel subtypes that respond to neurotransmitters are found throughout the nervous system and come in several different varieties specifically activated by certain neurotransmitters. These include the subfamilies of ligand-gated receptor channels similar to the nicotinic acetylcholine receptors (nAChRs), the ionotropic glutamate receptors, and the ionotropic ATP receptors (see Lovinger, 1997 for review). The functional hallmarks of these proteins

159

are their ability to bind neurotransmitter and to open an ion channel that is an intrinsic part of the receptor protein complex. Normally, these proteins reside at synapses where concentrations of neurotransmitter can reach near-millimolar levels within microseconds and dissipate nearly as rapidly. Thus, the ligand-gated ion channels normally operate on a very fast time scale dictated by the rapid transient rises in synaptic neurotransmitter concentrations. By necessity, the kinetic state transitions within the receptor channels must occur very rapidly. This presents special challenges for the study of ligand-gated ion channel function because the investigator must take into account the normal pattern of neurotransmitter release, as well as the rapid state transitions of the channels when designing such experiments.

A large literature has now accumulated indicating that ethanol alters the function of ligand-gated ion channels (Forman et al., 1995; Harris et al., 1995; Lovinger, 1997; Mihic et al., 1997). Clinically relevant concentrations of ethanol have been demonstrated to significantly potentiate the function of muscle nAChR, γ-aminobutyric acid type A (GABA$_A$), glycine, and 5-hydroxytrypyamine-3 (5-HT$_3$) receptors and inhibit the function of N-methyl-D-aspartate (NMDA) and non-NMDA glutamate receptors (Diamond and Gordon, 1997; Lovinger, 1997; Grant, 1995). Evidence that some or all of these channels contribute to alcohol's intoxicating effects (Grant, 1995; Diamond and Gordon, 1997; Lovinger, 1997) is mounting. Thus, studies of acute effects of ethanol on the function of these channels have become increasingly common. A variety of techniques have been developed for drug application to receptor-containing cells, as well as analysis of the resultant data both in terms of pharmacological properties of the receptor channels and the kinetics of these channels. With this explosion of studies in the area, it is a good time to consider optimal techniques for measuring pharmacology and function of these receptors.

This chapter will focus on methods for experimental observation and analysis of the kinetics of ligand-gated ion channels. Methodology for examining channel function on the proper time scale will be discussed. In addition, measurement of ion current kinetics, as well as methods for simulation and model fitting for currents, will be considered. We will focus on whole-cell currents because the majority of work with alcohols has used whole-cell recording preparations. Much of the information presented herein is applicable to the analysis of single-channel data. However, there are several specialized aspects of single-channel analysis, and thus the reader is referred to other sources for in-depth consideration of single-channel measurement and kinetic analysis (e.g., Sakmann and Neher, 1995).

7.2 PATCH-CLAMP RECORDING WITH RAPID DRUG SUPERFUSION

7.2.1 Brief Primer on Patch-Clamp Electrophysiology

Patch-clamp methodology has become the preferred approach for measuring ionic current in small mammalian cells, including most neurons. This methodology has been described in detail in several volumes (e.g., Sakmann and Neher, 1995; Hille, 1992), and thus, we will not present a detailed description of patch-clamp techniques in this chapter. However, a few features of the technique that make it advantageous for transmembrane current measurement must be emphasized. Patch-clamp recordings are made by forming a GΩ resistance seal between the tip of a blunt-ended glass microelectrode and the membrane of a cell. This high-resistance seal allows the investigator to record either from a tiny patch of membrane on the cell or excised from the cell. Alternatively, a recording can be made in the "whole-cell" mode, in which the investigator can break a small hole in the cell membrane and record current passing through the entire cell membrane. This "tight-seal" recording not only allows for ultra-low noise recording and low-resistance current-passing capabilities for faithful voltage control, but has the added advantage of allowing the investigator to move cells or pieces of cells away from a surface such as the bottom of a culture dish. This advantage of the recording approach will become more apparent as this chapter continues.

7.2.2 RAPID SUPERFUSION EXPERIMENTS: SPECIAL CIRCUMSTANCES AND SYSTEM DESIGN

Accurate measurement of the rapid kinetics of ligand-gated ion channels presents an especially vexing problem to the experimenter because a very rapid fluid exchange system must be used. Ordinarily, if one wishes to accurately measure agonist-induced current through a ligand-gated ion channel, a solution exchange system with the capability of exchange in μs-ms is needed. This will vary, of course, with the kinetics of the receptor under study. There are a number of approaches to achieving this rapid exchange time, but some factors are constant.

First, it is necessary to lift cells clear of any surface, be it a cover slip or culture dish bottom. Solution currents and eddy currents tend to occur as solution is superfused over a cell sitting in an otherwise static solution near an impermeable surface such as a dish bottom. Thus, a certain amount of local recirculation will allow persistent encounters between agonist and cell even when applying an agonist-free solution. In addition, solution will be trapped underneath cell elements, and this will slow solution exchange. Lifting the cell free of these surfaces eliminates such barriers to fast exchange.

Second, one must minimize the size of the cell or patch preparation under study. Solution exchange will be more rapid around a patch of membrane than around a cell (Figure 7.1), and will be faster for smaller cells than for larger cells. Thus, it is desirable to pull patches or "macropatches" from cells, whenever possible, to obtain optimal solution exchange. In deciding whether recordings from membrane patches yield meaningful data, it must be determined if a current of sufficient amplitude can be recorded in the patch, or if single channels can be measured for the receptor or channel in question. Often, to allow examination of the relationship between agonist concentration and channel kinetics, macroscopic currents must be measured at agonist concentrations that produce fairly low occupancy of a receptor. In such cases, channel density must be high enough to allow for measurement of current at low receptor occupancy. For this reason, over-expression of receptors in a heterologous system may be very useful in some instances.

A third consideration is the lower limit of the timing of solution exchange with the agonist application system itself. There are several ways to achieve the μs-ms

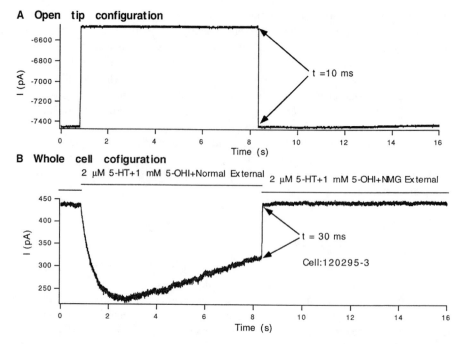

FIGURE 7.1 Solution exchange times measured for an open pipette tip vs. a whole cell. Note that exchange is ~ 3 × faster around the open tip in comparison with the whole cell. Note also that much faster exchange times than shown in this figure can be achieved in the open-tip configuration.

exchange rate that is needed. For example, patches can be inserted into the end of a Y-shaped tube and solutions switched in the short arms of the tube (Brett et al., 1986). One approach used for very rapid exchange is to arrange two or more micropipettes with tip sizes much larger than the patch or cell under study, such that the tips of both pipettes face the cell and solution streams from the pipettes produce superfusion. The strength of solution flow through the different pipettes is then varied and the stronger-flowing solution will dominate application to the cell or patch (Maconochie and Knight, 1989). If the faster-flowing pipette contains agonist-free solution, then the cell is constantly washed. Drugs can then be applied to the preparation by switching to faster superfusion via a pipette containing the appropriate compound(s). This system works well for producing solution exchange times of <1 ms once it is optimized for speed of change of the two solution streams (Maconochie and Steinbach, 1998). An added advantage is that the cell does not move.

An alternative approach is to rapidly move the cell between two or more continuously flowing solution streams or to rapidly move the pipettes producing these streams such that the stream bathing the cell is switched on the desired time scale. Theoretically, this can be accomplished by moving the micromanipulator that holds the cell-attached micropipette or by moving the micromanipulator that holds the drug application micropipettes. In practice, the latter approach is almost always preferred, because it minimizes mechanical disturbance of the cell and loss of the GΩ seal.

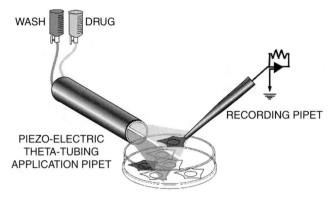

FIGURE 7.2 Schematic illustration of a rapid superfusion system using θ tubing to apply agonist to a single cell.

Figure 7.2 schematically illustrates a typical arrangement of a solution application micropipette system in relation to the cell under study. The drug delivery micropipette is placed within tens of μm from the cell, with the solution stream that contains no agonist superfusing the cell. To initiate drug application, the micromanipulator holding the drug delivery micropipettes is moved 20–200 μm such that the pipette or pipette array is moved, placing the agonist-containing pipette in front of the cell, rapidly exposing the cell to agonist. At the end of drug application, the manipulator returns to its original position and the cell is rapidly washed. Two types of micropipette arrays are commonly used in this system. Single glass or fused silica tubes can be fashioned into a lateral array of up to several pipettes in width. This design has the advantage of allowing complete separation of solutions in the different channels ending in each pipette tip. However, the disadvantage of this system is that movement between different pipettes in the array will cover a greater distance the farther apart the pipettes are on the array. Movement between more distant pipette tips will add unwanted extra time to the solution exchange time. If only two micropipette channels are used, one can attach a separate manifold to each micropipette. This allows the investigator to run several solutions through both the wash and agonist application pipettes. However, there is some mixing of solutions when switching between channels running into the manifold, and this must be factored into the timing of the experiment. Ultimately, the use of separate micropipettes for "wash" solution and agonist application is limited in time by the solution streams' being separated by the combined thickness of the walls of both pipettes. The second type of micropipette that is often used is made from θ–tubing that has a thin septum of down to 0.3 mm separating two different solution-containing chambers within the pipette (e.g., Jonas, 1995). This tubing can be pulled by hand or using a micropipette puller to yield a tip of 300 μm outer diameter and 50–100 μm total inner diameter. Pulling the electrode thins the septum that separates the two chambers at the tip of the pipette. After the glass is pulled to the desired width, it is scored and broken. A neat break is desirable, as this will minimize flow artifacts. When properly constructed, these θ-tubing applicators will have a septum of a few μm or less in width. An even thinner septum width can be achieved by acid etching of the glass.

Extensive polishing of the glass is not recommended because the inner diameters of each chamber in the applicator must remain larger than the largest cells to be examined. The thinner the septum, the faster solution can be exchanged by lateral displacement of the application pipette. Exchange times as fast as 100 μs can be achieved with this system. An additional consideration is the length of the shank of the glass. Longer shanks will tend to oscillate or "ring" during rapid movement, and thus, it is best to keep the glass as short as possible. The shape of the narrow region of the applicator tube is also important, because an excessive taper of this part of the glass can increase resistance to flow.

There are two predominant mechanisms used to achieve rapid lateral movement of the solution application pipette. A stepper motor is an electrically driven motor that, for solution application purposes, is attached to a rod. Application of a gating voltage pulse to the motor will displace the rod by a distance dependent on the voltage applied to the system and the mechanical systems attached to the rod. The range of such movements can be set on the order of tens to hundreds of μms. The initiation of rod movement is usually controlled by a standard TTL pulse or a command voltage delivered from a computer or other gating device. Thus, once the size of the rod displacement step is set and the command signal is sent to initiate movement, the rod will move the prescribed distance and the pipette will move accordingly. Upon discontinuation of the command signal, the rod and micropipette will be returned to the original position. The stepper motor is mounted on a standard micromanipulator for use during a patch-clamp experiment. One advantage of the stepper motor system is that multiple steps of different distances and durations can be pre-programmed and this allows precise timing of multiple solution exchanges if a multi-barreled drug delivery system is used (e.g., Benveniste and Mayer, 1991).

The most popular alternative is a piezoelectric translator-driven micromanipulator (Jonas, 1995). The piezoelectric device contains piezoelectric crystals or strips of this material that expand and contract in response to electrical charging and discharging of the material. The piezoelectric translator can be used to move a micropipette mounted on a manipulator in distances of up to hundreds of μm. The distance traveled will depend on the amplitude of the voltage pulse applied to the material, and the rate of movement will depend, among other things, on the slope of the leading and trailing edges of this command potential as well as the length of the glass that is to be moved. The relationship between command potential characteristics and displacement of the pipette tip must be determined for the particular equipment employed. The most common use of this system is to produce a single fixed movement of the application pipette that is initiated at the beginning of the command potential, with the pipette returning to its original position at the cessation of the command step. However, multiple-step movements are possible with multipipette arrays and interleaved piezoelectric strips (Jones et al., 1998). Usually, initiation of the command step is gated by computer output through a standard digital-to-analog converter port. Most commercially available piezoelectric-driven manipulators are capable only of moving a standard micropipette laterally by a few hundred μm and such systems are generally used for exchange between two solution streams only. However, piezoelectric bimorph strips can be purchased from commercial vendors and used to construct systems capable of multiple movements, and, with sufficiently short applicator micropipettes, movement of larger distances is possible.

It is worth noting that the solution exchange systems designed to optimize study of ligand-gated channels are also useful for other pharmacological applications such as precise timing of drug delivery while studying pharmacology of voltage-gated channels or drug effects on cellular signaling measured using imaging techniques. In the remainder of this chapter, we will describe experimental procedures using the θ-tubing application system mounted on a piezoelectric bimorph-driven micromanipulator as described in our publications (Zhou et al., 1998; Lovinger et al., 2000).

One method that has not been discussed in the foregoing section is uncaging of agonist molecules. In this approach, a neurotransmitter or receptor agonist is chemically bonded within a light-sensitive "cage" molecule. Upon exposure to a brief flash of light of a particular wavelength (often UV or IR), the molecule is released from the cage. This uncaging can occur within microseconds to milliseconds (Adams and Tsien, 1993), making this technique quite useful for rapid agonist delivery, and can be combined with patch-clamp recording (e.g., Grewer et al., 2000). This makes the technique quite useful for accurately measuring current onset rates. However, one needs to combine this approach with a technique for rapid agonist removal if deactivation is to be studied, and it is necessary to have a caged version of the agonist of interest to implement this technique.

7.2.3 TESTING THE SPEED OF SOLUTION EXCHANGE

Prior to initiating an experiment, it is necessary to establish the solution exchange time for the applicator in use. Normally, this is done in the "open-tip" mode. That is, a standard patch-clamp recording micropipette with an open tip is placed in the solution stream coming from one of the applicator channels, and the solution is switched to another stream in which a solution of a different ionic strength is flowing (usually a HEPES-buffered saline solution with a lower concentration of NaCl relative to the normal recording solution). When the solution facing the pipette is fully exchanged, the potential at the interface of the pipette tip and solution (the liquid junction potential or simply the junction potential) will change. The switching time can then be calculated as the time necessary to achieve a new steady-state junction potential. The design of the experiment must be considered at this point. If experiments are to be performed using membrane patches only, the open-tip switching time will accurately reflect the switching time during the experiment, since the patch will not likely be much bigger than the tip itself. However, if experiments on large-membrane macropatches of whole cells are planned, a different approach to solution switching measurement must be used. This is because solution exchange around the entire diameter of a macropatch or whole cell will necessarily be slower than that at the pipette tip because of the larger surface area and curvature of the macropatch or cell.

The best method for estimating solution exchange for a whole cell is to create a steady-state current in the cell by activating some ion channel and then rapidly inactivating current flow through the channel in a manner that depends only on solution exchange and diffusion (and not on any slower steps such as ligand unbinding or channel closing kinetics). The following is a standard protocol we have used to estimate exchange time for whole-cell recordings in NCB-20 neuroblastoma cells that naturally express $5\text{-}HT_3$ receptors (see Figure 7.1). After establishing a whole-cell recording and voltage-clamping the membrane potential at a very hyperpolarized level, cells or membrane patches are lifted clear of the bottom of the culture dish and placed in front of one side of the θ-tubing pipette that extrudes a control HEPES-buffered saline solution. This solution is then exchanged within the same stream for a solution containing 2 μM 5-HT and 1 mM 5-OH indole, a compound that greatly slows desensitization of the $5\text{-}HT_3$ receptor-channel. This produces a large, long-lasting current mainly carried by Na^+ ions. The other chamber of the tubing contains a solution with the same concentrations of 5-HT and 5-OH indole but with complete substitution of N-methyl-D-glucamine for NaCl. Thus, when the application pipette is moved from the original solution stream to the stream containing no NaCl, almost all of the charge carrier is washed away rapidly and the current will fall to near zero even though ligand is still bound and the channels remain open. The investigator can then measure the exchange time as the time needed to reach a new steady-state current level in the absence of NaCl. Using this approach, we have consistently observed exchange times 3–10 times slower for whole-cells of 20–30 μm in diameter relative to open pipette tips. The time to completely exchange solution around whole cells was estimated as in Zhou et al. (1998), and exchange times ranged from 20–40 ms depending on the cell diameter. This finding underscores the need to accurately measure exchange times using a system appropriate for the recording configuration.

It is important that the exchange time be measured under conditions as nearly the same as the experimental conditions as possible. Changes in, for example, the orientation of the solution application pipette and the cell-attached recording pipette, as well as the distance between the cell or patch and the solution interface can have large effects on exchange times. Thus, it is best to test solution exchange just before or after a recording with all of the pieces in the same configuration. One other useful tip for keeping the exchange times consistent from day to day, as well as for avoiding other artifacts, is to keep solution lines clean during and between experiments. Filtering solutions that go through the lines, and thoroughly cleaning the lines after use, are important steps in this process. One should be very careful to clean out any particulate matter, such as dust, that accumulates in the solution application pipette.

7.2.4 DATA COLLECTION

During the performance of the actual experiment, the cell is continuously superfused with normal external solution (usually HEPES-buffered saline) from one side of the θ-tubing, while from the other side of the θ-tubing pipette flows agonist-containing solution with or without alcohols or other drugs. The solution bathing the cell is exchanged by rapidly moving the θ-tubing pipette laterally using the piezoelectric manipulator so that the agonist-containing side faces the cell. Solutions flowing through either the control or agonist side of the θ-tubing pipette can be changed upstream by opening or closing stopcocks, solenoid valves, or any one of a number of solution switching devices.

Whenever experiments are performed with alcohols or other hydrophobic or amphipilic liquids, it is important to keep in mind that these agents can and will act as solvents. Ethanol and other alcohols will dissolve hydrophobic agents in tubing and this can cause artifactual effects on ion channel function. The alcohol-soluble compounds range from molecules within plastic tubing itself to molecules that adhere to tubing or other plastic elements when applied via a standard solution application system. For example, benzodiazepines that are near the limit of solubility in a predominantly aqueous solution will come out of solution and dissolve in the tubing and may remain there until a suitable solvent is perfused through the system. Thus, alcohols applied through the same tubing as such a compound will dissolve the leeched drug and the experimenter will observe an effect that appears to be due to the alcohol, but is actually due to the compound dissolved by the alcohol. There are several ways to avoid such artifactual alcohol effects. These include using Teflon tubing to avoid molecular leeching, washing tubing with alcohols or other hydrophobic solvents after use, and never applying alcohol via tubing that has been exposed to hydrophobic channel-acting compounds. It is more difficult to prevent effects due to agents within the tubing itself. In general, if an alcohol effect can be observed with a variety of solution application systems with different components, it is safe to assume the effect is due to the alcohol and not to other compounds dissolved in the alcohol solution.

Ion currents are measured using standard patch-clamp recording techniques. In our experiments we use an Axopatch-200 amplifier (Axon Instruments), with low-pass filtering using an eight-pole Bessel filter (Frequency Devices). Data are collected through online digitization at 1 to 5 kHz. We have used an ITC-16 analog/digital interface and a Macintosh-IIfx computer (Verdoorn, 1993). However, most commercially available patch-clamp recording software (e.g., pClamp) will work provided DAC output is sufficient to gate the command potential for movement of the piezoelectric manipulator. We fashion our standard recording electrodes from thin-walled Borosilicate glass (1.5 mm diameter, World Precision Instruments) using a horizontal puller (Sutter Instruments model p-87) or a vertical puller (LIST-MEDICAL model LIM-3P-A). For NCB-20 cells or HEK 293 cells of 20-30 μm diameter we can obtain >10 GΩ seals with patch-pipettes having tip resistances of ~1 MΩ. This allows us to obtain series resistances during recordings ranging from 2 to 5 MΩ that can then be reduced by compensation using the prediction and correction circuits of the amplifier (setting 65–75%). Figure 7.3 shows currents activated by different concentrations of agonist at different ligand-gated ion channels expressed in HEK 293 cells. All of the experiments we will describe in this chapter were performed under voltage-clamp with the membrane potential held at –60 mV.

7.2.5   EXAMPLES OF FAST APPLICATION ACTIVATION AND ANALYSIS OF LIGAND-GATED ION CHANNEL FUNCTION: THE 5-HT$_3$ RECEPTOR CHANNEL

5-HT$_3$ receptors, found exclusively on neurons, belong to a large subfamily of ligand-gated ion channels including nACh, GABA$_A$, and glycine receptors (Jackson and Yakel, 1995). Despite their distinguishing ligand specificity, protein distribution and

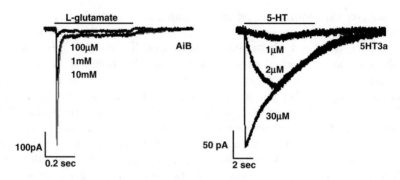

FIGURE 7.3 Ion currents activated by rapid agonist application to HEK 293 cells expressing either GluRA flip + GluRB flip-type AMPA/glutamate receptors (Left), or homomeric 5-HT$_{3A}$ receptors (Right). Current activated by different agonist concentrations are shown. Note the rapid activation and desensitization of GluR-mediated current, and the slower kinetics of 5-HT$_3$ receptor-mediated current.

channel kinetics, these receptors share many structural features with other members of the nACh-related subfamily of ligand-gated ion channels. These features include the presence of four transmembrane domains and a highly conserved cys-cys loop at the N-terminus, as well as a pentameric holoprotein structure (Maricq et al., 1991; Green et al., 1995). As the 5-HT$_3$ receptor can form a functional homomultimeric channel, and can be expressed at high levels in many native and heterologous systems, it provides a very attractive model for the study of this protein family (Maricq et. al., 1991). The 5-HT$_3$ receptor also plays many important roles in normal brain function. Some of these key functions include modulation of neurotransmitter release, a role in the sensation of pain, and involvement in chemically induced nausea and vomiting (Costall and Naylor, 1992). Extracellular dopamine levels in nucleus accumbens, which are thought to be critical for drug addiction, have been shown to increase upon activation of the 5-HT$_3$ receptor/channel complex (Chen et al., 1991; Costall et al., 1987; Hagan et al., 1987; Jiang et al., 1990). Because ethanol enhances the function of the 5-HT$_3$ receptor (Barann et al., 1995; Lovinger, 1991; Machu and Harris, 1994), it was hypothesized that ethanol may increase brain dopamine levels through 5-HT$_3$ receptors and this action might contribute to the reinforcing effects of the drug (Grant, 1995). There is also evidence for 5-HT$_3$ receptor involvement in alcohol drinking behavior and the discriminative stimulus effects of acute ethanol (Grant, 1995; Fadda et al., 1991; Johnson et al., 1993; LeMarquand et al., 1994a,b; Sellers et al., 1994; Tomkins et al., 1995). Hence, it is important to study alcohol actions on the 5-HT$_3$ receptor-channel complex.

One feature of the 5-HT$_3$ receptor in its homomultimeric form, is that the channel conductance is quite small (<1 pS). Thus, receptor-channel kinetics and alcohol effects on kinetics are best studied using whole-cell recording. We will focus on whole-cell recording as a means to provide the data needed for kinetic analysis of the function of this receptor and alcohol effects on the receptor. However, many of the techniques we use are also applicable to single-channel analysis, as we will note periodically in this chapter. For a more thorough description of single channel recording and analysis of single channel kinetics, we refer the reader to Sakmann and Neher (1995).

In this chapter, we will describe receptor function and the effects of alcohols examined using rapid drug application techniques and rigorous kinetic modeling based on a standard kinetic scheme for ligand-gated ion channels. Some of the primary data presented in the chapter and a preliminary kinetic analysis of these data were presented in a previous publication (Zhou et al., 1998). The methods for cell preparation, electrophysiological recording, rapid drug application and measurement of current are the same as those used in Zhou et al. (1998), and we refer the reader to that paper for a complete description of methodology.

7.3 ANALYSIS OF WHOLE-CELL DATA

Analysis of channel kinetics can be performed for several reasons:

1. To gain a first estimate of a particular component of channel function (e.g., fast desensitization or deactivation)
2. To examine effects of a particular agonist or pharmacological agent on channel behavior
3. To relate channel kinetics to channel structure
4. To try to obtain a reasonable description of channel function by finding the simplest kinetic model that can account for channel behavior.

Depending on the object of the study, the analysis may range from simply fitting particular current components to simulation of current and fitting current traces using well-defined kinetic models. We will consider each of these levels of analysis in the following sections of this chapter.

It must be stressed that only a limited amount of information can be gained from analysis of whole-cell kinetic data. Ligand-gated ion channels undergo very rapid transitions between agonist-bound open and closed configurations (Colquhoun and Hawkes, 1995a), and these transitions are not easily measured from macroscopic currents. Thus, it is often necessary to examine single-channel behavior to obtain a full description of channel behavior. Single ligand-gated ion channels open in "bursts" that exhibit rapid open-closed transitions. These transitions are believed to reflect transitions involving agonist bound states. Single channel burst kinetic analysis is necessary to estimate the rate constants governing these fast transitions (Colquhoun and Hawkes, 1995a).

7.3.1 MEASUREMENT OF WHOLE-CELL KINETICS

One way to analyze channel function and alcohol effects on ion channel kinetics is to measure particular current components that provide information about kinetic transitions. This can be done by measuring various parameters such as current amplitude and slope or by estimating the rates of particular channel transitions using exponential curve fitting with standard non-linear least-squares routines (Levenberg, 1944; Marquardt, 1963). The choice of components may seem at first to be intuitively quite obvious. The initial current rising phase has something to do with channel activation, the peak amplitude of the current has something to do with the number of channels that have been activated, and the decay of the current in the continuous presence of agonist provides some information about "fast desensitization" of the receptor-channel. However, extracting the best parameters for measurement is complicated by the fact that all channels will not be in the same state at the same time, owing to the stochastic nature of channel opening and closing. Thus, transitions in whole-cell current that appear to be straightforward (e.g., movement from no current to peak current after agonist application) will often be the result of interactions between several kinetic processes (e.g., channel activation, fast desensitization and deactivation). Likewise, realization of a steady-state current in the presence of agonist does not indicate a lack of desensitization or a cessation of the desensitization process, but for most channels, is indicative of a balance among all of the rate constants driving the channel in and out of the open state at a given time and agonist concentration. In light of these sorts of considerations, it is a good idea to compare whole-cell data to simulations of channel function to estimate the best parameters to measure in comparing channel kinetics under different experimental conditions. In Zhou et al. (1998), we provide an example of the derivation of parameter measures from kinetic simulations. However, these simulations will be only as good as the models from which they are constructed.

Measurement of whole-cell kinetic data can provide a first indication of the effect of drugs, such as alcohols, on receptor-channel rate constants. In the case of the homomeric 5-HT$_3$ receptor, for example, a change in the initial slope of current activation is most likely an indication of altered channel activation rate, while a change in current decay after a brief, non-desensitizing agonist application indicates a change in channel deactivation (often called agonist unbinding). Alteration in the amplitude of a true steady-state current during prolonged agonist application is indicative of a change in the balance of the rates of desensitization and resensitization.

Other measurements are more difficult to assign to a change in any one kinetic rate constant or set of rate constants. A case in point is current decay in the continuous presence of agonist. This phenomenon involves the fast desensitization process, but is sensitive to changes in other rate constants as well. For example, if an increase in this parameter is observed in the presence of a concentration of agonist producing low-to-moderate receptor occupancy, increased decay could be the ultimate result of a drug-induced increase in agonist affinity or current activation. Such a case is described in Zhou et al. (1998). When alcohols were applied along with 5-HT both the initial slope and decay rate of current were increased. Using a simple model of receptor-channel kinetics to simulate behavior of ligand-gated channels the increase in decay rate would be expected to be proportional to the increase in initial current slope because as more channels enter the open state, more will enter the desensitized state even if the rate constant for desensitization does not increase (see Figure 7.5 from Zhou et al., 1998). This is most easily seen in the case of increasing agonist concentration. According to all current theories of ligand-gated ion channel function, only rate constants involved in channel activation are dependent on ligand concentration, while rate constants governing desensitization and resensitization are agonist concentration-independent (see Chapter 6 in Hille, 1992). However, one observes a clear increase in current decay rate as the agonist concentration is increased at the $5\text{-}HT_3$ receptor (Figure 7.3). The increase in current decay is normally proportional to the increased initial slope of current, which is our best estimate of channel activation rate (Zhou et al., 1998). Thus, it appears that the agonist-dependent increase in current decay reflects faster entry of channels into the open state that makes more channels available for entry into the desensitized state.

Examining the effect of alcohols on this decay rate is instructive. We consistently observe that alcohols increase the decay of current in the presence of a low concentration of 5-HT. This is due, in part, to an inhibitory effect of some alcohols on the receptor, but also reflects increased desensitization of the receptors. However, the proportionality of the initial current slope to the decay rate is altered in the presence of trichloroethanol (TCEt) (Zhou et al., 1998), indicating that the channels actually desensitize at a slower rate for a given activation rate. Thus, the increased current decay rate in the presence of alcohols is somewhat misleading when trying to interpret the actual changes in receptor channel kinetic rate constants. Indeed, when a higher agonist concentration is employed, then one does not observe enhanced current decay rate, but rather, a larger steady-state current is observed which is most likely a reflection of a change in the desensitization/resensitization rate constants.

7.3.2 BUILDING KINETIC MODELS

If measurement of the transitions of whole-cell currents does not provide a reliable indicator of changes in the rate constants governing channel transitions, then what can be done to more accurately estimate these changes in intrinsic receptor-channel kinetics? To better understand kinetics of biomolecular reactions, including ion channel function, investigators have long relied on models of molecular state transitions that posit reasonably stable states through which the molecule moves via transitions that are governed by rate constants that are thought to be intrinsic to the protein under study. The functional states are thought to reflect particular global protein conformations while the state transitions represent changes in these conformations. For ion channels, this sort of kinetic analysis works quite well because channels do appear to adopt discrete functional states such as an open state associated with relatively invariant channel conductance and well defined lifetime. Thus, application of kinetic modeling to analysis of ion channel function is a well-developed sub-field of ion channel biophysics and pharmacology.

The first step in kinetic analysis of a channel is to develop a reasonable multiple state model of the channel of interest. To do this, it is usually necessary to have some information about the function of the channel in question (in our case, this information comes from whole-cell recording data). The minimal number of states that must be postulated in order to explain channel behavior can be estimated from close inspection of the data and fitting individual kinetic parameters. The first rule to remember is that one can only measure function in the open channel state as well as transitions to the open channel state. Once the channel is open, it will transition to inactive or closed states via one of several mechanisms, deactivation (e.g., channel closing and unbinding of neurotransmitter reversing the binding-opening process) and fast desensitization (similar to inactivation of voltage-gated channels), in which the channel becomes trapped in a non-conducting but ligand-bound state. A first glance at whole-cell data derived from a full agonist concentration-response analysis can yield valuable clues as to the presence, number or potential rate constant values for these transitions. If the current shows no sign of decay during prolonged agonist exposure, then it is likely that the receptor-channel does not enter desensitized states, or that entry into these states is so slow as to be negligible in modeling channel behavior. However, if decay is observed, particularly at high agonist concentrations, at least one desensitized state must be added to the kinetic model to adequately describe channel behavior. The lifetime of any particular channel state can be described by a single exponential function (Hille, 1992). The transitions between states in whole-cell currents appear as "relaxations" that can be fit by one or a number of exponential functions. Based on the first-order exponential nature of the elementary channel transitions, it follows that a given number of channel states (N) will exhibit up to N-1 components in the exponential relaxations governing state transitions. Inversely, based on the number of identifiable relaxation times (M), one can infer that the channel has at least M+1 states. These basic tenets of kinetic behavior tell us that, if there is only a single open channel state and if the channel sojourns from that state predominantly to only one nonconducting state, then decay of current will be described by a single exponential function. However, if a current transition has a multi-exponential time course, there are at least three distinct states, and possibly more, that the channel can enter. To continue with the example of fast desensitization, if the decay of current in the presence of agonist is monoexponential, it is possible that a transition from a single open state to a single desensitized state is taking place (although one cannot exclude the presence of other desensitized states). A minimal model of channel kinetics might then assume a single desensitized state. However, the presence of additional exponential components indicates the presence of more than two channel states. Thus, at a minimum, a channel can enter into additional alternative desensitized states, or possibly enter into the desensitized state from more than one open state. Indeed, the situation might be even more complex. Thus, fitting the exponential current decay in the presence of agonist yields important information about how many states may be needed to minimally describe channel behavior, but a complete description of channel behavior is not forthcoming from such an analysis.

Estimating rate constant values is more difficult and is subject to the sorts of problems mentioned in the preceding discussion of measurement of whole-cell current kinetics. The intrinsic channel rate constant (usually expressed in transitions per second or s^{-1}) is the inverse of the exponential time constant of a single exponential function in a simple unidirectional two-state system. Theoretically, then, it is possible to derive a reasonable rate constant estimate from measured time constants. However, it is often difficult to isolate a single monoexponential state transition from whole-cell data, and, thus, great care must be taken in applying this approach. The rate constant estimate can be greatly improved under conditions where the forward or reverse rate constant can be set at a very low value. For example, when agonist is applied rapidly at a high concentration, the forward rate constants (i.e., binding and opening rate constants) that contribute to channel opening will greatly exceed the unbinding and deactivation rate constants (e.g., Maconochie and Steinbach, 1998). If desensitization does not have a very rapid onset, the initial slope of agonist-activated current will yield a reasonable estimate of the limit of the rate constants governing channel activation. A similar situation occurs when agonist application is rapidly terminated. In this case, the rate constants driving the forward reactions toward agonist binding become minimal and the backward rate constants that lead to unbinding predominate. If the agonist application was sufficiently short to prevent any desensitization, then the decay of current after agonist removal can provide a good estimate of the rate constants that predominate in the deactivation or unbinding process (Jones et al., 1998; Orser et al., 1994; Zhou et al., 1998).

In most instances, however, it is difficult to obtain good rate constant estimates from measurement of current alone. Thus, it is desirable to use simulation of current to obtain reasonable estimates. Kinetic simulation does not depend on any physiological data's actually being collected. One initially constructs a kinetic model, such as that shown below (Model 7.1), and then enters appropriate rate constants for the different kinetic transitions according to the best estimates that can be gained from data in the literature or theories of channel function. Using an appropriate computer algorithm, such as those supplied commercially in MATLAB the estimated current level can be generated on a time point-by-time point basis as governed by the entered rate constants. For example, a deactivation/unbinding rate constant of 3.17 s^{-1} will give rise to a current that decays with a τ of ~315 ms. One can vary the rate constant estimates and view the resultant current kinetic changes. These data can be compared to experimentally collected data to obtain an idea of what kinetic changes might underlie certain drug actions, for example.

7.3.3 SIMULATING CHANNEL KINETICS AND FITTING WHOLE-CELL DATA WITH A DEFINED MODEL

Once the investigator has a reasonable idea of the possible range of rate constant values for a particular state transition, then estimation of kinetic parameters can proceed along well-established lines. One approach is to continue simulations of current to find a set of rate constants that describe channel behavior reasonably well. This can be accomplished using the so-called Q-matrix method in which one constructs a matrix containing the rate constants that adequately describe channel behavior under a particular condition and assuming a defined kinetic model (Colquhoun and Hawkes, 1995b). This approach provides reasonable estimates of kinetic parameters that give a quantitative picture of channel behavior. However, the simulation approach may be quite time consuming and, ultimately, better parameter estimates can be obtained using a model-fitting approach.

The fitting procedure is different from current simulation in several respects. The goal is to evaluate whether a particular kinetic model can account for channel behavior under different conditions, and to obtain reasonable estimates of rate constant values. This approach proceeds from the collection of data. Once the results are in hand and a reasonable kinetic model is constructed, the data and model, along with best-guess estimates of the rate constant for each state transition and the starting condition are entered into a program that has the capability of generating non-linear least-squares fits to a multi-component kinetic model. We will not describe the mathematics underlying either of these procedures in any detail. We refer the reader to several previous publications describing these procedures in greater detail (Beechem, et al. 1991; Beechem, 1992; Levenberg, 1944; Marquardt, 1963; Press et al., 1992). We have used software that was originally developed for global analysis of the adequacy of a single physical model for description of time-resolved fluorescence data (Beechem et al., 1991). This program is more than adequate for fitting data to channel kinetic models when properly adapted (Zhou et al., 1998). The essential statistical procedure is similar to that used for any nonlinear fitting procedure in that a comparison is made between the observed data and the data expected from the model using the $\chi 2$ statistic according to the standard iterative method of Marquardt and Levenberg (Levenberg, 1944; Marquardt, 1963; Press et al., 1992). The nonlinear fitting routines use matrix algebra to generate an overall solution matrix from combinations of rate constant values (Press et al., 1992). The standard software routines used for such fitting will generally have the matrix manipulations already available as subroutines, and, thus, it is necessary only to enter matrix elements and call up the desired algebraic manipulations to start the process. However, procedures for convergence on a good fit require some sort of software optimization engine. In a standard fitting routine, the kinetic model is constrained by the experimenter as to number of states and state transitions based on measurements of the data as described above. The experimenter enters initial estimates of rate constants generated by data measurement and simulations. In the straightforward fitting routine, the rate constants are unconstrained and, thus, can be freely varied. The fitting algorithm will calculate a model ion current based on the initial rate constants. This model current will then be compared with the actual data for goodness of fit. In the modified Globals (tm) procedure that we have used, a χ^2 value is generated for the model current in comparison to the actual data. If the value is not sufficiently low (criteria are defined by the experimenter), all of the rate constants are varied in directions that will lead to a closer fit to the data. The size and direction of the change in the model parameters is calculated using a parameter improvement vector defined by Marquardt (Beechem et al., 1991; Marquardt, 1963). A new model current is generated and the χ^2 values are again calculated. This process is repeated with constant adjustment of the parameters until an acceptably low χ^2 value is obtained. The parameters that provide a "good fit" to the data are then taken as the estimate of the rate constant under that condition. Using the Globals program it is also possible to perform actual Global Analysis. This is a method in which simultaneous fitting of data under several different conditions (e.g., full concentration–response curves) is used to provide very precise estimates of rate constant values (Beechem et al., 1991). However, we have not yet used the program in this way. A number of commercially available programs, such as ScoP and MATLAB, contain matrix algebraic routines that can be used in current modeling and simulation, and, thus, it is possible to put together fitting routines using this type of software (e.g., Jones and Westbrook, 1995).

We have applied both current simulation and model fitting to gain a better understanding of drug effects on the 5-HT$_3$ receptor (Zhou et al., 1998). In our initial study, we described the results of fitting a fairly simple kinetic model to data obtained

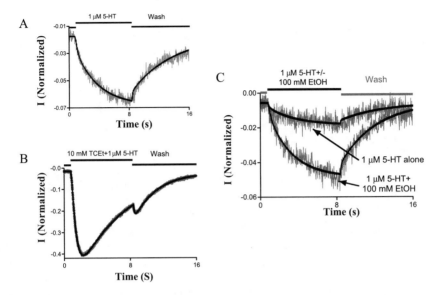

FIGURE 7.4 Fits of 5-HT-induced current from five cells. (Left) Normalized average current sweeps generated by 1 μM 5-HT (TOP) and 1 μM 5-HT+10 mM TCEt (bottom) were fit using kinetic Model 7.1. Gray lines are averaged real current traces and black lines are fitting traces. (Right) Normalized average current sweeps generated by 1 μM 5-HT and 1 μM 5-HT+100 mM ethanol (EtOH). All currents are normalized to current induced by 10 μM 5-HT, a saturating concentration of agonist.

from activation of the receptor in the absence and presence of ethanol and TCEt. It is well documented that alcohols, such as ethanol and TCEt, enhance ion current mediated by the 5-HT$_3$ receptor-channel at low agonist concentrations (Lovinger and White, 1991; Zhou et al., 1998). Figure 7.4 shows examples of the type of data we collected in our studies. Current traces in this figure show 5-HT$_3$ receptor-mediated currents (noisy gray lines) activated by 1 μM 5-HT in the absence or presence of 10 mM TCEt, a TCEt concentration that produces maximal current potentiation. The current records shown in this figure are averages of responses from five different NCB-20 cells, which were normalized to the peak amplitude of the current activated by 10 μM 5-HT (a saturating concentration of agonist) from the same cells, to provide values for percent of maximal current. A simple kinetic model described below, and the modified Globals program was used to fit the averaged data (Figure 7.4, black traces). As in previous studies, we observed that this concentration of TCEt increased the amplitude of the agonist-induced current. A small tail or "after-response" current (Orser et al., 1994) was observed upon washing away the agonist + TCEt solution, as previously noted (Lovinger and Zhou, 1993). This feature of the current is believed to represent unblocking of channels following removal of TCEt when the drug is applied at high concentrations (Lovinger and Zhou, 1994; Zhou et al., 1998). The model used for this fit is shown below. It was constructed based on the well-known nAChR channel kinetic scheme modified to accommodate experimental data from studies of the 5-HT$_3$ receptor.

$$R + A \underset{k_{-1}}{\overset{2*k_1}{\rightleftarrows}} AR + A \underset{2*k_{-1}}{\overset{k_1}{\rightleftarrows}} A_2R \underset{\alpha}{\overset{\beta}{\rightleftarrows}} A_2O \underset{k_{-d}}{\overset{k_d}{\rightleftarrows}} A_2D$$

$$+$$

$$\text{TCEt}$$

$$k_b \updownarrow k_{-b}$$

$$A_2TB$$

MODEL 7.1

Because the Hill coefficient of the 5-HT$_3$ receptor for serotonin is generally greater than 1 and close to 2 in tissues and clonal cell lines, (Neijt et. al., 1988; Yakel et. al., 1991), it is assumed that at least two ligand molecules bind to the receptor before the channel can open. For simplicity, the two binding sites for agonist in this model are assumed to be independent with identical association and dissociation rate constants (k_1 and k_{-1}). R, AR, A$_2$R and A are the free, monoliganded, and doubly liganded closed states, and agonist respectively. A decay phase of 5-HT-activated current was consistently observed during continuous agonist application, indicating the existence of at least one desensitized state. Also, considering the evidence for a blocked state induced by high concentrations of TCEt, a kinetic model consisting of two identical binding steps, one isomerization step, one desensitization step and one blocking step for TCEt was used in the fitting. A$_2$O, A$_2$D, and A$_2$TBlock (A$_2$TB) represent the open, desensitized and blocked states, respectively. The constants β and α represent the receptor opening and closing rate constants. It is known from work on single ligand-gated ion channels that fast transitions between ligand-bound open and closed states occur (Colquhoun and Hawkes, 1995a), and these rate constants govern those transitions. The symbols k_d and k_{-d} refer to the desensitization and resensitization rate constants. Finally, k_b and k_{-b} are the TCEt blocking and unblocking rate constants, which were introduced to account for an after-response or "tail" current upon washout of high concentrations of TCEt (Zhou et al., 1998). All receptors are assumed to reside in the resting closed state (R, no conductance) prior to agonist application.

Excellent fits were obtained to the averaged current records induced by 1 μM 5-HT and 1 μM 5-HT + 10 mM TCEt, suggesting that this simple kinetic model is sufficient to describe the behavior of the 5-HT$_3$ receptor activated by a low agonist concentration in the absence or presence of TCEt. Consistent with our earlier observations (Zhou et. al., 1998), TCEt increases the agonist association rate constant by 218.7%, while dramatically decreasing the agonist dissociation rate constant (only 6% of the control). Hence, agonist will have a higher apparent affinity (k_1/k_{-1}) for the receptor in the presence of TCEt than in its absence, and this effect will result in a leftward shift of the agonist concentration–response curve, as previously described for TCEt (Lovinger and Zhou, 1993; Zhou et al., 1998). These estimated kinetic parameters also suggest that TCEt drastically reduces the rate constant for receptor desensitization. The desensitization rate constant is only 3.1% of control in the presence of 10 mM TCEt. TCEt apparently reduces the resensitization rate constant to such a degree that during the time period of the drug application (8 sec) the degree of resensitization is negligible. A somewhat surprising finding is that TCEt appears to have very little, if any, effect on the receptor opening and closing rate constants (β and α).

Similar results were obtained when examining the effects of ethanol (Zhou et al., 1998). Like TCEt, ethanol also potentiates the function of 5-HT$_3$ receptors. As is also shown in Figure 7.4C, ethanol (100 mM) clearly increased the amplitude of peak current (average current from four different cells, the noisy gray lines represent the actual data). Reasonable fits (dark smooth lines) to the average currents induced by 1 µM 5-HT and 1 mM 5-HT + 100 µM ethanol were obtained using the same kinetic model without the A$_2$TB state, suggesting that this simple model is sufficient to describe the behavior of the 5-HT$_3$ receptor/channel at low concentrations of agonist with or without ethanol.

7.3.4 Estimating Error of Fits

A caveat of model fitting is that the fitting algorithms will blindly return the best fit based solely on a statistical criterion such as the minimization of χ^2 However, the χ^2 value for the "best fit" may not be statistically significantly different from another fit with drastically different values for some parameters. Once an adequate fit to the data is found, the goodness of this fit needs to be evaluated. There are several ways to do this.

In our experiments, we wished to determine whether alcohols altered particular rate constants within the kinetic scheme. To determine if a change in a particular rate constant is important for the effect of such a manipulation, one can set the value of that rate constant at an invariant value during the fitting procedure. In this way, one can determine if an adequate fit depends on a particular value of that rate constant. An example of this sort of analysis is given in the next few paragraphs, which describe our fitting of 5-HT$_3$ receptor-mediated ion current in the presence and absence of TCEt. To further understand the contribution of each pair of transition rate constants to the effects of 10 mM TCEt, we examined whether it was necessary for a particular pair of rate constants to change to obtain adequate fitting of the data. This analysis was carried out one by one for receptor association and dissociation, opening and closing, desensitization and resensitization rate constants, as well as blocking and unblocking rate constants. In this procedure, the rate constant values were fixed to those obtained from fitting responses to 1 µM 5-HT alone. As shown in Figure 7.5A, when the values of receptor association (k_1) and dissociation (k_{-1}) rate constants were fixed to those of current activated by 1 µM 5-HT alone, the Globals program was unable to find a set of parameters to mimic the average current sweep induced by 1 µM 5-HT + 10 mM TCEt (the small noisy gray line in this figure is the 1 µM 5-HT generated current sweep and the dark smooth line is the fit to those data). Although the current activation rate of the best fit sweep was increased, most likely due to an increase in the channel opening rate constant (β), this is clearly not sufficient to give rise to the full peak current observed in the presence of 10 mM TCEt. This observation indicates that an increase in the receptor

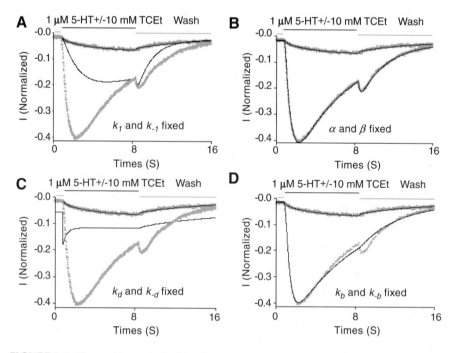

FIGURE 7.5 Changes in particular kinetic parameters are necessary to generate a proper fit to TCEt data. Pairs of parameters were fixed to values obtained from the best fit for the averaged current activated by 1 μM 5-HT. A: The receptor activation (k_1) and deactivation (k_{-1}) rate constants were fixed to those estimated for 1 μM 5-HT. B: The channel opening (β) and closing (α) rate constants were fixed to those estimated for 1 μM 5-HT. C: The channel desensitization (k_d) and resensitization rate constants (k_{-d}) were fixed to those estimated for 1 μM 5-HT. D: The TCEt blocking (k_b) and unblocking (k_{-b}) rate constants were fixed to zero. In each section of the figure, the gray lines are averaged, normalized current traces, while the black lines are best fits generated using Model 7.1.

association rate constant in combination with a decrease in the receptor dissociation rate constant is required to fully account for the potentiating effect of 10 mM TCEt.

In agreement with the observation that TCEt induces no change in the channel opening (β) and closing (α) rate constants, fixing the values of these two constants to the values obtained when fitting current generated by 1 μM 5-HT alone did not alter the goodness of the fit for the 1 μM 5-HT + 10 mM TCEt generated current (Figure 7.5B). This result suggests that modifications of transition rate constants other than the opening and closing rate constants by TCEt are sufficient to explain the observed potentiating effects of TCEt on the function of 5-HT$_3$ receptors.

Experimental results from measured desensitization rate and steady-state current studies suggest that TCEt may slow the desensitization process of the 5-HT$_3$ receptor-channel (Zhou et al., 1998). Consistent with this idea, current fitting also indicated a slowing of the receptor desensitization by TCEt (Figure 7.5C). The Global Analysis program failed to obtain a reasonable fit of the current generated by 1 μM 5-HT +

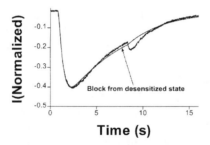

FIGURE 7.6 Blocking from the desensitized state fails to account for the "tail" current. Black line is 1 μM 5-HT+10 mM TCEt-induced current. The smooth line is the fit to these data with entry into the blocked state from the desensitized state as in Model 7.2.

10 mM TCEt when the values of desensitization and resensitization rate constants were fixed to those of 1 μM 5-HT-activated current. The "best" fit in Figure 7.5C suggests that receptor desensitization is simply too fast to reconstitute the average current sweep generated by 5-HT + TCEt.

Also in agreement with previous observations, the after-response "tail" current occurring upon the removal of 10 mM TCEt was abolished by forcing the blocking and unblocking rate constants to zero, which in turn effectively eliminated any contribution to the fitting from the TCEt-induced blocking effect (Figure 7.5D). Again, this observation is consistent with the notion that TCEt binds to the open channel and thus reduces current conduction. A potential alternative mechanism for the TCEt removal-induced after-response current is the presence of a second, "deeper" desensitized channel state in the presence of high TCEt concentrations. Kinetic Model 7.2 illustrates this situation. In this model, all of the channel states are those described in Model 7.1. The only difference is that the A_2TB state is now entered through a reversible step from the desensitized rather than the open channel state. However, an adequate fit to the current observed in the presence of 5-HT + 10 mM TCEt cannot be obtained using this model (Figure 7.6). This is mainly because receptor resensitization appears to be much slower than channel closing. The inadequacy of this alternative model supports the idea that high concentrations of TCEt produce a slowly developing inhibitory effect by interacting with the open channel state.

$$R + A \underset{k_{-1}}{\overset{2*k_1}{\rightleftarrows}} AR + A \underset{2*k_{-1}}{\overset{k_1}{\rightleftarrows}} A_2R \underset{\alpha}{\overset{\beta}{\rightleftarrows}} A_2O \underset{k_{-d}}{\overset{k_d}{\rightleftarrows}} A_2D$$

$$+$$

$$TCEt$$

$$k_b \updownarrow k_{-b}$$

$$A_2TB$$

MODEL 7.2

Analysis of the goodness of fit of data in the presence and absence of ethanol was also performed in which each pair of important transition rate constants was fixed to the 1 μM 5-HT values as described for the TCEt data above. The χ^2 value for the original fit was 0.00033 and the 95% confidence range of the χ^2 value in this experiment was 0.00034 (Figure 7.7). Fixing the β and α values of the receptor-channel to those of 1 μM 5-HT activated current, again, did not affect the χ^2 of the fit, consistent with the analysis of TCEt effects (Figure 7.7B). However, fixing either the receptor association and dissociation rate constants or the receptor desensitization and resensitization rate constants resulted in a much worse fit than the original one (Figures 7.7C and D); $\chi^2 = 0.0094$ for fixing the activation and deactivation rate constants, and $\chi^2 = 0.00156$ for fixing the desensitization and resensitization rate constants). These observations are consistent with the idea that ethanol potentiates 5-HT$_3$ receptor function via a mechanism similar to TCEt, involving changes in agonist association/dissociation and receptor desensitization/resensitization rate constants but not opening or closing rate constants. Actually, 100 mM ethanol appears to increase the receptor association rate constant and decrease both the receptor dissociation rate constant and the receptor desensitization rate constant, again consistent with observations from the study of TCEt actions (Zhou et al., 1998).

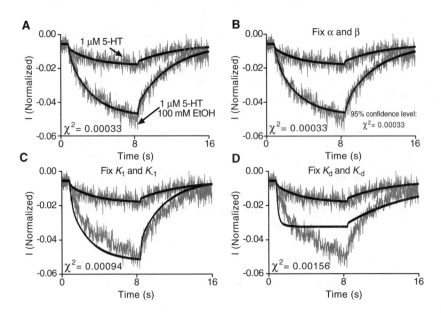

FIGURE 7.7 Changes in particular kinetic parameters are necessary to generate a proper fit to EtOH data. One pair of parameters was fixed to values estimated from the best fit for 1 μM 5-HT activated average current traces (n = 4). A: The control fits for the currents activated by 1 μM 5-HT alone and 1 μM 5-HT + 100 mM ethanol. B: The channel opening (β) and closing (α) rate constants were fixed to those of 1 μM 5-HT. C: The receptor activation (k_1) and deactivation (k_{-1}) rate constants were fixed to those of 1 μM 5-HT. D: The channel desensitization (k_d) and resensitization rate constants (k_{-d}) were fixed to those of 1 μM 5-HT.

A more thorough way to test the adequacy of the fit is by performing a rigorous error analysis for each of the transition rate constants to define the range of rate constant values needed to obtain a reasonable fit with the confidence level set at 95%. In the rigorous error analysis, a tested parameter is forced to walk through a preset range of values centered around the best fit value of that particular parameter, at a predetermined step size. For each value of the tested parameter, the Globals program will find a set of values for all other variable parameters to obtain the best possible fit. The χ^2 values of these best fits are then plotted against the corresponding value of the tested parameter. Theoretically, this numerical well of χ^2 values should minimize at the best-fit value of the tested parameter itself. Then the 95% confidence interval of the χ^2 value is calculated and plotted as a horizontal line when plotting the χ^2 minimum or well. We have performed this type of analysis for the data we have obtained in the presence of 5-HT alone and 5-HT + 10 mM TCEt (Figures 7.8A, B, and C). It is evident from the rigorous error analysis that fits (as judged by statistically indistinguishable χ^2 values) for the receptor association, dissociation and resensitization rate constants for responses generated by either 1 μM 5-HT alone or 1 μM 5-HT + 10 mM TCEt were obtained only over a relatively narrow range of values. Furthermore, the range of values that gave adequate fits in the absence of TCEt were well separated from those rendering adequate fits in the presence of TCEt. Although the range of desensitization rate constants that yielded an adequate fit for the average 1 μM 5-HT activated current is a little broad, the χ^2 wells for the 1 μM 5-HT and 1 μM 5-HT + 10 mM TCEt are still well separated. These results suggest the conclusion that at the 95% confidence level 10 mM TCEt increased the receptor association rate constant and decreased the receptor dissociation, desensitization and resensitization rate constants for channels activated by 1 μM 5-HT. This conclusion is consistent with the analysis presented above and the previously observed experimental data (Zhou et al., 1998). Due to a lower signal to noise ratio for the effects of ethanol in comparison to those of TCEt, a meaningful rigorous error analysis was unachievable for our ethanol data.

Again, the opening (α) and closing (β) rate constants for the channel behaved differently from other parameters (Figure 7.8B). The χ^2 wells for the opening or closing rate constants estimated from fitting 1 μM 5-HT and 1 μM 5-HT + 10 mM TCEt-induced currents were ill defined and the estimates for current in the absence or presence of TCEt overlapped each other at the 95% confidence level. A broad and flat χ^2 well for a parameter generally indicates that there is redundancy in the kinetic model and that proportional changes over a wide range of values of particular rate constants in opposite directions can produce an adequate fit. At this level of analysis, the impact of the β and α rate constants on channel gating cannot be completely identified by the Globals program. After closely examining how other parameters were adjusted, it became clear that it was the opening rate constant and closing rate constant themselves that counteracted each other. While one rate constant increased within a certain range, the other would increase proportionally and the χ^2 of the fit remained approximately the same. To further elucidate this point, the average current sweep generated by 1 μM 5-HT + 10 mM TCEt was fit with the closing rate constant preset at 500 s^{-1} (the best fit value) or 150 s^{-1}, respectively (Figures 7.9A and B). The best-fit values of parameters other than the opening rate

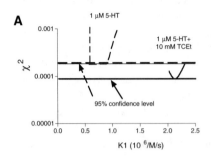
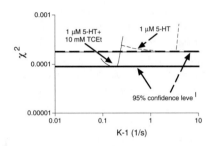

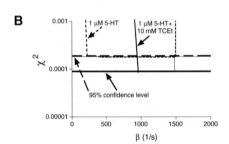
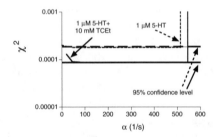

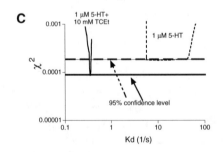
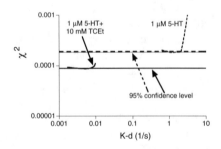

FIGURE 7.8 Rigorous error analysis for the estimated rate constant values. The minimum or "well" represents the range of χ^2 values that provide statistically indistinguishable values for a given rate constant from the fits generated using Model 7.1. The horizontal line is the χ^2 value at the 95% confidence level. The dotted and solid lines represent the rate constant estimates for 1 μM 5-HT alone and 1 μM 5-HT + 10 mM TCEt, respectively. A: TCEt significantly increases the estimated receptor association rate constant and decreases the dissociation rate constant at the 95% confidence level. B: TCEt effects on estimated channel opening and closing rate constants are not significantly altered at the 95% confidence level. C: TCEt significantly decreases both the estimated desensitization and resensitization rate constants at the 95% confidence level.

constant were almost identical in these two fits and yielded similar χ^2 values. Indeed, the ratios of the opening and closing rate constants were almost identical for the two fits (β/α = 2.82 for a = 500 s^{-1} and β/α = 2.86 for a = 150 s^{-1}) suggesting a linear relationship between these two parameters.

Current simulations generated using the Q-matrix approach (Colquhoun and Hawkes, 1995b) also support this interpretation. Four simulations with the same values of parameters other than the opening and closing rate constants were plotted on the same graph (Figure 7.9c). Simulation one was generated using the original best fit values for all parameters (β = 1452 s^{-1} and α = 488 s^{-1}). In simulation two, only the opening and closing rate constants were altered, and both were reduced to half of their initial values (β = 726 s^{-1} and α = 244 s^{-1}). In simulation three, the values of these two rate constants were further reduced to one third of the original values (β = 484 s^{-1} and α = 163 s^{-1}). The model currents generated by simulations using these three different sets of parameters were indistinguishable. Only when the ratio of the opening and closing rate constants was altered in simulation four (β = 1452 s^{-1} and α = 244 s^{-1}), did the simulation yield a different result from the original (compare simulations one, three and four). This result suggests that the opening and closing rate constants can compensate for each other. As long as the ratio of the two is not altered, the macroscopic current should remain the same.

Rigorous error analysis for β/α was carried out to more closely examine the conclusion reached above. The results of four such analyses of β/α are shown in Figure 7.9D. The corresponding χ^2 value was plotted against β/α in each case. The dashed and solid horizontal lines are the 95% confidence χ^2 level for the current induced by 1 μM 5-HT in the absence and presence of 10 mM TCEt, respectively. The two dashed curves represent the results of two rigorous error analyses of β/α for 1 μM 5-HT-activated current. For the dark dashed curve, α was fixed at 300 s^{-1}, while β was changed stepwise from 100 to 1650 s^{-1} so that the β/α ratio walked through a range from 0.333 to 5.50. For the gray dashed curve, β was fixed at 873 s^{-1}, while α was changed stepwise from 25 to 565 so that the value of β/α walked through a range from 34.92 to 1.55. Both curves gave almost identical results at the overlapping region, despite the fact that, at only one point (β/α =2.91) these two curves had the same value (β =873 s^{-1}, α =300 s^{-1}). This observation supports the idea that only the ratio of the channel opening (β) and closing (α) rate constants is important for fitting the macroscopic current mediated by the 5-HT$_3$ receptor/channel in the framework of the current model. The two solid curves represent the results of rigorous error analysis performed on β/α for 1 μM 5-HT + 10 mM TCEt activated current while holding constant the closing (α = 300 s^{-1}) or opening (β = 873 s^{-1}) rate constant, respectively. Again, the two curves overlap. The two χ^2 wells for 1 μM 5-HT-generated current and the two χ^2 wells for 1 μM 5-HT + 10 mM TCEt-generated current are well defined and overlapping. These four χ^2 wells are all minimized at the point at which β/α equals 2.91, giving an intrinsic P_{open} of 0.74. Therefore, TCEt does not appear to alter the ratio between the channel opening (β) and closing (α) rate constants, which is consistent with the conclusions from fitting and simulation analyses. In the final analysis, the kinetic model we have used likely contains more parameters than are strictly necessary to adequately describe our whole-cell currents. However, from analysis of single ligand-gated channels, we

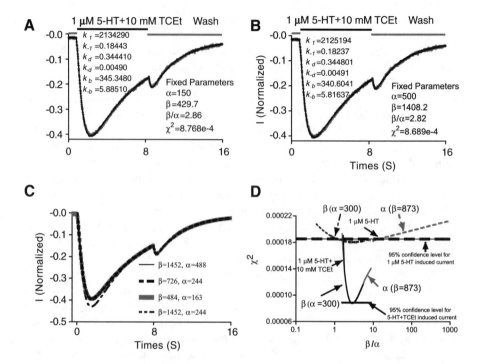

FIGURE 7.9 TCEt does not appear to alter the value of β/α. (A) and (B): Average current generated by 1 μM 5-HT + 10 mM TCEt are the same as shown in Figure 7.1 with the exception that the closing rate constant is preset at 150 s⁻¹ in A and 500 s⁻¹ in B. The values of other parameters derived from the best fit are almost identical except for the opening rate constant. However, the ratio of opening and closing rate constants are similar in A and B. C: Four current simulations generated using the Q-matrix approach as described in the text. The opening and closing rate constants were varied in the different simulations, while the values of all other parameters were set to those that yielded the best fit to current generated by 1 μM 5-HT + 10 mM TCEt in Figure 7.1. Values of opening and closing rate constants for the first simulation were also set to those of best fit. In simulations 2 and 3, values of these two rate constants were reduced by one half and two thirds respectively, but the ratio of these two parameters was not altered. In simulation 4, the ratio of those two rate constants intentionally altered as indicated. (d) χ^2 for four separate rigorous error analyses of β/α are plotted as a function of β/α. Dotted and black horizontal lines = 95% confidence χ^2 values for the current induced by 1 μM 5-HT in absence or presence of 10 mM TCEt, respectively. Black and gray dotted curves = results of rigorous error analysis of β/α for 1 μ5HT-activated current while holding (α = 300 s⁻¹) or (β = 873 s⁻¹), respectively. Black and gray solid curves = the results of rigorous error analysis of β/α for 1 μM 5-HT + 10 mM TCEt-activated current while holding (α = 300 s⁻¹) or (β = 873 s⁻¹) constant, respectively. Two χ^2 wells for 1 μM 5-HT-generated current and two χ^2 wells for 1 μM 5-HT + 10 mM TCEt-generated current are well defined and overlap. These four χ^2 wells are all minimized where β/α equals 2.91.

know that separate binding and opening or closing steps are necessary to fully describe channel behavior (Colquhoun and Hawkes, 1995a), and, thus, we have kept these steps and rate constants in the present model.

7.4 FUTURE DIRECTIONS

Kinetic analysis coupled with accurate measurement of whole-cell and single-channel currents provides useful descriptions of ligand-gated ion channel function and insight into the mechanisms by which pharmacological agents including alcohols alter channel function. However, this approach alone cannot yield much information about the molecular mechanisms involved in channel function and modulation. Molecular biological and biophysical techniques hold great promise for providing information about these molecular mechanisms. However, channel function must always be monitored as molecular-level studies proceed, and kinetic analysis provides an in-depth portrait of channel behavior that cannot be obtained from simple pharmacological and electrophysiological analyses. Good examples can be found in studies of the effects of amino acid mutation. Mutation of a single amino acid within a channel can alter channel function in many ways. If the function of that amino acid is not suspected with any certainty prior to mutagenesis, then kinetic analysis can help to generate hypotheses about the functional role of that particular residue. For example, an amino acid deep within the membrane-spanning domain of a channel would not be likely to directly interact with agonist or antagonist binding in a classical ligand-gated neurotransmitter receptor. However, mutations near the middle of the second transmembrane spanning domain of both the nAChR and the 5-HT$_3$ receptor have been shown to alter agonist potency at these receptors (Yakel et al., 1993; Labarca et al., 1995). While these findings might seem paradoxical, it must be remembered that the kinetics of the channel in the agonist-bound state strongly influences the apparent affinity of the agonist (Colquhoun, 1998).

A single-point mutation of an amino acid in a ligand-gated ion channel may affect both channel function and pharmacology. It is quite possible that effects on pharmacology may result from changes in receptor function, and not through disruption of a "binding site" for a particular drug (e.g., Boileau and Czajkowski, 1999). For example, a mutation may eliminate amino acids with key roles in transduction of binding into channel activation or modulation; or the mutation may alter the receptor-channel kinetics in a manner that mimics the effect of the drug of interest. In the latter case, drug actions may be prevented simply because the kinetic alterations normally produced by the drug are occluded by those produced by the mutation itself. This could happen because of a change in the drug binding site itself or because of a change at a site distant from the site of drug interaction with the channel. In cases where drug binding to the channel cannot be directly measured (as is the case of most alcohols), then one must rely on detailed kinetic information to provide clues as to the reason for a loss of drug effect. If kinetic effects produced by a mutation appear to occlude drug actions, then one must view the possibility of having identified a drug-binding site with some skepticism until a more reliable way is found to demonstrate direct interactions between the drug and a particular amino acid residue within the receptor.

7.5 CONCLUSIONS

Examination of the kinetics of ligand-gated ion channels provides information about mechanisms of channel gating, as well as detailed information about pharmacological effects of drugs, such as alcohols on these channel function. Kinetic analysis requires a combination of electrophysiological techniques and kinetic simulation and/or model fitting. Fast solution exchange is necessary to resolve the rapid activation and other kinetic transitions of ligand-gated ion channels. This can be accomplished using either whole-cell or excised-patch recording techniques, but a system capable of micro- to millisecond time-scale repositioning of cells or solution application streams is needed.

Measurement and non-linear fitting of different current components can provide initial information about receptor-channel kinetics. However, a more complete picture of receptor-channel kinetics can be gained by simulation of currents using different kinetic models, and fitting experimental data directly using a defined kinetic model is a rigorous way to examine drug-induced changes in channel kinetics. Additional techniques, such as rigorous error analysis, can be used to estimate the precision of estimates of kinetic parameters. The analysis presented in this chapter suggests that alcohols potentiate the function of the 5-HT_3 receptor by favoring and stabilizing the open-state configuration of the receptor. This conclusion agrees well with our early observations of alcohol effects on this receptor-channel using more descriptive methods and kinetic simulations (Zhou et al., 1998).

The simple model that we have presented (Model 7.1) adequately describes currents generated by a low concentration of agonist in the absence or presence of alcohols (Figure 7.5). This observation again supports the notion that the 5-HT_3 receptor is very similar to the well-studied nACh receptor both in its structure and kinetics. It is important to point out that the success of this model in fitting the experimental data only indicates that this particular model is sufficient to account for the kinetics of the 5-HT_3 receptor and alcohol effects at the macroscopic current level. Other models incorporating the major features of this one while having more kinetic steps, such as opening from the mono-liganded receptor or two distinct desensitized states, may also be able to fit the data well.

ACKNOWLEDGMENTS

We thank Drs. Todd A. Verdoorn and Dirk J. Snyders for helpful advice on receptor kinetics and Drs. Feng Li and David Weaver for technical assistance with the fast-application system. We also thank Dr. Joseph M. Beechem for providing the Global Analysis program and especially Dr. Eric Hustedt for help with use of Globals program and for reading and providing helpful comments on the chapter. This work was supported by grants from the National Institute on Alcohol Abuse and Alcoholism (AA08986 and AA11542).

REFERENCES

Adams SR and Tsien RY (1993) Controlling cell chemistry with caged compounds. *Annu Rev Physiol* 55:755-784

Barann M et al.(1995) Increasing effect of ethanol on 5-HT$_3$ receptor-mediated 14C-guanidinium influx in N1E-115 neuroblastoma cells. *Naunyn-Schmiedebergs Archives of Pharmacology* 352(2):149-156.

Beechem JM (1992). Global analysis of biochemical and biophysical data. *Meth Enzymol.* 210:37-54.

Beechem JM et al. (1991) The global analysis of fluorescence intensity and anisotropy decay data: Second-generation theory and programs. In *Topics in Fluorescence Spectroscopy, Vol. 2: Principles* (Lakowicz JR Ed.) New York: Plenum Press.

Benveniste M and Mayer ML (1991) Kinetic analysis of antagonist action at N-methyl-D-aspartic acid receptors. Two binding sites each for glutamate and glycine. *Biophys J* 59(3):560-573.

Brett RS, Dilger JP, Adams PR and Lancaster B (1986) A method for the rapid exchange of solutions bathing excised membrane patches. *Biophys J* 50:987-991.

Boileau AJ and Czajkowski C (1999) Identification of transduction elements for benzodiazepine modulation of the GABA(A) receptor: three residues are required for allosteric coupling. *J Neurosci* 19(23):10213-10220.

Chen JP, Van Praag HM and Gardner EL (1991) Activation of 5-HT$_3$ receptors by 1-phenylbiguanide increases dopamine release in the rat nucleus accumbens. *Brain Res* 543:354-357.

Colquhoun D (1998) Binding, gating, affinity and efficacy: the interpretation of structure-activity relationships for agonists and of the effects of mutating receptors. *Br J Pharmacol* 125(5):924-947.

Colquhoun D and Hawkes AG (1995a) The principles of stochastic interpretation of ion-channel mechanisms. in *Single Channel Recording* (Sakmann B and Neher E, Eds.) pp 397-482. New York: Plenum Press.

Colquhoun D and Hawkes AG (1995b). A Q-matrix cookbook: How to write only one program to calculate the single-channel and macroscopic predictions for any kinetic mechanism, in *Single Channel Recording* (Sakmann B and Neher E, Eds) pp 589-636. New York: Plenum Press.

Costall B and Naylor RJ (1992) The psychopharmacology of 5-HT$_3$ receptors. *Pharmacol and Toxicol* 71:401-415.

Costall B et al. (1987) Effects of the 5-HT$_3$ receptor antagonist GR38032F, on raised dopaminergic activity in the mesolimbic system of the rat and marmoset brain. *Br J Pharmacol* 92:881-894.

Diamond I and Gordon AS (1997) Cellular and molecular neuroscience of alcoholism. *Physiol Rev* 77(1):1-20.

Fadda F et al. (1991) MDL 72222, a selective 5-HT$_3$ receptor antagonist, suppresses voluntary ethanol consumption in alcohol-preferring rats. *Alcohol and Alcoholism* 26:107-110.

Forman SA, Miller KW and Yellen G. (1995) A discrete site for general anesthetics on a postsynaptic receptor. *Mol Pharm* 48:574-581.

Grant KA (1995) The role of 5-HT$_3$ receptors in drug dependence. *Drug and Alcohol Dependence* 38:155-171.

Green T, Stauffer KA and Lummis SCR (1995) Expression of recombinant homo-oligomeric 5-hydroxytryptamine3 receptors provides new insights into their maturation and structure. *J Biol Chem* 270:6056-6061.

Grewer C et al. (2000) Glutamate translocation of the neuronal glutamate transporter EAAC1 occurs within milliseconds. *Proc Natl Acad Sci USA* 97(17):9706-11.

Hagan RM et al. (1987) Effect of the 5-HT$_3$ receptor antagonist, GR38032F, on responses to injection of a neurokinin agonist into the ventral tegmental area of the rat brain. *Eur J Pharmacol* 138:303-305.

Harris RA et al. (1995) Actions of anesthetics on ligand-gated ion channels: role of receptor subunit composition. *FASEB J* 9:1454-1462.

Hille (1992) *Ionic Channels of Excitable Membranes*. Sunderland MA: Sinauer Associates.

Jackson M and Yakel JL (1995) The 5-HT$_3$ receptor channel. *Ann. Rev. Physiol.* 57:447-468.

Jiang LH et al. (1990) The effect of intraventricular administration of the 5-HT$_3$ receptor agonist 2-methylserotonin on the release of dopamine in the nucleus accumbens: an *in vivo* chronocoulometric study. *Brain Res* 513:156-160.

Johnson BA et al. (1993) Attenuation of some alcohol-induced mood changes and the desire to drink by 5-HT3 receptor blockade: A preliminary study in healthy male volunteers. *Psychopharm* 112:142-144.

Jonas P (1995) Fast application of agonists to isolated membrane patches, in *Single Channel Recording* (Sakmann B and Neher E, Eds.) pp 231-243. New York: Plenum Press.

Jones MV and Westbrook GL (1995) Desensitized states prolong GABAA channel responses to brief agonist pulses. *Neuron* 15(1):181-191.

Jones MV et al. (1998) Defining affinity with the GABAA receptor. *J Neurosci* 18(21):8590-8604.

Labarca C et al. (1995) Channel gating governed symmetrically by conserved leucine residues in the M2 domain of nicotinic receptors. *Nature* 376(6540):514-516.

LeMarquand D, Phil RO and Benkelfat C (1994a) Serontonin and alcohol intake, abuse and dependence: clinical evidence. *Biol Psych* 36(5):326-337.

LeMarquand D, Pihl RO and Benkelfat C (1994b) Serotonin and alcohol intake, abuse, and dependence: findings of animal studies. *Biol Psych* 36(6):395-421.

Levenberg K. (1944) A method for the solution of certain nonlinear problems in least squares. *Quart Appl Math* 2: 164-168.

Lovinger DM (1991) Ethanol potentiates 5-HT$_3$ receptor-mediated ion current in NCB-20 neuroblastoma cells. *Neurosci Lett* 122:54-56.

Lovinger DM, Sung KW and Zhou Q (2000) ethanol and trichloroethanol increase the probability of opening of 5-HT$_3$ receptor-channels in NCB-20 neuroblastoma cells. *Neuropharm* 39:561-570.

Lovinger DM and White GG (1991) Ethanol potentiation of the 5-hydroxytryptamine$_3$ receptor-mediated ion current in neuroblastoma cells and isolated mammalian neurons. *Mol Pharm* 40:263-270.

Lovinger DM and Zhou Q (1993) Trichloroethanol potentiation of 5-hydroxytryptamine$_3$ receptor mediated current in nodose ganglion neurons from adult rat. *J. Pharm Exper Therap* 265:771-777.

Lovinger DM and Zhou Q (1994) Alcohols potentiate ion current mediated by recombinant 5-HT3RA receptors expressed in a mammalian cell line. *Neuropharm* 33:1567-1572.

Lovinger DM (1997) Alcohols and neurotransmitter gated ion channels: past, present and future. *Naunyn-Schmiedeberg's Arch Pharmacol.* 356:267-282.

Machu TK and Harris RA (1994) Alcohols and anesthetics enhance the function of 5-hydroxytryptamine$_3$ receptors expressed in Xenopus laevis ooctyes. *J Pharm Exper Therap* 271(2):898-905.

Maconochie DJ and Knight DE (1989) A method for making solution changes in the sub-millisecond range at the tip of the patch pipette. *Pflugers Arch* 414:589-596.

Maconochie DJ and Steinbach JH (1998) The channel opening rate of adult- and fetal-type mouse muscle nicotinic receptors activated by acetylcholine. *J Physiol* 506 (Pt 1):53-72.

Maricq AV, Peterson AS, Brake AJ, Myers RM, Julius D (1991). Primary structure and functional expression of the 5-HT$_3$ receptor, a serontonin-gated ion channel. *Science* 254:432-437.

Marquardt DW (1963) An algorithm for least squares estimation of nonlinear parameters. Society for Industrial and Applied Mathematics 11:431-441.

Mihic SJ et al. (1997) Sites of alcohol and volatile anaesthetic action on GABA$_A$ and glycine receptors. *Nature* 389:385-389.

Neijt HC et al. (1988). Pharmacological characterization of serotonin 5-HT$_3$ receptor-mediated electrical response in cultured mouse neuroblastoma cells. *Neuropharm* 27:301-307.

Orser SA et al. (1994). Propofol modulates activation and desensitization of GABA$_A$ receptors in culture murine hippocampal neurons. *J Neurosci* 14(12):7747-7760.

Press WH et al. (1992) *Numerical Recipes in Fortran: The Art of Scientific Computing. 2nd ed.*, Cambridge University Press, Cambridge UK.

Sakmann B and Neher E (1995) *Single-Channel Recording.* New York: Plenum Press.

Sellers EM et al. (1994) Clinical efficacy of the 5-HT$_3$ antagonist ondansetron in alcohol abuse and dependence. *Alcohol Clin Exper Res* 18:879-885.

Tomkins DM, Le AD and Sellers EM (1995) Effect of the 5-HT$_3$ antagonist ondansetron on voluntary ethanol intake in rats and mice maintained on a limited access procedure. *Psychopharm* 117(4):479-485.

Verdoorn TA (1993) Formation of heteromeric γ-aminobutyric acid type A receptors containing two different subunits. *Mol Pharmacol.* 45:475-480.

Yakel JL, Shao XM and Jackson MB (1991). Activation and desensitization of the 5-HT$_3$ receptor in a rat glioma x neuroblastoma hybrid cell. *J. Physiol* 436:293-308.

Yakel JL et al. (1993) Single amino acid substitution affects desensitization of the 5-hydroxytryptamine type 3 receptor expressed in Xenopus oocytes. *Proc Natl Acad Sci USA* 90(11):5030-5033.

Zhou Q, Verdoorn TA and Lovinger DM (1998). Alcohols potentiate the function of 5-HT$_3$ receptor-channels on NCB-20 neuroblastoma cells by favouring and stabilizing the open channel state. *J. Physiol* (London) 507:335-352.

8 Electrophysiological Assessment of Synaptic Transmission in Brain Slices

Jeff L. Weiner

CONTENTS

8.1 INTRODUCTION

Over the last century, many advances have been made in elucidating the physiological mechanisms underlying the behavioral and cognitive effects of alcohol (ethanol). Early theories suggested that ethanol partitioned into neuronal membranes in a rather nonspecific manner, thereby causing a generalized disruption of central nervous system (CNS) function (Seeman, 1974; Goldstein, 1986). Over the past 30 years, it has become increasingly clear that ethanol perturbs neuronal function in a far more selective manner than originally thought. It is now widely appreciated that acute exposure to alcohol directly alters the activity of some of the ligand-gated ion channels that mediate excitatory and inhibitory synaptic transmission (Faingold et al., 1998; Tsai and Coyle, 1998). Alcohol also disrupts the function of a select group of voltage-gated ion channels that regulate neurotransmitter release and neuronal firing (Brodie and Appel, 1998; Dopico et al, 1999; Walter and Messing, 1999). Moreover, these same synaptic proteins appear to undergo extensive neuroadaptive changes in response to chronic alcohol exposure and withdrawal.

A number of methodological advances have played a central role in facilitating the transition in our current understanding of the mechanisms underlying alcohol's effects on CNS function. One key has been the development of electrophysiological techniques that enable the study of synaptic transmission in acute brain slice preparations. These methods were instrumental in first identifying the synapse as a central target of alcohol action. Moreover, these methods continue to be used by many investigators to unravel the complex mechanisms through which alcohol disrupts interneuronal communication.

The purpose of this chapter is to describe the methods involved in preparing brain slices for electrophysiological recording as well as some of the more common recording techniques in use today. The focus will be on providing practical insight into the application of these methods. Where possible, examples of how these techniques have been used in alcohol research will also be provided.

The final section of this chapter will focus on methodological approaches that can be used to distinguish between pre- and postsynaptic mechanisms of alcohol action. As our understanding of how alcohol interacts with synapses progresses, it is becoming increasingly apparent that alcohol acts at a number of pre- and postsynaptic loci to alter CNS activity. A variety of electrophysiological approaches have been developed to differentiate between these different mechanisms of synaptic modulation. The use of three common methods employed in alcohol research will be discussed.

8.2   THE *IN VITRO* TISSUE SLICE PREPARATION

Thanks to the pioneering work of Henry McIlwain and his colleagues, the acute brain slice preparation has become a mainstay in electrophysiological research. Ironically, this preparation was originally developed for use in biochemical studies characterizing energy metabolism in neuronal tissue. These early studies demonstrated that, under appropriate experimental conditions, acutely prepared slices of brain tissue could maintain normal levels of ATP and phosphocreatine for several hours (McIlwain et al., 1951). Soon after, McIlwain and his colleagues discovered that, in addition to demonstrating biochemical markers of viability, neurons in brain slices could actually maintain viable membrane potentials (Li and McIlwain, 1957). In addition, synaptic responses could be recorded in these preparations in response to electrical stimulation (Yamamoto and McIlwain, 1966). These early groundbreaking studies laid the foundation for countless numbers of investigators who have adapted and extended these *in vitro* methods to make seminal contributions to our understanding of the complex processes that underlie synaptic transmission.

In vitro tissue slice preparations have proven particularly useful in alcohol research. Previous investigators have employed these methods in the initial discovery and subsequent characterization of the effects of ethanol on excitatory and inhibitory synaptic transmission. Tissue slices have been used to investigate ethanol modulation of synaptic transmission in areas ranging from the spinal cord (Wang et al., 1999) and brainstem (Eggers et al., 2000) to the cerebral cortex (Soldo et al., 1994; Sessler et al., 1998). In addition, synaptic effects of ethanol have been characterized in slices from a wide variety of species ranging from mice (Whittington and Little, 1990) and rats (Carlen et al., 1982) to monkeys (Ariwodola et al., 2000).

A number of features of the brain slice preparation make it an ideal model system with which to investigate the synaptic effects of drugs like ethanol. First, unlike most cell culture models, many of the interneuronal connections that form between different cell types are preserved. While not all afferent and efferent projections are maintained in brain slices, the synapses that are preserved have developed in their native environment and are representative of the anatomical relationships that are observed *in vivo*. These features provide the opportunity to characterize the effects of ethanol on native synaptic receptors. The importance of this feature is underscored by reports highlighting differences between the effects of ethanol on native and recombinant receptors (e.g., Sapp and Yeh, 1998) and even between the effects of ethanol on native receptors characterized in slices and in culture (Weiner et al., 1999; Costa et al., 2000). Second, the stability and experimental access afforded by the brain slice preparation greatly facilitates the application of techniques like single cell recording that are extremely difficult to carry out in intact animals. These same features also allow for much greater experimental regulation of the time course of drug applications as well as precise control of drug concentrations. Finally, since brain slices can be prepared from animals following chronic exposure and withdrawal of alcohol, these preparations provide the opportunity to assess *in vitro*, the neuroadapative changes associated with alcohol abuse that develop *in vivo*.

Although brain slices can be prepared from virtually any area of the brain, the hippocampal slice has been, and continues to be, the most popular *in vitro* slice preparation. Therefore, many of the slice methods discussed in this chapter were originally developed for use in hippocampal tissue. Nevertheless, these methods can generally be adapted for use in preparing slices from other brain regions. The popularity of the hippocampal slice rests largely on its lamellar and laminar neuronal organization. Neurons in many brain regions are either sparsely distributed or localized in clusters containing many different cell types. In addition, most brain regions receive afferent inputs from a complex and dizzying array of sources, making it difficult to preserve and subsequently identify synaptic inputs in slices of these regions. In contrast, the hippocampus contains a number of large, homogenous populations of densely packed cells (Schwartzkroin, 1975; Dingledine et al., 1980). These cells form a well-described trisynaptic circuit that is essentially preserved in transverse sections made along its longitudinal axis (Andersen et al., 2000). In addition, much of the inhibitory control of this brain region arises from a diverse array of local circuit interneurons and this inhibitory circuitry is also well preserved in acutely prepared slices (Freund and Buzsaki, 1996).

8.2.1 Preparation of Brain Slices

Regardless of the brain region to be studied, the first steps in preparing brain tissue for electrophysiological investigation are similar. The animal to be used is sacrificed by decapitation and the brain is rapidly removed (no longer than 2 min) into a well-oxygenated ice-cold artificial cerebrospinal fluid (aCSF) solution. Animals of any age can be used. However, we and others have generally found that brain slices prepared from younger rats (2–6 weeks old) tend to be more viable and easier to record from (Lipton et al., 1995). We have found that, if slices are prepared from aged animals or from species in which the removal of brain tissue may take longer than a couple of minutes, transcardial perfusion with ice-cold aCSF prior to removal of the brain enhances slice viability.

Most studies use an aCSF that is some modification of a "Krebs" or "Ringer" type saline solution. The specific ingredients are often customized to the particular needs of the experiment (e.g., omitting Mg^{++} to record N-methyl-D-aspartate (NMDA) receptor-mediated responses). We have used the same basic aCSF recipe for many years with generally good results (Weiner et al., 1994; 1997a,b; 1999). This solution contains (in mM): 126 NaCl, 3 KCl, 1.5 $MgCl_2$, 2.4 $CaCl_2$, 1.2 NaH_2PO_4, 11 glucose, and 26 $NaHCO_3$. The aCSF is then saturated with 95% O_2-5% CO_2 using a gas dispersion stone. This mixture ensures adequate oxygenation of the tissue and maintains the pH within a physiological range (7.3–7.5).

One variable that has been reported to alter some of the acute effects of alcohol in *in vitro* studies is the use of an anesthetic agent prior to sacrifice. For example, methoxyflurane and CO_2 anesthesia have been shown to disrupt the allosteric modulation of γ-amino-butyric acid$_A$ (GABA$_A$) receptors by flunitrazepam and ethanol (Engel et al., 1996). Although rapid decapitation without anesthesia is permissible under the guidelines of the American Veterinary Medical Association (2001), the use of an anesthetic agent is recommended when it will not interfere with experimental outcomes. To date, the acute effects of ethanol on excitatory (Weiner et al., 1999) and inhibitory (Weiner et al., 1994; 1997a, b; 1999) synaptic transmission in brain slices that we have characterized do not appear to be influenced by the use of halothane anesthesia prior to decapitation. However, whenever characterizing novel effects of ethanol, it is often necessary to carry out an empirical evaluation to ensure that the use of an anesthetic does not interfere with the ethanol effects being investigated.

Once the brain has been removed and incubated in chilled, aerated aCSF for no more than 1 to 2 min, the tissue is then prepared for slicing. Two methods are commonly employed in the preparation of brain slices for electrophysiological recording. The first method involves using a manual (Figure 8.1A) or automated tissue chopper to section the tissue. In this method, the brain is hemisected and the region of interest is dissected free of surrounding tissue. The dissected tissue block is then secured on a piece of filter paper and positioned perpendicular to the blade of the chopper. The blade is usually a thin, double-edged carbon or steel razor adjusted such that it cuts into, but not through, the filter paper securing the tissue. These choppers are typically some variation of the first device developed for tissue slice preparation

A   B

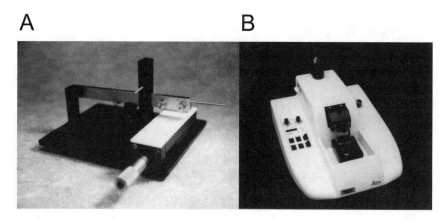

FIGURE 8.1 Photographs of instruments commonly used in the preparation of tissue slices for electrophysiological recording. (A) A manual tissue chopper (Stoelting, Wood Dale, IL) with an analog micrometer that is frequently used for preparing hippocampal slices for extracellular or "Blind" patch-clamp recordings. (B) A motorized, oscillating tissue microtome with digital section thickness controller (Leica Microsystems) commonly employed when preparing relatively thin (100–300 µm) sections for visualized cell recording. This model is designed to allow ice to be packed around the inner cutting chamber to keep the aCSF at a cold temperature during tissue sectioning.

by Henry McIlwain. The advantages of this method are that it is inexpensive, relatively fast (total slicing time < 5 minutes) and generally yields excellent slices. The disadvantages of these McIlwain-type choppers are that they are not practical for cutting uniform, thin tissue sections (< 300 µm), which are often necessary for visualized cell recordings (described in Section 3.3). Chopper methods may also not be suitable for slicing brain regions that are difficult to dissect free of surrounding tissue, such as the thalamus or striatum. Finally, we have found that this method, which requires considerable handling of the tissue, requires more training than using a vibrating tissue slicer (see below). However, for a well-trained investigator on a budget, tissue choppers produce excellent quality slices, particularly from hippocampus, that are suitable for most *in vitro* electrophysiological experiments.

The second option for slicing tissue is to use a motorized, vibrating tissue slicer (Figure 8.1B). In this method, the brain is rapidly removed as described above and a block containing the region of interest is fixed to a cutting stage with cyanoacrylate glue. After the tissue is secured, the stage is fastened to the base of a chamber filled with chilled, oxygenated aCSF and the tissue is sectioned using the vibrating, oscillating blade of the slicer. There are a number of makes and models of tissue slicers to choose from. All models provide some control of the cutting angle, speed, and oscillation frequency of the blade. In our experience, a blade angle of 10–15 degrees, a slow cutting speed and high oscillation frequency produce the best results. Of course, a slower cutting speed will prolong the slicing process. Therefore, a balance must be reached such that the blade advances slowly enough that it does not push the tissue and the slicing process takes no longer than 10–15 min. Another helpful tip is to keep the aCSF bathing the tissue as cool as possible (by surrounding the cutting chamber with ice) and well aerated throughout this procedure.

There are several advantages of a vibrating tissue slicer. Most investigators report that these instruments provide better-quality tissue sections than the classical tissue choppers (Lipton et al., 1995). Vibrating slicers appear to cause less damage to the surface of the tissue, a factor that is particularly important when preparing thin sections for visualized cell recording (see section 3.3). As noted above, vibrating slicers are also better suited for preparing tissue sections from brain regions that are not readily dissected free from the whole brain. The disadvantages of tissue slicers are that they can be considerably more expensive than choppers and the slicing procedure usually takes longer than the chopper method.

8.2.2 SLICE INCUBATION AND RECORDING CHAMBERS

Once the tissue has been sectioned using either a chopper or vibrating slicer, it is immediately transferred to an incubation chamber. Most studies suggest that brain slices be allowed to equilibrate for a minimum of 1 h before recording (Lipton et al., 1995). The two main options for storing and recording from brain slices are interface and submersion chambers. In an interface chamber, the slices rest on a nylon mesh and are slowly perfused with oxygenated aCSF with the fluid level adjusted such that it just covers the surface of the tissue. The chamber is usually maintained between 32–35° C and warmed, moisturized O_2/CO_2 mixture is blown over the tissue. Some investigators have suggested that interface chambers provide the optimal method for storing and recording from acutely prepared brain slices (Lipton et al., 1995). Moreover, since the aCSF level of the bath is set to be just above the surface of the tissue, the tissue surface can be detected electrically once the recording electrode makes contact with the bath solution. This is particularly useful when using sharp intracellular electrodes as their tips can be very difficult to visualize. The only real disadvantage of interface chambers relates to their use as recording chambers in pharmacological studies, as the exchange of the bath solution can be extremely slow. Therefore, the use of interface chambers is not recommended in studies that require the frequent introduction and washout of drugs into the recording chamber.

The second option for storing and recording from brain slices is to use a submersion chamber. Submersion storage chambers are typically constructed from a simple beaker fitted with a mesh-covered grid (Figure 8.2). The beaker is filled with aCSF and continuously aerated with 95% O_2/5% CO_2. Slices are gently transferred to individual wells of the chamber using a polished Pasteur pipette and stored for the day's experiment. We have used submersion chambers such as the one illustrated in Figure 8.2 for many years and have found them to be very reliable for storing tissue sections. In a recent study on monkey brain slices (Ariwodola et al., 2000), viable recordings of dentate granule neurons, judged by an initial membrane potential more negative than –65 mV, were obtained from slices stored in a submersion chamber for up to 15 h. The primary advantage of submersion chambers relates mainly to their use as recording chambers. Submersion recording chambers usually have a relatively small volume (150–500 µL) and a faster aCSF perfusion rate (2 ml/min) than that of interface chambers. These

FIGURE 8.2 Submersion storage chamber for incubation of acutely prepared tissue slices. The chamber is constructed from a 500 mL beaker and is filled with aCSF that is continuously aerated with 95% O_2/5% CO_2 delivered via a gas diffusion stone. The chamber is fitted with a mesh-lined grid constructed from polyethylene filters (540μm thickness, 526μm openings) that serves to keep tissue sections separated during storage. The temperature of the chamber can be adjusted using a heated water-bath.

A   **B**

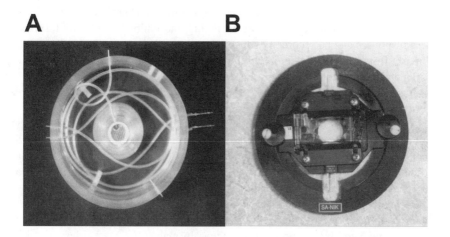

FIGURE 8.3 Common submersion recording chambers used in tissue slice electrophysiological studies. (A) A custom-made Haas-type chamber in which the tissue rests on a nylon-mesh that is fitted into a raised chamber in the center of the bath. The tissue is perfused from above and below with aerated aCSF. The outer chamber is equipped with a nichrome wire that can be used to heat the aCSF perfusing the recording chamber. (B) Glass-bottom chamber used in visualized slice recording (Warner Instruments, Hamden, CT). The tissue is weighted down using small platinum wire weights and the chamber can be heated by warming the perfusate through an external water bath prior to perfusion of slices.

factors greatly facilitate the application and washout of drugs. Therefore, most electro-physiological studies that focus on characterizing the effects of drugs, like ethanol, on synaptic transmission generally employ a submersion chamber, at least for recording. Two common styles of submersion chambers are illustrated in Figure 8.3.

One important variable regarding the storage and recording of brain slices involves the temperature at which slices are maintained. It may seem intuitive that storing and recording from slices at, or near, physiological temperatures would be desirable for most experiments. However, for practical reasons, many electrophysi-ological studies are carried out at room temperature. In fact, even when trying to carry out experiments at physiological temperatures, most investigators opt to record at temperatures slightly below those observed *in vivo* (32–35° C). We have found that, regardless of the temperature at which experiments will be carried out, it is best to maintain the tissue slices at a constant temperature throughout storage and recording. In rat hippocampal slices, gradually increasing the temperature following the dissection from 13° C to ambient levels can result in profound changes in a variety of electrophysiological parameters (Watson et al., 1997) and in basal protein kinase C activity (Weiner et al., 1997b).

8.2.3 STIMULATION AND PHARMACOLOGICAL ISOLATION OF SYNAPTIC RESPONSES

As mentioned earlier, one of the most powerful advantages of brain slices is that functional synapses are preserved in these preparations. Therefore, electrical stim-ulation of afferent inputs can generate excitatory and inhibitory synaptic responses. These responses can be recorded extracellularly from large groups of cells or intra-cellularly from individual neurons. Stimulation is usually delivered using relatively small metal electrodes. We employ concentric (Figure 8.4A) or bipolar (Figure 8.4B) tungsten wire electrodes that provide a relatively discrete stimulation focus. Bipolar electrodes can also be fashioned by pulling glass theta tubing to a relatively thin tip (1–2 μm). Both sides of the electrode are then filled with a concentrated electrolyte solution (e.g., 3M NaCl) and connected to a stimulus isolation unit with thin wire connectors (Banks et al., 1998; Jensen and Mody, 2001). These glass electrodes are particularly useful for triggering monosynaptic responses from defined morpholog-ical zones (e.g., somatic vs. dendritic synapses).

Because most slices contain a variety of functional excitatory and inhibitory synaptic inputs, electrical stimulation typically generates complex synaptic responses mediated by multiple ion channels. In many studies, the desired goal is to characterize the effects of a drug on an individual element of synaptic transmis-sion, such as NMDA receptor-mediated synaptic excitation or GABA_A receptor-mediated synaptic inhibition. This can usually be achieved by using selective receptor antagonists to "pharmacologically isolate" the synaptic response of interest. For example, GABA_A receptor-mediated inhibitory synaptic responses in most brain regions can be effectively isolated using a cocktail of GABA_B, NMDA, and AMPA/kainate receptor antagonists. Unfortunately, the identity of all synaptic ele-ments may not always be known and appropriate pharmacological tools may not be readily available. For example, it has been known for many years that kainate

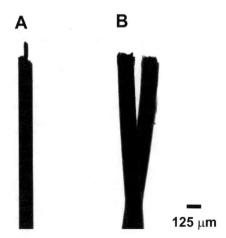

FIGURE 8.4 Stimulating electrodes used to evoke synaptic responses in brain slices. (A). Concentric bipolar stimulating electrode constructed from a 125 μm stainless steel outer pole and a 25 μm platinum/iridium inner pole (FHC, Bowdoinham, ME). (B) Twisted tungsten wire bipolar electrode (Plastics1, Roanoke, VA) with a slightly larger tip (~200 μm).

receptors are highly expressed in the rat hippocampus (Foster et al., 1981). However, the lack of an antagonist that can distinguish between kainate receptors and their more prominent relatives, the AMPA receptors, precluded any assessment of the physiological role or pharmacological profile of these receptors in tissue slices. The relatively recent development of AMPA receptor-selective antagonists (Bleakman et al., 1995) sparked a flurry of studies demonstrating a number of novel pre- and postsynaptic roles for kainate receptors in the hippocampus and other brain regions (see Ben-Ari and Cossart, 2000 for review). We used one of these selective AMPA receptor antagonists, LY 303070, to pharmacologically isolate a kainate receptor-mediated synaptic current in hippocampal CA3 pyramidal neurons and characterized its sensitivity to ethanol (Figure 8.5). Surprisingly, this study revealed that, although ethanol had no effect on AMPA excitatory postsynaptic currents (EPSCs), as shown by many others (Faingold et al., 1998; Tsai and Coyle, 1998), it potently inhibited kainate receptor-mediated synaptic responses in these cells (Weiner et al., 1999). We have since identified a number of other kainate receptor-mediated responses that are also inhibited by ethanol in the rat hippocampus (Crowder et al., 2000) and nucleus accumbens (Crowder and Weiner, 2001).

Because the majority of slice electrophysiology experiments require the intro-duction of at least some drugs into the bath, a number of methods have been developed to accomplish this process. The most common approach is simply to exchange the solution bathing the tissue with one containing the desired drug(s). A number of reservoirs can be set up with different combinations of drugs or varying concentrations of a single drug, like ethanol. Each reservoir must be equipped with a gas diffusion stone to provide continuous aeration with the O_2/CO_2 mixture. Solution flow is usually driven by gravity and switching between reser-voirs can be accomplished by manual or electrically controlled valves. When several reservoirs are used, their fluid levels and height should all be similar to

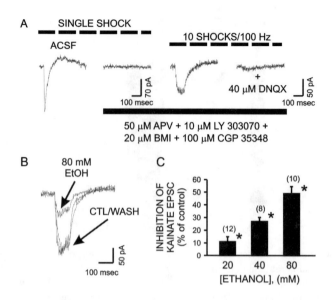

FIGURE 8.5 (A) Pharmacological isolation of slow kainate EPSCs in rat hippocampal CA3 pyramidal neurons. Note that in normal aCSF, individual stimuli evoke a biphasic response that is completely blocked a cocktail containing an NMDA receptor antagonist (APV), an AMPA receptor antagonist (LY303070), a $GABA_A$ antagonist, and a $GABA_B$ antagonist. In the presence of this cocktail, a stimulus train of 10 pulses delivered at 100Hz evokes a slow inward synaptic current that is then blocked by a mixed AMPA/KA receptor antagonist (DNQX). (B) Averages of four to seven pharmacologically isolated KA EPSCs evoked prior to, during, and after a 6-min application of 80 mM ethanol. (C) Bar graph summarizing the concentration dependence of ethanol inhibition of KA EPSCs. Numbers in brackets indicate the number of cells tested under each condition; *, $p < 0.05$. (Modified with permission from Weiner et al., 1999.)

ensure a constant flow rate into the bath. Roller bar perfusion pumps can be used instead of gravity to control the flow rate of the different solutions. Care must be taken to eliminate any pulsation introduced by these pumps as the mechanical interference can disrupt the recording.

We employ an alternative method that involves introducing drugs into the recording chamber via calibrated syringe pumps. In this method, adapted from Tom Dunwiddie's laboratory, an enclosed perfusion system is set up to provide a continuous, regulated flow of aCSF into the recording chamber. The perfusion is driven by pressure created by the O_2/CO_2 gas mixture aerating the aCSF. The flow rate is adjusted to 2 ml/min and this rate is continuously monitored using a calibrated flowmeter (Barnant Company, Barrington, IL). Concentrates of drugs (50–200X) to be introduced into the bath are mixed with the aCSF in a small port near the recording chamber via calibrated syringe pumps. The final bath concentration of each drug is controlled by the flow rate of each syringe pump. The main advantage of this perfusion method is that multiple concentrations of a drug can be introduced into the recording chamber from a single syringe, simply by altering the speed of the syringe pump. Also, by placing the mixing port close to the recording chamber, drug exchanges can be achieved relatively quickly.

8.3 RECORDING TECHNIQUES

A number of electrophysiological methods have been developed for the study of synaptic transmission in brain slices. The first techniques enabled the detection of extracellular electrical activity from large numbers of neurons. Subsequent advances permitted recording synaptic potentials from individual cells. All of these techniques are still in use today and have a broad number of applications in alcohol research.

8.3.1 EXTRACELLULAR RECORDING

This technique involves recording the electrical activity of large populations of neurons by placing recording electrodes in the extracellular space adjacent to the target cells of interest. Extracellular recording is usually carried out with low resistance glass microelectrodes filled with concentrated NaCl or NaAcetate (3–10 mM; 5–10 MΩ tip resistance). The shape of these electrodes is similar to that of patch-clamp electrodes (Figure 8.6A), although extracellular electrode tip diameters may be slightly smaller to reduce leakage of the pipette contents. Extracellular signals are generally very small as they reflect the flow of ionic current through the relatively low resistance of the extracellular fluid. Because these responses are typically on the order of 10–500 μV, they require fairly extensive amplification to be detected, making low instrument noise a critical technical consideration in the successful application of this technique.

Extracellular recording offers several advantages over intracellular recording methods. First, this technique is relatively simple and inexpensive to carry out. Because electrodes are placed in the extracellular space and not onto individual neurons, relatively inexpensive manipulators and stereomicroscopes provide adequate stability and visualization for these kinds of experiments. Moreover, although extracellular signals require extensive amplification, amplifiers that can perform these functions are generally much less expensive than those required for single cell recording. Other important advantages of this technique are that it can be carried out for relatively long periods of time and extracellular responses tend to be more stable and less variable than those recorded from individual cells, as they represent the summed activity of large populations of neurons. The stability and relatively low variability of these recordings makes them particularly well suited for studies that require long-term monitoring of a synaptic responses, such as during alcohol withdrawal (Thomas et al., 1998B) or when carrying out inter-subject comparisons (Nelson et al., 1999). Finally, some extracellular recording methods have been adapted to record extracellular activity *in vivo* in anesthetized and awake, freely moving animals (see Chapters 9 and 10).

However, several limitations of extracellular recording methods exist. Typically extracellular signals can be detected only in brain regions in which the activity of large populations of neurons is highly synchronized and the dipole orientation of the cells is uniform. Therefore, compound action potentials (population spikes) or field excitatory postsynaptic potentials (fEPSPs) from cortical or hippocampal

A   **B**

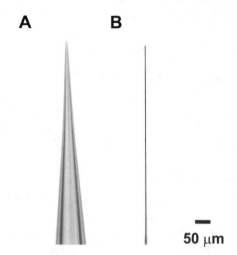

50 μm

FIGURE 8.6 Glass micropipettes used for single-cell recording. (A) Patch microelectrode designed to record whole-cell responses using "blind" or visualized cell patch-clamp recording. (B) Microelectrode fashioned for intracellular current clamp recording. Note that the intracellular electrode tip diameter is only slightly smaller than that of the patch electrode, however the extended narrow taper results in a much higher tip resistance (Patch electrode = 8 MΩ; Sharp electrode = 135 MΩ, both measured with a 150 mM Kgluconate filling solution). Both electrodes were pulled using a horizontal, Flaming-Brown type puller using filamented, borosilicate glass (outer diameter = 1.5 mm, inner diameter = 0.86 mm) (both from Sutter, Novato, CA).

pyramidal cells are readily detected using extracellular recording methods. However, recording GABAergic inhibitory synaptic potentials through extracellular methods is considerably more difficult due to the diffuse density of these currents and their short dipoles. Extracellular methods can be used to assess GABAergic inhibition indirectly in some brain regions like the hippocampus where GABAergic synapses mediate a potent feed-forward inhibition of excitatory synaptic responses. This inhibition can be readily detected by measuring the inhibition of the second of two excitatory responses evoked at short interstimulus intervals (known as paired-pulse inhibition).

Extracellular recordings were used in one of the first demonstrations that intoxicating concentrations of ethanol did, in fact, alter synaptic transmission in the mammalian CNS. Durand and co-workers (1981) demonstrated that ethanol significantly attenuated orthodromically evoked population spikes in the rat hippocampal CA1 region. Using extracellular techniques, they showed that this inhibition was not due to a direct effect of ethanol on the voltage-gated channels that underlie action potentials, since antidromically evoked spikes, triggered by directly stimulating the axons of the CA1 pyramidal cells, were unaffected by ethanol. Extracellular recording methods were later used in the first demonstration that ethanol specifically inhibited the NMDA receptor-mediated component of excitatory synaptic transmission in the rat hippocampus (Lovinger et al., 1990). These techniques have also

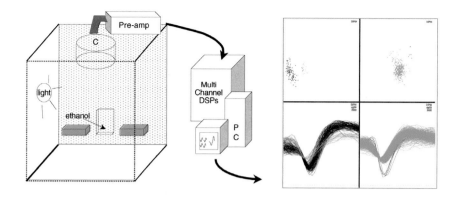

FIGURE 10.1

In vivo multichannel recording setup. Neural signals are acquired from rats during alcohol self-administration within a customized operant behavior chamber. Commutator (C) installed in center of chamber ceiling turns as the animal turns, and is designed to preserve integrity of signals from each channel while allowing free movement of subject throughout experimental chamber. C is connected to a preamplifier that amplifies and filters signals. Neural signals are then sent through analog-to-digital boards, where additional amplification or filtering may take place, then to digital signal processing (DSP) boards under the control of a personal computer (PC). Signals from each channel are sorted into waveforms that correspond to action potentials from individual neurons using specialized data acquisition software running on the PC. Examples of waveforms visible on two individual channels (wires) are shown on right. Two distinct waveforms are visible on each wire. Directly above waveforms is principal component analysis for each wire, which aids in sorting and confirms separation of the waveforms into two separate units because two separate clusters are visible. Digitized and sorted waveforms and timestamps (time of occurrence of each waveform) are then saved to PC along with times of occurrence of any behavioral events.

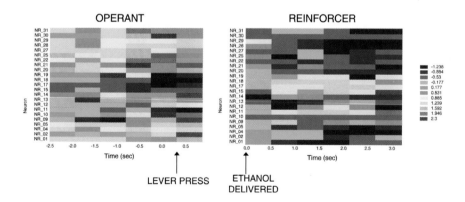

FIGURE 10.7

Patterns of neuronal activity within NAC are different during appetitive and consummatory phases of self-administration trial. Standardized firing rates of 21 simultaneously recorded NAC neurons are depicted relative to lever press response and alcohol receipt and consumption periods for a single alcohol self-administration session. Each neuron is represented in one row of contour plots, while time, in 500-msec blocks, is arranged across abscissa. There are more excitatory responses during appetitive phase of trial.

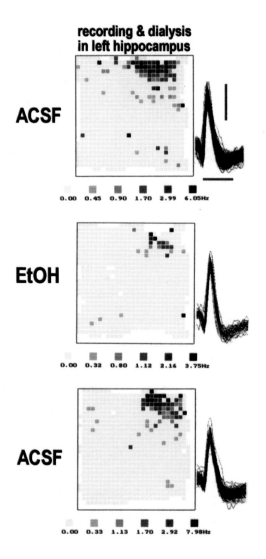

recording & dialysis in left hippocampus

ACSF

0.00 0.45 0.90 1.70 2.99 6.05Hz

EtOH

0.00 0.32 0.80 1.12 2.16 3.75Hz

ACSF

0.00 0.33 1.13 1.70 2.92 7.98Hz

FIGURE 11.8

The effect of ethanol, reverse-dialyzed in the hippocampus, on the location-specific firing of a place cell in the microdialysis area. Firing rate distribution maps are shown, generated with the MapMaker software. The maps indicate the average firing rate of the place cell within the areas of pixels during 15-min data collection periods, as the rat moved around chasing food pellets on the floor of a rectangular test chamber. Black pixels represent areas where the cell fired with the highest firing rates; yellow pixels represent areas where the cell did not fire at all, and white areas represent areas not visited by the animal. During microdialysis with ACSF (upper panel), the cell displayed a clear location-specific firing (6.05 Hz) in the northern part of the chamber. When ethanol (1 M for 30 min) was perfused (middle panel), the location-specific firing rate of the cell dropped to 3.75 Hz. After washing out ethanol from the recording site, full recovery of the original location-specific firing (7.90 Hz) was observed (lower panel). The similarity of the corresponding overlaid action potential waveforms on the right of each panel indicate that the data were collected from the same place cell throughout the recording and dialysis session. Calibrations: horizontal bar: 0.5 msec; vertical bar: 0.1 mV.

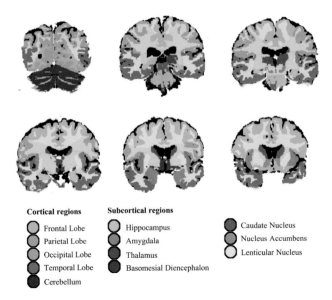

Cortical regions

- ⬤ Frontal Lobe
- ⬤ Parietal Lobe
- ⬤ Occipital Lobe
- ⬤ Temporal Lobe
- ⬤ Cerebellum

Subcortical regions

- ⬤ Hippocampus
- ⬤ Amygdala
- ⬤ Thalamus
- ⬤ Basomesial Diencephalon

- ⬤ Caudate Nucleus
- ⬤ Nucleus Accumbens
- ⬤ Lenticular Nucleus

FIGURE 13.4

A set of "processed" MR images from a single individual studied in an MR morphometry study performed in the author's laboratory. The different regions and structures examined are color-coded to show the structural boundaries.

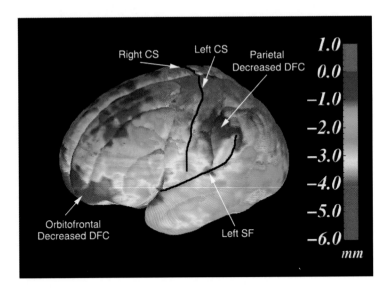

FIGURE 13.5

DFC group-difference maps showing differences in DFC (in mm) between the alcohol-exposed and control subjects according to the color bar on the right. Note negative effects in nearly all regions, most prominent in parietal and orbital frontal regions, i.e., DFC is greater in controls than in alcohol-exposed patients, with regional patterns of DFC reduction up to 8 mm. While only left hemisphere is fully visualized here, DFC reduction in both parietal and orbital frontal regions is bilateral. Left and right central sulcus (CS) and left hemisphere Sylvian fissure (SF) are highlighted in black.

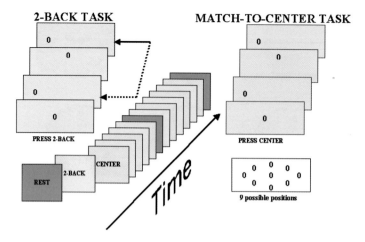

FIGURE 14.7
A pictorial representation of a block design, spatial working memory experiment and its three conditions: rest (green), 2-back task (yellow), and match-to-center attentional task (aqua). In the 2-back task, subjects were instructed to press a response key when a 0 appeared in the same position as one presented two presentations back. In the match-to-center task, subjects pressed a response key when a 0 appeared in the middle.

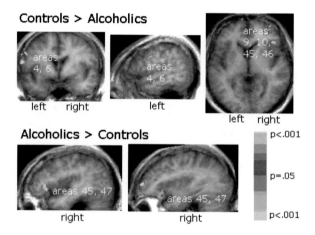

FIGURE 14.8
An example of SPM-derived significant activation clusters of differences between the attention task and a rest condition. Loci where the activation of the control group was significantly greater than that of the alcoholic group are depicted in the red to yellow tones. Loci where the activation of the alcoholic group was significantly greater than that of the control group are depicted in the blue to purple tones. Activation of control group was greater than that of alcoholic group in anterior medial and superior prefrontal cortex, where a large region of activation included Brodmann areas 9, 10, 45, and 46 bilaterally, although more widespread in the right than left hemisphere; the left motor cortex (areas 4 and 6) was also significantly more activated. By contrast, activation of the alcoholic group was greater than that of the control group in the right prefrontal cortex (areas 45 and 47), positioned more inferiorly and posteriorly than prefrontal regions significant in the controls. The differences in activations between the groups were due to greater activation in one group rather than deactivation in the other group (Figure 4 in Pfefferbaum et al., *Neuroimage* 14: 7-20, 2001).

proven invaluable in studies investigating the effects of fetal alcohol exposure (Bellinger et al., 1999) and chronic alcohol exposure and withdrawal on excitatory (Morrisett, 1994; Nelson et al., 1999) and inhibitory (Rogers and Hunter, 1992; Kang et al., 1996) synaptic transmission.

8.3.2 INTRACELLULAR (SHARP) ELECTRODE RECORDING

With the landmark demonstration that synaptic responses could be evoked in acutely prepared brain slices and recorded with extracellular techniques (Yamamoto and McIlwain, 1966), many investigators adopted these preparations to characterize the physiology and pharmacology of synaptic transmission. The subsequent demonstration that intracellular recording techniques, initially developed to study invertebrate neurophysiology and mammalian peripheral neurons, could be applied in brain slices to study the excitability and synaptic regulation of individual neurons, solidified the role of the brain slice as an invaluable tool in electrophysiology.

While classical "sharp electrode" intracellular recording is less frequently used today since the development of the patch-clamp technique, a number of important applications for this technique still exist. In this method, a glass microelectrode is used to impale an individual neuron, thereby providing a direct measurement of the electrical potential across the cell membrane. Intracellular electrodes are usually fashioned from borosilicate glass using methods similar to those used to prepare extracellular electrodes. However, thinner walled glass is usually employed and the shape of intracellular electrodes must be adjusted so that they can penetrate neurons without causing excess mechanical damage. Intracellular electrodes must have a homogenously tapered shank that ends in a tip that is usually less than 1 µm (Figure 8.6B). Electrodes are filled with a concentrated electrolyte solution (usually 3–10 M KCl or KAcetate) and have a tip resistance between 60–120 MΩ. The electrode is advanced through an area of tissue with a high density of cells of interest (e.g., hippocampal pyramidal layer, cortical layer, dentate gyrus granule cell layer). The electrode movement is controlled using a manipulator that allows for high velocity micro-steps with minimal lateral oscillations. This is best achieved with stepping motor driven devices or piezoelectric manipulators. This rapid acceleration has been found to be necessary for successful penetration of the cell membrane (Sonnhof et al., 1982). While the electrode is advanced through the tissue, a small current pulse is injected into the electrode to monitor the tip resistance. Cells are "detected" electrically as a small increase in the resistance of the electrode. Once a cell is detected, the electrode is then advanced further, puncturing the membrane of the cell. Successful membrane penetration is immediately evidenced by a rapid drop in the electrode voltage, as the cytosolic potential is significantly more negative than that of the extracellular space. As this technique requires that the electrode actually puncture the cell membrane, it is often necessary to wait several minutes following membrane impalement while the membrane seals up around the intracellular electrode. This can be monitored by measuring the input resistance of the cell. Once the input resistance has stabilized, stable intracellular recordings can often be maintained for several hours.

One important feature of the intracellular recording configuration is that the cytosolic contents of the cell being recorded are largely preserved. This contrasts with the whole cell patch-clamp recording method discussed below, in which the contents of the cytosol may be extensively dialyzed. As a result, the use of intracellular electrodes may be preferable when characterizing the activity of ion channels that are particularly sensitive to modulation by second messenger systems, as their activity could be compromised during whole cell patch-clamp recording.

The major disadvantage of this technique is that, for cells to survive impalement, intracellular electrodes must have a small tip diameter and an extended, tapered shank. This electrode profile results in a relatively high tip resistance, making it impractical to measure the actual currents underlying synaptic transmission (voltage-clamp configuration), particularly in highly arborized neurons in brain slices. However, when used as a "voltage follower," intracellular electrodes perform very well.

The fine, tapered shape of intracellular electrodes does offer one important advantage over patch-clamp electrodes. Under appropriate experimental conditions, these electrodes can be used to carry out intracellular recordings in anesthetized animals *in vivo* (O'Donnell and Grace, 1995; Reynolds and Wickens, 2000; Aksay et al., 2001; Goto and O'Donnell, 2001). Such experiments, while technically demanding, provide one of the most sensitive ways to characterize synaptic transmission under conditions where all afferent and efferent inputs are preserved. Although patch-clamp recordings have been achieved in intact animals (Light and Willcockson, 1999), the shape of these electrodes limits their utility *in vivo*.

The intracellular recording technique was used to carry out the first detailed analysis of the effects of ethanol on excitatory and inhibitory components of synaptic transmission in the mammalian CNS. Carlen and co-workers (1982) used this method to characterize the acute effects of ethanol on synaptic transmission in rat hippocampal CA1 pyramidal neurons. In their studies, 10-20 mM ethanol, applied by superfusion or local microdrops, significantly enhanced excitatory and inhibitory synaptic potentials in the majority of cells recorded. Importantly, ethanol had minimal effects on the membrane potential and firing properties of these cells. Five years later, Siggins and co-workers (1987) carried out a similar study and confirmed that the predominant effect of ethanol was on synaptic transmission rather than on the passive or active membrane properties of the CA1 pyramidal cells. Unfortunately, they observed that ethanol, more often than not, inhibited rather than potentiated EPSPs and inhibitory postsynaptic potentials (IPSPs). Although these two studies differed in some of their conclusions regarding the effects of ethanol on synaptic transmission, they did agree that ethanol produced its most significant effects at the level of the synapse, rather than by directly altering the passive and active membrane properties of the principal neurons.

Intracellular recording techniques continue to be used today to study the effects of ethanol on neuronal activity. For example, this method was recently used to demonstrate that, in locus coeruleus neurons recorded in pontine slices, ethanol significantly inhibits EPSPs while having no significant effects on IPSPs or membrane properties of these cells (Nieber et al., 1998). Evans and colleagues (2000) recently used intracellular recordings in inferior colliculus neurons to characterize hyperexcitability associated with ethanol withdrawal. Finally, although ethanol does not seem to have pronounced effects on the resting membrane potential of most neurons, there are a few exceptions. Brodie and Appel (1998) have used intracellular recordings to demonstrate that ethanol can directly depolarize and increase the firing rate of dopamine cells in the ventral tegmental area, an effect that may play an integral role in the rewarding properties of this drug.

8.3.3 WHOLE-CELL PATCH-CLAMP RECORDING

The patch-clamp recording technique was originally developed by Sakmann and Neher to study the biophysical properties of individual ion channels in small patches of muscle fiber (Neher et al., 1978). Soon this method was used to obtain high-resolution whole-cell recordings from isolated or cultured cells (Hamill et al., 1981). The major advance for synaptic electrophysiologists came with the application of this method to the study of individual neurons in brain slices (Edwards et al., 1989). Since these initial studies, the use of the patch-clamp technique has revolutionized the way that ion channel activity is studied. Although many of the details of patch-clamp recording have already been discussed in Chapters 4 and 7, several methodological considerations related to the application of this recording technique in brain slices warrant some discussion.

Each of the two general methods of obtaining whole-cell patch-clamp recordings in brain slices has its own advantages and disadvantages. In the "Blind" method, a patch electrode is advanced through a brain region in which there is a relatively dense, homogenous population of cells. Positive pressure is applied to the patch electrode to prevent membrane debris from blocking the electrode tip as it is advanced through the tissue. The electrode resistance is monitored in current-clamp mode by passing a small current step across the tip of the electrode. Much like in intracellular recording, cells are detected as an increase in the resistance across the tip of the electrode. Once a cell is detected, the positive pressure is removed, often resulting in a rapid increase in tip resistance due to the sealing of the cell membrane onto the electrode tip. If this does not occur spontaneously, a small amount of negative pressure is gently applied to the tip to facilitate seal formation. Injecting a small amount of negative DC current through the patch electrode seems to facilitate seal formation. Once a seal of at least 1 GΩ (preferably 3-5 GΩs) is formed, an additional application of negative pressure is applied, resulting in the sudden rupture of the membrane between the electrode tip and the cell.

There is considerable debate among "patch-clampers" about whether the various pressure applications are best administered by mouth or via a syringe. Through many years of trial and error, we can unquestionably state that, when recording in brain slices, a syringe is adequate for maintaining pressure while in the "cell search" mode. However, seal formation and "going whole-cell" are more consistently achieved using experimenter-controlled oral pressure application.

The primary advantage of the "Blind" patch method is that it is relatively inexpensive to set up, as a simple dissecting microscope provides adequate magnification (Figure 8.7a). In addition, a recording chamber that provides superfusion of both sides of the tissue slice can be used (Figure 8.4A). We have generally found that tissue slices seem to remain viable for longer periods of time in these kinds of

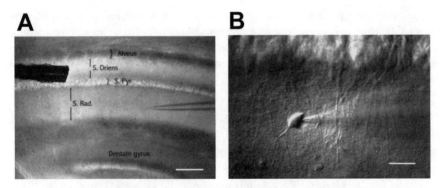

FIGURE 8.7 (A) Low-power view of hippocampal slice preparation showing location of a bipolar stimulation electrode (top left) as well as a patch-clamp recording electrode located in the stratum radiatum of the CA1 region. Scale bar, 100 μm. (B) A high-power DIC image of a single stratum radiatum interneuron, located ~100 μm below the surface of the slice, during recording with a patch-clamp electrode. The pyramidal cell layer is visible at the top of the view. Scale bar, 10 μm. (Reproduced with permission from Frazier et al., 1998; Copyright 1998 by the Society for Neuroscience.)

slice chambers compared with the glass-bottomed submersion chambers used in visualized cell recording (Figure 8.4B). This is particularly noticeable when recording at more physiological temperatures (32–35° C). The main disadvantage of the "Blind" patch method is that, as the name implies, visualization of the cell being recorded is not achieved. This is not a significant problem when recording from a densely packed, homogenous cell population like CA1 pyramidal neurons. This method would not, however, be very effective for recording from a sparsely distributed neuronal type such as hippocampal GABAergic interneurons.

The alternative method of whole-cell patch-clamp recording in brain slices involves directly visualizing the target neuron prior to recording. This can now be readily achieved by using a number of image enhancement techniques that compensate for the poor imaging qualities of brain slices. The first requirement is a microscope that is equipped with differential interference contrast (DIC) optics (Dodt and Zeiglgansberger, 1990). The DIC objective converts phase gradients created by the light-scattering properties of tissue slices into amplitude gradients and essentially allows for optical sectioning of relatively thick specimens (100–300 μm). Optical sectioning permits visualization of cells that may be located relatively deep within a tissue slice (Figure 8.7B). Additional improvement of image quality can be achieved using video enhancement technology (Allen and Allen, 1983) and by using infrared illumination. Infrared light penetrates through thick tissue sections better than visible light, serving to further aid in the visualization of individual cells (Dodt and Zeiglgansberger, 1990). In the DIC patch-clamp method, the recording electrode is positioned above the tissue and a cell of interest is visually identified. The recording electrode is then visually guided to the surface of the neuron using a mechanical or motorized micromanipulator. The remaining steps involved in forming a seal and gaining electrical access to the cell are identical to those described above for "Blind" patch-clamp recording.

The obvious advantage of this technique is that it enables the application of the whole-cell patch-clamp recording technique to be applied on virtually any type of neuron, no matter how sparsely distributed (e.g., hippocampal GABAergic interneurons; Figure 8.7B). Moreover, this technique has made it possible to stimulate (Banks et al., 1998) and actually record from (Stuart et al., 1997) dendritic processes in tissue slices. While DIC-video enhanced microscopy is unquestionably revolutionizing the use of brain slices in electrophysiology, there are several disadvantages of this method compared with "Blind" patch-clamp recording. The quality of DIC images is considerably better in tissue slices prepared from relatively young animals, possibly due to a lesser degree of myelination (Dodt and Zeiglgansberger, 1990). Therefore, the application of this technique on slices prepared from adult animals can be problematic. In addition, as already mentioned, tissue slices are less viable in the DIC recording chambers, particularly when recording at, or near, physiological temperatures. This may be because only the top surface of the tissue is perfused with oxygenated aCSF in DIC recording chambers as the tissue rests on a glass bottom that is required for DIC imaging. Nevertheless, DIC-video-enhanced microscopy has made possible experiments that previously could only be hypothesized, and technical improvements will no doubt soon overcome some of these methodological limitations.

Many investigators have employed the whole-cell patch-clamp technique to study the effects of ethanol on synaptic transmission in brain slices. This recording method was used in the first demonstration that ethanol significantly inhibited the currents underlying NMDA receptor-mediated synaptic transmission in the mammalian CNS (Morrisett and Swartzwelder, 1993). Similarly, we used this technique in the first demonstration that ethanol could potentiate pharmacologically isolated $GABA_A$ receptor-gated synaptic currents in rat hippocampal neurons (Weiner et al., 1994). Recently, we have used the patch-clamp technique in brain slices to uncover novel presynaptic effects of ethanol at GABAergic (Crowder et al., 2000; Ariwodola et al., 2001) and glutamatergic (Crowder and Weiner, 2001) synapses in the rat hippocampus and nucleus accumbens. We have also used this technique to characterize the acute effects of ethanol on excitatory and inhibitory synaptic transmission in monkey hippocampal slices (Ariwodola et al., 2000). Many other investigators continue to use these recording techniques to study the acute effects of ethanol on synaptic transmission (Soldo et al., 1998; Eggers et al., 2000; Poelchen et al., 2000) as well as the synaptic consequences of chronic ethanol exposure and withdrawal (Thomas et al., 1998b).

8.4 DISTINGUISHING BETWEEN PRESYNAPTIC AND POSTSYNAPTIC MECHANISMS OF ACTION OF ETHANOL

As already discussed, ethanol interacts with a number of targets within the central nervous system that participate, in a variety of ways, in synaptic transmission. One of the major advantages of brain slice preparations is that they retain intact synapses that can be triggered by electrical stimulation (Dingledine et al., 1980). This feature makes brain slices an ideal model system to elucidate the complex effects of ethanol on synaptic transmission. However, because both pre- and postsynaptic changes can contribute to any observed effects of ethanol, identifying the specific mechanisms underlying modulatory effects of ethanol on synaptic transmission in brain slices can be difficult. For example, many studies have demonstrated that, at least under some conditions, ethanol enhances $GABA_A$ receptor-mediated synaptic transmission in the rat hippocampal CA1 region (Weiner et al., 1994; Wan et al., 1996; Weiner et al., 1997a,b; Poelchen et al., 2000). However, despite much effort, the synaptic mechanism(s) underlying this effect have yet to be conclusively resolved. The quest to identify the specific loci that mediate various forms of synaptic plasticity has also proven to be a major challenge to slice electrophysiologists (e.g., Manabe et al., 1993; Nicoll and Malenka, 1999). A variety of methodological approaches have been developed to distinguish between pre- and postsynaptic sites of synaptic modulation. Three of the most common methods will be discussed and the advantages and disadvantages of each will be reviewed.

8.4.1 FOCAL APPLICATION OF AGONIST

One of the most straightforward approaches to delineate the locus of a synaptic modulation is to activate the postsynaptic ion channels directly by exogenous application of an appropriate agonist. The effect of the synaptic modulator can then be tested on these agonist-evoked responses. Since these agonist-evoked responses are generally devoid of any presynaptic influences, any observed effects of the modulator can be taken as evidence of a postsynaptic mechanism of action. This approach is frequently used in the study of recombinant ligand-gated ion channels expressed in cell lines or when studying the pharmacology of native ion channels in cells grown in culture. Methods for such preparations are described in Chapter 7 in this volume and typically involve rapid exchange of the solution bathing the cell being recorded. Rapid agonist application is necessary as most ligand-gated ion channels that underlie excitatory and inhibitory synaptic transmission undergo rapid desensitization when exposed to an agonist for prolonged intervals. While it is relatively straightforward to effect rapid solution exchanges across an isolated or cultured cell, such rapid bath exchanges are not generally possible in brain slice preparations. Therefore, alternative approaches have been developed that permit localized application of receptor agonists directly onto individual neurons in tissue slices. These methods involve filling a glass microelectrode, similar in shape and size to a patch electrode, with a receptor agonist of choice (e.g., GABA or glutamate). The electrode is then placed in proximity to the cell being recorded and agonist is released for brief intervals (usually 1–10 ms) using current injection (iontophoresis) or a brief pressure application.

Focal agonist application has been used in a number of brain slice electrophys-iological studies to demonstrate that inhibitory effects of ethanol on NMDA- (Proctor et al., 1992) and kainate receptor-evoked EPSCs (Weiner et al., 1999) are mediated predominantly via postsynaptic mechanisms. These methods have also been used to determine the synaptic loci that mediate potentiation by ethanol of $GABA_A$ receptor-mediated synaptic inhibition in rat hippocampal neurons. Initial studies reported that ethanol had no effect on GABA-evoked responses in hippocampal neurons (Proctor et al., 1992), even under conditions where ethanol did enhance $GABA_A$ IPSP/Cs

(Wan et al., 1996). We have previously shown that ethanol enhances somatic $GABA_A$ IPSCs to a greater extent than dendritic $GABA_A$ synaptic responses (Weiner et al., 1997a). Therefore, in preliminary experiments, we used DIC-video enhanced microscopy to directly test the ethanol sensitivity of currents evoked by focal application of GABA directly onto the soma of CA1 pyramidal cells. In these experiments, 80 mM ethanol, which produced significant potentiation of somatic $GABA_A$ IPSCs, had no effect on somatically evoked GABA currents. In contrast, GABA-evoked currents were potentiated by the $GABA_A$ receptor modulator flunitrazepam and were completely abolished by the $GABA_A$ receptor antagonist bicuculline methiodide (Figure 8.8). These experiments suggest that ethanol may not act solely via a postsynaptic mechanism. However, these negative results also highlight some of the limitations of this method. Even when carried out under direct visual guidance, it is not possible to precisely control the concentration of agonist that actually reaches the cell. This may be problematic when testing the effects of competitive agonists like ethanol that have been shown, using other methods, to potentiate $GABA_A$ receptor function only at low GABA concentrations. Furthermore, it is not possible to ensure that the receptors being activated by focal agonist application are actually the same as those underlying the synaptic responses. Differences have been noted between the physiological and pharmacological properties of synaptic and extrasynaptic $GABA_A$ receptors (Brickley et al., 1999; Bai et al., 2001). For these reasons, negative results obtained with focal agonist application in slices may be difficult to interpret.

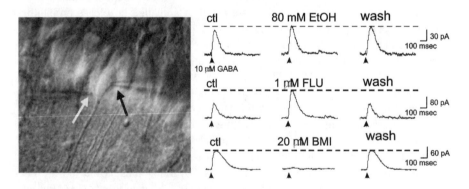

FIGURE 8.8 Ethanol has no effect on currents evoked by focal application of GABA in rat hippocampal CA1 pyramidal neurons. Photograph is a video-enhanced DIC image of the CA1 pyramidal layer of 300 μm rat hippocampal slice. Illustrated are the position of a patch-clamp recording electrode (right) sealed onto a CA1 pyramidal neuron and a pressure ejection electrode (left) filled with 10 μM GABA. Traces to the right of the photograph are averages of 3–5 GABA evoked currents recorded prior to, during, and after bath application of ethanol (EtOH), flunitrazepam (FLU), and bicuculline methiodide (BMI). Note that 1 μM FLU potentiates and 20 μM BMI completely inhibits GABA-evoked currents. However, 80 mM EtOH has no effect on these GABA-mediated responses.

8.4.2 PAIRED-PULSE FACILITATION RATIO

Because of the limitations associated with focal application of agonist in brain slices, a number of methods have been devised to evaluate the synaptic locus of a drug's action that rely solely on evaluating the effect(s) of the modulator on synaptic responses. One commonly employed method is to examine the effects of the modulator on paired-pulse facilitation (PPF), a form of short-term synaptic plasticity that is present at many synapses in the mammalian central nervous system (McNaughton, 1982; Manabe et al., 1993). PPF is induced by pairing two stimuli, typically with an interstimulus interval of 5–50 ms. This pairing results in a facilitation of the second response relative to the first that is thought to result from an increase in the probability of neurotransmitter release (Zucker, 1989). The ratio of Peak 2 to Peak 1 has been shown to be sensitive to a broad range of presynaptic experimental manipulations that alter the probability of neurotransmitter release (e.g., changing extracellular Ca^{++} concentration). However, this ratio is generally unaffected by postsynaptic manipulations, such as administration of a submaximal concentration of an antagonist of the synaptic receptor being studied (Creager et al., 1980; Dunwiddie and Haas, 1985; Manabe et al., 1993; Brundege and Dunwiddie, 1996).

The advantages of PPF in elucidating the synaptic locus of a drug's action are that it is relatively simple to carry out and has been subjected to extensive empirical validation. Moreover, this method can be employed with either single cell recordings or extracellular measures of synaptic transmission. In fact, PPF has even been used to probe the synaptic mechanisms underlying long-term synaptic plasticity in freely moving rats (Li et al., 2000). The main disadvantage of this method is that presynaptic manipulations that do not arise as a result of a change in release probability may not affect the PPF ratio, even though they are presynaptic in origin. For example, increasing the stimulation intensity used to evoke synaptic responses, a presynaptic manipulation, will increase the magnitude of these responses by recruiting more fibers but will not change the PPF ratio (Manabe et al., 1993).

Nevertheless, because many neuromodulators that act presynaptically seem to do so by changing release probability, PPF is still a commonly employed electrophysiological assay. For example, several recent studies have used PPF to support the hypothesis that cannabinoids reduce excitatory synaptic transmission via a presynaptic reduction in transmitter release at glutamatergic synapses (Gerdeman and Lovinger, 2001; Hoffman and Lupica, 2001; Huang et al., 2001). Several studies have also used PPF to characterize the synaptic mechanisms associated with ethanol exposure. Tan and colleagues (1990) showed an increase in PPF at glutamatergic synapses in the hippocampal CA1 region of rats that had been exposed to prenatal ethanol, consistent with a reduction in the probability of glutamate release at these synapses. Nelson and coworkers (1999) used PPF, along with a number of other measures of synaptic transmission, to elucidate the synaptic loci associated with the complex neuroadaptive changes associated with chronic ethanol exposure. In their study, rats were exposed to chronic intermittent ethanol for 2 weeks. One of their many interesting findings was that chronic ethanol exposure resulted in a significant reduction in somatic and dendritic excitatory synaptic responses in the hippocampal CA1 region when recorded in the continued presence of ethanol or within an hour of ethanol withdrawal. In contrast,

PPF was not significantly affected by this ethanol treatment, suggesting that a presynaptic reduction in the probability of glutamate release was unlikely to account for the reduction in glutamatergic neurotransmission associated with chronic ethanol exposure. Interestingly, although acute exposure to ethanol had no effect on PPF of the somatic population spike in slices recorded from control animals, ethanol did increase PPF in slices recorded from chronically treated subjects, revealing perhaps, that there are some changes in presynaptic function that occur following chronic intermittent ethanol exposure.

8.4.3 MINIATURE SYNAPTIC CURRENTS

As already described, synaptic responses can be triggered in acutely prepared brain slices by electrical stimulation of presynaptic afferents. Another feature of brain slice preparations is that synaptic responses can often be detected in the absence of external stimulation. With the enhanced signal-to-noise ratio afforded by whole-cell patch-clamp recording, spontaneously occurring excitatory and inhibitory synaptic currents can often be detected in whole-cell patch-clamp recordings. In some cases, these spontaneous synaptic currents arise from the firing of presynaptic cells within the slice whose afferent projections impinge upon the cell being recorded. However, synaptic responses can also result from the random fusion of neurotransmitter vesicles at presynaptic terminals. These latter responses, termed miniature postsynaptic currents, persist in the presence of drugs that block action potentials, like tetrodotoxin. Since these miniature synaptic currents typically reflect the activation of individual synapses, or at most, the synaptic contacts of a single presynaptic cell, these responses tend to be much smaller than electrically evoked synaptic responses. However, these spontaneous events are a powerful tool to characterize the mechanisms underlying the effects of synaptic modulators. Under most experimental conditions, the frequency of these miniature synaptic currents is extremely sensitive to presynaptic manipulations, thus providing an excellent index of presynaptic function. In contrast, the kinetics of the miniature synaptic currents are generally only altered by postsynaptic manipulations. Not surprisingly, miniature synaptic currents have been used extensively in many physiological and pharmacological studies of synaptic transmission (Mody et al., 1994). The main disadvantage of this method is that miniature synaptic currents can be very small, sometimes making it problematic to distinguish real synaptic events from the baseline noise. Therefore, care must be taken to ensure that any changes in the amplitude of miniature synaptic events do not result in an artifactual increase in their frequency, as more events may simply be detected from the baseline noise. Excellent software programs are available that facilitate event detection and can significantly reduce these problems (e.g., Mini-Analysis, Synaptosoft Inc, Decatur, GA).

Because the application of the whole-cell patch-clamp technique in brain slices has been in common use for less than 10 years, relatively few studies have employed the analysis of miniature synaptic currents to characterize possible presynaptic effects of ethanol. One recent study used these methods, along with several other approaches, to demonstrate that ethanol potentiation of glycinergic synaptic transmission in rat spinal cord slices involves both pre- and postsynaptic mechanisms

(Eggers et al., 2000). These authors noted that ethanol significantly increased the frequency and amplitude of TTX-resistant miniature glycinergic currents in hypoglossal motoneurons in spinal cord slices (Figure 8.9). Additional experiments, including an analysis of the effects of ethanol on currents evoked by focal application of glycine directly onto HM neurons, demonstrated that ethanol acts both pre- and postsynaptically to facilitate glycinergic synaptic transmission in the rat spinal cord. Interestingly, the postsynaptic ethanol sensitivity of these synapses changed during development, with glycine-evoked currents recorded from neonatal rats (P1–3) being significantly less sensitive to ethanol than those recorded from juveniles (P9–13). In contrast, the presynaptic effects of ethanol at these synapses were not developmentally regulated.

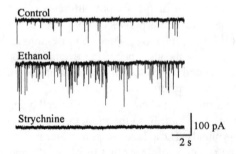

FIGURE 8.9 Glycinergic miniature inhibitory postsynaptic currents (mIPSCs) in hypoglossal motoneurons (HMs) are enhanced by ethanol and blocked by strychnine. Representative traces from a voltage-clamp recording of glycinergic mIPSCs in a HM from a juvenile rat [*postnatal day 9 (P9)*] are shown. Action potentials and nonglycinergic currents were blocked by tetrodotoxin (TTX), Cd^{++}, 6,7-dinitro-quinoxaline (DNQX), D(—)-2-amino-5-phosphonopentanoic acid (APV), and bicuculline methiodide (BMI). In this cell, adding 100 mM ethanol to the bath increased the average mIPSC amplitude by 32% and decreased the interval between successive mIPSCs by 58% (distributions significantly different by Kolmogorov-Smirnov test, $P<0.001$). Glycinergic mIPSCs were blocked by strychnine (2 μM). (Reproduced with permission from Eggers et al., 2000.)

A number of additional methods are often employed to tease out the mechanisms underlying the effects of synaptic modulators. Measuring the frequency of failures in transmission in response to minimal stimulation of an individual synapse provides an alternative method to assess changes in release probability (e.g., Nicoll and Malenka, 1999; Cossart et al, 2001). In addition, methods have been developed to differentiate between pre- and postsynaptic mechanisms that take advantage of the quantal nature of synaptic transmission. These methods involve a statistical analysis of the variance in the amplitude of a synaptic response evoked repeatedly in the absence and presence of a synaptic modulator (Clements, 1990). In theory, the amplitude of an individual synaptic response will fluctuate between a finite numbers of equally spaced quantal levels (Fatt and Katz, 1951). Empirical studies have demonstrated that manipulations that are purely presynaptic will not affect the individual quantal levels but will alter their probabilities of occurrence. In contrast, postsynaptic manipulations may alter the quantal amplitudes but will not affect their relative probabilities of occurrence.

While empirical studies have provided some validation for all of the methods discussed, no single approach is likely to provide unequivocal proof of a synaptic mechanism of action. The best approach is to employ several different methods and, whenever possible, to empirically validate each method with known pre- and postsynaptic manipulations. This is particularly important when investigating a complex synaptic modulator like ethanol that may well have a variety of pre- and postsynaptic effects.

8.5 CONCLUSIONS AND FUTURE DIRECTIONS

The use of electrophysiological methods in acutely prepared brain slices has contributed immeasurably to our current understanding of how ethanol alters synaptic transmission in the mammalian CNS. The presence of functional synapses that are representative of the interneuronal connections observed *in vivo*, the ease of recording synaptic responses from individual neurons, and the ability to precisely control the time course and concentration of drug applications are just some of the features that make brain slice preparations a valuable tool for investigating the complex synaptic mechanisms underlying actions of ethanol.

A number of recent scientific and technical advances are providing even more applications for the use of brain slices in alcohol research. For example, over the last several years, a number of transgenic mouse models have been developed that display marked differences in their preference for or sensitivity to ethanol. Interestingly, many of the targeted genes are either involved in neurotransmitter synthesis (e.g., Weinshenker et al., 2000), encode for neurotransmitter receptors (e.g., Hall et al., 2001) or for signaling molecules that regulate the activity of synaptic ion channels (e.g., Miyakawa et al., 1997; Hodge et al., 1999; Thiele et al., 2000). It will be interesting in the coming years to delineate the synaptic correlates that underlie the behavioral differences in ethanol sensitivity in these various transgenic models using electrophysiological techniques in brain slices.

Another important technical advance has been the development of methods for culturing brain slices (Gahwiler, 1981). In this technique, brain slices are prepared from rat pups and grown for several weeks in culture media. These brain slice explant cultures maintain the cytoarchitectural organization of the tissue of origin and flatten out into layers of only 1-3 cell diameters, providing excellent visualization thus facilitating the use of patch-clamp and optical imaging techniques. Moreover, since these slices remain viable for several weeks, they provide an ideal model system for the investigation of the synaptic consequences associated with chronic alcohol exposure and withdrawal. Several studies have already employed slice culture methods to demonstrate upregulation of NMDA receptor function during alcohol withdrawal (Thomas et al., 1998a) and the possible cytotoxic sequelae associated with this upregulation in excitatory synaptic transmission (Prendergast et al., 2000; Thomas and Morrisett, 2000).

Finally, electrophysiological recording methods are but one of many experimental techniques that can be applied in brain slices. High resolution imaging of intracellular calcium levels (Conner et al., 1994) and electrochemical detection of neurotransmitters such as dopamine (Jones et al., 1999) are just two examples of other methods that can be used in brain slices to provide additional insight into the physiology and pharmacology of synaptic transmission. Technical advances are rapidly making it possible to apply these, along with many other methods (e.g., RT-PCR, gene arrays), in concert with electrophysiological recordings of synaptic transmission in brain slices. Such multidisciplinary studies will likely yield important new insights into the complex mechanisms underlying the acute and chronic effects of ethanol.

ACKNOWLEDGMENTS

I would like to thank Dr. Kristin Anstrom and Mr. Doug Byrd for their assistance with digital imaging and photography. This work was supported by grants from the National Institute on Alcoholism and Alcohol Abuse (AA12251 and AA11997), the Alcoholic Beverage Medical Research Foundation, and the U.S. Army (grant DAMD17-00-1-0579).

REFERENCES

Report of the AVMA Panel on Euthanasia (2001) *J Am Vet Med Assoc* 218:669-696.

Aksay E et al. (2001) *In vivo* intracellular recording and perturbation of persistent activity in a neural integrator. *Nat Neurosci* 4:184-193.

Allen RD and Allen NS (1983) Video-enhanced microscopy with a computer frame memory. *J Microsc* 129:3-17.

Andersen P, Soleng AF and Raastad M (2000) The hippocampal lamella hypothesis revisited. *Brain Res* 886:165-171.

Ariwodola OJ et al. (2000) Ethanol modulation of excitatory and inhibitory synaptic transmission in monkey hippocampal neurons. *Alcohol Clin Exp Res* 24S:8A.

Ariwodola OJ and Weiner JL (2001) Ethanol potentiation of GABAergic synaptic transmission in the rat hippocampus may be self-limiting. *Alcohol Clin Exp Res* 25S:11A.

Bai D et al. (2001) Distinct functional and pharmacological properties of tonic and quantal inhibitory postsynaptic currents mediated by gamma-aminobutyric acid(A) receptors in hippocampal neurons. *Mol Pharmacol* 59:814-824.

Banks MI, Li TB and Pearce RA (1998) The synaptic basis of $GABA_A$ slow. *J Neurosci* 18:1305-1317.

Bellinger FP et al. (1999) Ethanol exposure during the third trimester equivalent results in long-lasting decreased synaptic efficacy but not plasticity in the CA1 region of the rat hippocampus. *Synapse* 31:51-58.

Ben-Ari Y and Cossart R (2000) Kainate, a double agent that generates seizures: two decades of progress. *Trends Neurosci* 23:580-587.

Bleakman D et al. (1995) Activity of 2,3-benzodiazepines at native rat and recombinant human glutamate receptors in vitro: stereospecificity and selectivity profiles. *Neuropharmacol* 35:1689-1702.

Brickley SG, Cull-Candy SG and Farrant M (1999) Single-channel properties of synaptic and extrasynaptic $GABA_A$ receptors suggest differential targeting of receptor subtypes. *J Neurosci* 19:2960-2973.

Brodie MS and Appel SB (1998) The effects of ethanol on dopaminergic neurons of the ventral tegmental area studied with intracellular recording in brain slices. *Alcohol Clin Exp Res* 22:236-244.

Brundege JM and Dunwiddie TV (1996) Modulation of excitatory synaptic transmission by adenosine released from single hippocampal pyramidal neurons. *J Neurosci* 16:5603-5612.

Carlen PL, Gurevich N and Durand D (1982) Ethanol in low doses augments calcium-mediated mechanisms measured intracellularly in hippocampal neurons. *Science* 215:306-309.

Clements JD (1990) A statistical test for demonstrating a presynaptic site of action for a modulator of synaptic amplitude. *J Neurosci Methods* 31:75-88.

Connor JA et al. (1994) Calcium signaling in dendritic spines of hippocampal neurons. *J Neurobiol* 25:234-242.

Cossart R et al. (2001) Presynaptic kainate receptors that enhance the release of GABA on CA1 hippocampal interneurons. *Neuron* 29:497-508.

Costa ET et al. (2000) Acute effects of ethanol on kainate receptors in cultured hippocampal neurons. *Alcohol Clin Exp Res* 24:20-225.

Creager R, Dunwiddie T and Lynch G. (1980) Paired-pulse and frequency facilitation in the CA1 region of the in vitro rat hippocampus. *J Physiol* 299:409-24.

Crowder TL., Ariwodola OJ and Weiner JL (2000) Ethanol sensitivity of pre- and postsynaptic kainate receptors in the CA1 region of the rat hippocampus. *Alcohol Clin Exp Res* 24S:9A.

Crowder TL and Weiner JL (2001) Ethanol inhibits presynaptic kainate receptors at glutamatergic synapses in the rat nucleus accumbens. *Alcohol Clin Exp Res* 25S:10A.

Dingledine R, Dodd, J and Kelly, JS (1980) The *in vitro* brain slice as a useful neurophysiological preparation for intracellular recording. *J Neurosci Meth* 2:323-362.

Dodt HU and Zieglgansberger W (1990) Visualizing unstained neurons in living brain slices by infrared DIC-videomicroscopy. *Brain Res* 537:333-336.

Dopico AM et al. (1999) Alcohol modulation of calcium-activated potassium channels. *Neurochem Int* 35:103-106.

Dunwiddie TV and Haas HL (1985) Adenosine increases synaptic facilitation in the *in vitro* rat hippocampus: evidence for a presynaptic site of action. *J Physiol* 369:365-377.

Durand D et al. (1981) Effect of low concentrations of ethanol on CA1 hippocampal neurons *in vitro*. *Can J Physiol Pharmacol* 59:979-984 .

Edwards FA et al. (1989) A thin slice preparation for patch clamp recordings from neurones of the mammalian central nervous system. *Pflugers Arch* 414:600-612.

Eggers ED, O'Brien JA and Berger AJ (2000) Developmental changes in the modulation of synaptic glycine receptors by ethanol. *J Neurophysiol* 84:2409-2416.

Engel SR et al. (1996) Effect of *in vivo* administration of anesthetics on $GABA_A$ receptor function. *Lab Anim Sci* 46:425-429.

Evans MS, Li Y and Faingold C (2000) Inferior colliculus intracellular response abnormalities in vitro associated with susceptibility to ethanol withdrawal seizures. *Alcohol Clin Exp Res*: 24:1180-1186.

Faingold CL, N'Gouemo P and Riaz A (1998) Ethanol and neurotransmitter interactions-from molecular to integrative effects. *Prog Neurobiol* 55:509-535.

Fatt P and Katz B (1951) An analysis of the end plate potential recorded with an intracellular electrode. *J Physiol* 115:320-370.

Foster AC et al. (1981) Synaptic localization of kainic acid binding sites. *Nature* 289:73-75.

Frazier CJ et al. (1998) Synaptic potentials mediated via alpha-bungarotoxin-sensitive nicotinic acetylcholine receptors in rat hippocampal interneurons. *J Neurosci* 18:8228-8235.

Freund TF and Buzsaki G (1996) Interneurons of the hippocampus. 6:347-470.

Gahwiler BH (1981) Organotypic monolayer cultures of nervous tissue. *J Neurosci Methods* 4:329-342.

Gerdeman G and Lovinger DM (2001) CB1 cannabinoid receptor inhibits synaptic release of glutamate in rat dorsolateral striatum. *J Neurophysiol* 85:468-471.

Goldstein DB (1986) Effect of alcohol on cellular membranes. *Ann Emerg Med* 15: 1013-1018.

Goto Y and O'Donnell P (2001) Synchronous activity in the hippocampus and nucleus accumbens *in vivo*. *J Neurosci* 21:RC131-133.

Hall FS, Sora I and Uhl GR (2001) Ethanol consumption and reward are decreased in mu-opiate receptor knockout mice. *Psychopharmacology* 154:43-49.

Hamill OP et al. (1981) Improved patch-clamp techniques for high-resolution current recording from cells and cell-free membrane patches. *Pflugers Arch* 391:85-100.

Hodge CW et al. (1999) Supersensitivity to allosteric GABA(A) receptor modulators and alcohol in mice lacking PKCepsilon. *Nat Neurosci* 2:997-1002.

Hoffman AF and Lupica CR (2001) Direct actions of cannabinoids on synaptic transmission in the nucleus accumbens: a comparison with opioids. *J Neurophysiol* 85:72-83.

Huang CC, Lo SW and Hsu KS (2001) Presynaptic mechanisms underlying cannabinoid inhibition of excitatory synaptic transmission in rat striatal neurons. *J Physiol* 532:731-748.

Jensen K and Mody I. (2001) GHB depresses fast excitatory and inhibitory synaptic transmission via GABA(B) receptors in mouse neocortical neurons. *Cereb Cortex* 11:424-429.

Jones SR, Gainetdinov RR and Caron MG (1999) Application of microdialysis and voltammetry to assess dopamine functions in genetically altered mice: correlation with locomotor activity. 147:30-42.

Kang M et al. (1996) Persistent reduction of GABA(A) receptor-mediated inhibition in rat hippocampus after chronic intermittent ethanol treatment. *Brain Res* 709:221-228.

Li, C-L and McIlwain H (1957) Maintenance of resting membrane potentials in slices of mammalian cerebral cortex and other tissues *in vitro*. *J Physiol* 139:178-190.

Li S, Anwyl R and Rowan MJ (2000) A persistent reduction in short-term facilitation accompanies long-term potentiation in the CA1 area in the intact hippocampus *Neuroscience* 100:213-220.

Light AR and Willcockson HH (1999) Spinal laminae I-II neurons in rat recorded *in vivo* in whole cell, tight seal configuration: properties and opioid responses. *J Neurophysiol* 82:3316-3326.

Lipton P et al. (1995) Making the best of brain slices: comparing preparative methods. *J Neurosci Methods* 59:151-156.

Lovinger DM, White G and Weight FF (1990) NMDA receptor-mediated synaptic excitation selectively inhibited by ethanol in hippocampal slice from adult rat. *J Neurosci* 10:1372-1379.

Manabe T et al. (1993) Modulation of synaptic transmission and long-term potentiation: effects on paired pulse facilitation and EPSC variance in the CA1 region of the hippocampus. *J Neurophysiol* 70:1451-1459.

McIlwain H, Buchel L and Cheshire, JD (1951) The inorganic phosphate and phosphocreatine of brain especially during metabolism *in vitro*. *Biochem J* 48:12-20.

McNaughton BL (1982) Long-term synaptic enhancement and short-term potentiation in rat fascia dentata act through different mechanisms. *J Physiol* 324:249-262.

Miyakawa T et al. (1997) Fyn-kinase as a determinant of ethanol sensitivity: relation to NMDA-receptor function. *Science* 278:698-701.

Mody I et al. (1994) Bridging the cleft at GABA synapses in the brain. *Trends Neurosci* 17:517-525.

Morrisett RA (1994) Potentiation of N-methyl-D-aspartate receptor-dependent after discharges in rat dentate gyrus following in vitro ethanol withdrawal. *Neurosci Lett* 167:175-178.

Morrisett RA and Swartzwelder HS (1993) Attenuation of hippocampal long-term potentiation by ethanol: a patch-clamp analysis of glutamatergic and GABAergic mechanisms. *J Neurosci* 13:2264-2272.

Nicoll RA and Malenka RC (1999) Expression mechanisms underlying NMDA receptor-dependent long-term potentiation. *Ann N Y Acad Sci* 868:515-525.

Nieber K et al. (1998) Inhibition by ethanol of excitatory amino acid receptors in rat locus coeruleus neurons *in vitro*. *Naunyn Schmiedebergs Arch Pharmacol* 357:299-308.

Neher E, Sakmann B and Steinbach JH (1978) The extracellular patch clamp: a method for resolving currents through individual open channels in biological membranes. *Pflugers Arch* 375:219-28.

Nelson TE, Ur CL and Gruol DL (1999) Chronic intermittent ethanol exposure alters CA1 synaptic transmission in rat hippocampal slices. *Neuroscience* 94:431-442.

O'Donnell P and Grace AA (1995) Synaptic interactions among excitatory afferents to nucleus accumbens neurons: hippocampal gating of prefrontal cortical input. *J Neurosci* 15:3622-3639.

Poelchen W, Proctor WR and Dunwiddie TV (2000) The *in vitro* ethanol sensitivity of hippocampal synaptic gamma-aminobutyric acid(A) responses differs in lines of mice and rats genetically selected for behavioral sensitivity or insensitivity to ethanol. *J Pharmacol Exp Ther* 295:741-746.

Prendergast MA et al. (2000) *In vitro* effects of ethanol withdrawal and spermidine on viability of hippocampus from male and female rat. *Alcohol Clin Exp Res* 24:1855-1861.

Proctor WR, Allan AM and Dunwiddie TV (1992) Brain region-dependent sensitivity of $GABA_A$ receptor-mediated responses to modulation by ethanol. *Alcohol Clin Exp Res* 16:480-489.

Reynolds JN and Wickens JR (2000) Substantia nigra dopamine regulates synaptic plasticity and membrane potential fluctuations in the rat neostriatum, *in vivo*. *Neuroscience* 99:199-203.

Rogers CJ and Hunter BE (1992) Chronic ethanol treatment reduces inhibition in CA1 of the rat hippocampus. *Brain Res Bull* 28:587-592.

Sapp DW and Yeh HH (1998) Ethanol-$GABA_A$ receptor interactions: a comparison between cell lines and cerebellar Purkinje cells. *J Pharmacol Exp Ther* 284:768-776.

Schwartzkroin PA (1975) Characteristics of CA1 neurons recorded intracellularly in the hippocampal in vitro slice preparation. *Brain Res* 85:423-436.

Sessler FM et al. (1998) Effects of ethanol on rat somatosensory cortical neurons. *Brain Res* 804:266-274.

Seeman P (1974) The membrane expansion theory of anesthesia: direct evidence using ethanol and a high-precision density meter. *Experientia* 30:759-760.

Siggins GR, Pittman QJ and French ED (1987) Effects of ethanol on CA1 and CA3 pyramidal cells in the hippocampal slice preparation: an intracellular study. *Brain Res* 414:22-34.

Soldo BL, Proctor WR and Dunwiddie TV (1994) Ethanol differentially modulates $GABA_A$ receptor-mediated chloride currents in hippocampal, cortical, and septal neurons in rat brain slices. *Synapse* 18:94-103.

Soldo BL, Proctor WR and Dunwiddie TV (1998) Ethanol selectively enhances the hyperpolarizing component of neocortical neuronal responses to locally applied GABA. *Brain Res* 800:187-197.

Sonnhof U et al. (1982) Cell puncturing with a step motor driven manipulator with simultaneous measurement of displacement. *Pflugers Arch* 392:295-300.

Stuart G, Schiller J and Sakmann B (1997) Action potential initiation and propagation in rat neocortical pyramidal neurons. *J Physiol* 505:617-632.

Tan SE et al. (1990) Prenatal alcohol exposure alters hippocampal slice electrophysiology. *Alcohol* 7:507-511.

Thiele TE et al. (2000) High ethanol consupmtion and low sensitivity to ethanol-induced sedation in protein kinase A-mutant mice. *J Neurosci* 20:RC75 1-6.

Thomas MP et al. (1998a) Organotypic brain slice cultures for functional analysis of alcohol-related disorders: novel vs. conventional preparations. *Alcohol Clin Exp Res* 22:51-59.

Thomas MP, Monaghan DT and Morrisett RA (1998b) Evidence for a causative role of N-methyl-D-aspartate receptors in an in vitro model of alcohol withdrawal hyperexcitability. *J Pharmacol Exp Ther* 287:87-97.

Thomas MP and Morrisett RA (2000) Dynamics of NMDAR-mediated neurotoxicity during chronic ethanol exposure and withdrawal. *Neuropharmacology* 39:218-226.

Tsai G and Coyle JT (1998) The role of glutamatergic neurotransmission in the pathophysiology of alcoholism. *Annu Rev Med* 49:173-184.

Walter HJ and Messing RO (1999) Regulation of neuronal voltage-gated calcium channels by ethanol. *Neurochem Int* 35:95-101.

Wan FJ et al. (1996) Low ethanol concentrations enhance GABAergic inhibitory postsynaptic potentials in hippocampal pyramidal neurons only after block of $GABA_B$ receptors. *Proc Natl Acad Sci* 93:5049-5054.

Wang MY, Rampil IJ and Kendig, JJ (1999) Ethanol directly depresses AMPA and NMDA glutamate currents in spinal cord motor neurons independent of actions on $GABA_A$ or glycine receptors. *J Pharmacol Exp Ther* 290:362-367.

Watson PL, Weiner JL and Carlen PL (1997) Effects of variations in hippocampal slice preparation protocol on the electrophysiological stability, epileptogenicity and graded hypoxia responses of CA1 neurons. *Brain Res* 775:134-143.

Weiner JL, Dunwiddie TV and Valenzuela CF (1999) Ethanol inhibition of synaptic kainate receptor function in rat hippocampal CA3 pyramidal neurons. *Mol Pharmacol* 56:85-90.

Weiner JL Gu C and Dunwiddie TV (1997a) Differential ethanol sensitivity of subpopulations of $GABA_A$ synapses onto rat CA1 pyramidal neurons. J Neurophys 77:1306-1312.

Weiner JL et al. (1997b) Elevation of basal PKC activity increases ethanol sensitivity of $GABA_A$ receptors in rat hippocampal CA1 neurons. *J Neurochem* 68:1949-1959.

Weiner JL, Zhang L and Carlen PL (1994) Ethanol modulation of $GABA_A$-mediated synaptic current in hippocampal CA1 neurons: possible role of protein kinase C. *J Pharmacol Exp Ther* 268:1388-1395.

Weinshenker D et al. (2000) Ethanol-associated behaviors of mice lacking norepinephrine. *J Neurosci* 20:3157-3164.

Whittington MA and Little HJ (1990) Patterns of changes in field potentials in the isolated hippocampal slice on withdrawal from chronic ethanol treatment of mice *in vivo*. *Brain Res* 523:237-244.

Yamamoto C and McIlwain H (1966) Electrical activities in thin sections from the mammalian brain maintained in chemically-defined media in vitro. *J Neurochem* 13:1333-1343.

Zucker RS (1989) Short-term synaptic plasticity. *Annu Rev Neurosci* 12:13-31.

9 Cognitive Correlates of Single Neuron Activity in Task-Performing Animals

Bennet Givens and T. Michael Gill

CONTENTS

9.1   INTRODUCTION

Cognitive neuroscience has enjoyed tremendous growth over the past decade due to the imaginative combination of computerized cognitive testing procedures and advanced neuroimaging and neurorecording techniques. In animals, the procedures now developed to assess cognitive functions have advanced tremendously, due, in part, to the use of highly automated operant procedures that can test many of the same cognitive functions that similar computerized procedures assess in humans (Buccafusco, 2001). Likewise, the availability of high-speed, high-capacity computers to neurophysiology labs has allowed for large scale neurophysiological acquisition of multiple single units (Nicoleis, 1999). The combination of large scale neurophysiological recording with sophisticated cognitive testing has opened up new opportunities for assessing the neural basis for ethanol's disruptive effects on cognition.

This chapter outlines one approach for assessing these effects, that of simulaneously recording multiple single neurons from rats that are performing an operant task. We will review the multiple neuron recording technique and operant behavioral procedures before describing some of the types of neurophysiological data that can be collected from behaving rats and various analytic tools for assessing physiological data with respect to behavioral events. We will then describe the application of these techniques to the assessment of the acute effects of ethanol on the neural circuits underlying sustained attention, and conclude the chapter with suggestions for future directions into which this research approach can lead.

9.2   BEHAVIORAL NEUROPHYSIOLOGY

9.2.1   SINGLE NEURON RECORDING

9.2.1.1   Advantages

Neurophysiological recording offers several distinct advantages over other techniques for investigating the neural basis of behavior. First, the temporal resolution of single unit recording is in the sub-millisecond range, thus allowing for exquisite moment-to-moment correlations with ongoing behavior. Second, recordings can be made over long periods of time. We routinely record from animals for 3–6 months following surgical implantation of the electrodes, and have recorded from some rats for 6-, 9-, and even 12-month periods. This longevity allows for experimental designs in which a variety of within-subject manipulations can be performed. For example, a series of drug studies can be performed in which multiple dose agonist or antagonist interactions can be determined, or a series of behavioral manipulations can be performed. In addition, in most cases, the same neurons can be held for a period of days to weeks, allowing not only within-animal comparisons, but also within-neuron comparisons following, for example, different doses of ethanol. An example of the power of this ability to follow a set of neurons over time is revealed in a study by Laubach et al. (2000), who recorded from a defined set of neurons in the motor cortex of a rat as it learned a simple reaction-time task. They found that the rate and pattern of activity, as well as the degree of correlated firing between neurons, changed as the rat learned the task, and began to predict the behavioral outcome of each trial.

A third major advantage of neurophysiological recording is that multiple single neurons can be recorded simultaneously. This can be accomplished either by recording single neurons from multiple electrodes, or by isolating multiple single neurons from a single electrode. Today, most behavioral neurophysiologists combine these techniques, and record multiple single neurons from multiple electrodes, yielding dozens of simultaneously recorded single neurons. Multiple neuron recording allows for higher-order population ensemble analysis with the use of multivariate statistics (see Chapin, 1999), that can reveal the relationship not only between neighboring neurons within a brain region, but can uncover the circuit properties and information flow between neural structures.

Another advantage of neurophysiological recording is that each recording session can last several hours. Thus, the same set of neurons can be recorded throughout protracted behavioral sessions. For operant procedures, unit responses can be averaged over many trials allowing one to gain an appreciation for the interplay between single unit activity and specific components of behavior. These statistically determined relationships, or behavioral correlates, are revealed in peri-event time histograms; when correlated with specific cognitive components of the task, they are called *cognitive correlates*. Another advantage of "holding" neurons throughout a behavioral session is that the time-dependent effects of drugs, given either systemically or by intracranial infusions, can be assessed, as well as time-dependent changes in behavior, such as a loss of vigilance over time in attention tasks (see below).

9.2.1.2 Electrodes

A small surface that electrically interfaces with the brain is necessary to record activity from single neurons. While rigid, etched metal electrodes and glass micropipettes are useful for acute and intracellular recordings, for stable unit recording from freely behaving animals, fine wire electrodes are essential. We use pairs of tungsten wires (20 mm diameter, California Fine Wire Co., Grover Beach, CA) that are immersed in epoxy and baked at 200° C for 1 h. Two pairs of fine wires are inserted into a 30-ga cannula and extended 2.0–2.5 mm beyond the distal end. The fine wire electrode cannula is affixed to a moveable microdrive that is constructed using a tapped carrier and threaded rods, which, when turned, lower the electrodes through the brain. The impedance of the electrodes ranges from 100 to 400 kΩ. The fine flexible wire allows the electrodes to move with the brain (during head movement or changes in blood pressure, for example) and to maintain stable recording of the same set of neurons throughout the recording session. In addition, the moveable drive allows for new sets of neurons to be isolated after one set has been thoroughly characterized, thus significantly enhancing the number of neurons that can be recorded from the same animal.

9.2.1.3 Electrode Implantation

Careful surgical implantation procedures are critical for successful unit recording. Surgery should always be performed under aseptic conditions. We use 0.7% isoflurane gas to achieve a level plane of anesthesia throughout surgery. It is also important to continuously monitor and maintain the body temperature of the animal within a normal range. Our surgical procedure is as follows: after initial anesthetic induction, the scalp is shaved and the head positioned in a small-animal stereotaxic apparatus equipped with an anesthesia mask to maintain surgical gas anesthesia. The scalp is cleansed with Betadine® scrub, incised along the midline, and the skin retracted and burr holes drilled through the skull. The electrode microdrive is implanted into the region of interest. A teflon-coated, stainless steel electrode (250 µm) is placed into superficial layers of a region of cortex near the recording site to serve as a reference. A second teflon-coated, stainless steel electrode (250 µm) is attached to a machine screw on the skull surface and serves as an animal ground. The recording electrode (attached to the microdrive), the reference electrode and the ground electrode are inserted into a plastic headstage connector and all three affixed to the skull surface with machine screws and dental acrylic. All rats receive an oral antibiotic twice daily for 3 days postoperatively to minimize the risk of infection.

9.2.1.4 Neurophysiological Recording

Neural activity is recorded using pairs of electrodes configured as stereotrodes (McNaughton et al., 1983). The stereotrode signals are recorded using an operational amplifier that is fitted onto a headstage connector and cabled to a commutator that relays the signals to a differential amplifier (A-M Systems, Inc., Everett, WA). The analog signals are amplified ($\times$ 10 k), band pass filtered (low pass–300 Hz, high pass–5 kHz), and then digitized by an analog-to-digital board (DT2821, 250 kHz; see Figure 9.1). Only those signals exhibiting a peak amplitude that exceeds a user-defined threshold on either electrode of the stereotrode are sampled (at 25kHz) and stored in 1.28 ms windows using Discovery® software (Datawave Technologies, Longmont, CO). Multiple unit activity on each stereotrode is separated into single units based on the clustering of signals after plotting the pair-wise relationships between parameters extracted from the waveform on both stereotrode electrodes as described in the next section. Single unit separation is performed prior to the start of each behavioral testing session. After thorough characterization of a set of single units which include testing throughout at least one behavioral session, the microdrive is incrementally advanced and on the following day another set of single units is isolated from the multiple unit activity.

9.2.1.5 Isolating Single Neurons

For each action potential, a neuron generates a field potential on the order of several hundred µV in amplitude, and approximately 1 ms in duration. These fields must be distinguished from neighboring neurons and noise for single units to be well isolated, and for the time of their discharge to be faithfully recorded. The isolation of action potentials from single neurons is done by template matching in which the

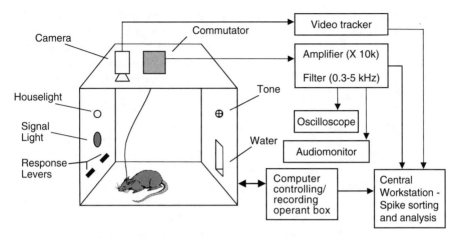

FIGURE 9.1 Schematic diagram of the operant chamber used to record single unit activity and assess behavioral peformance. Signals that arise from single neurons pass through a headstage pre-amplifier (not shown) and are sent through a commutator to rack-mounted amplifiers and filters. The neural activity is displayed on an oscilloscope and broadcast with an audiomonitor. Each unit is digitized by an A/D board within the central workstation, where on-line and off-line spike sorting and analysis take place. In addition, the computer for the behavioral system sends information about task events to the central workstation, which enters the timestamps for behavioral events into the physiological data stream.

waveform generated by each neuron is analyzed with respect to a variety of parameters, including peak and valley amplitudes and times, and transitional phase angles. In addition, the use of stereotrodes (McNaughton et al, 1983) or tetrodes (Gray et al., 1995), in which two or four closely spaced electrodes, respectively, enable signals from the same neuron to be "picked up" on multiple electrodes. Because the potentials from each neuron have slightly different amplitudes due to the different distances of the electrode tips from each neuron, the ratio of spike amplitudes from the same neuron across the different electrodes remains stable, and can be used as an additional parameter for sorting spikes. Two widely used software packages for isolating multiple single neurons in behaving animals are available from DataWave, Inc. (dwavetech.com), and Plexon, Inc. (plexoninc.com).

However, multiple single neuron recording techniques have several limitations. First, the so-called *sampling problem* tends to bias recordings toward larger neurons. Larger neurons are more easily and more frequently encountered, and also generate larger potentials that are more likely to be recorded. Second, *noise*, from a variety of electrical, thermal, magnetic, and mechanical sources, can obscure the recording of neurons. Most frequently these noises distort legitimate single units that are thus excluded from the analysis, but occasionally, certain types of noise match the waveform template, and thus are included as "real" spikes. Third, with multiple or ensemble unit recording, if two neurons both fire at or near the same time, one or both of the units will not be counted. The incidence of overlapping spikes increases (1) with high frequency activity, (2) as the number of units simultaneously recorded increases, and (3) when activity among neurons is driven or highly correlated as in event-related bursting. Additionally, multiple unit recording techniques are unable to individually locate and chemically identify neurons. While gross histological localization is possible with the use of small electrolytic lesions at the final site of recording or the bottom of the tract, morphological identification of the recorded neurons (e.g., Pinault, 1996) is not possible with current techniques for recording multiple units using fine wire electrodes.

9.2.2 OPERANT BEHAVIORAL PROCEDURES

We have found that operant tasks have several advantages for neurophysiological investigations of the cognitive effects of low doses of ethanol. Animals can be trained to perform a behavioral task with a high degree of proficiency. These well-trained animals are then used in ethanol testing. Typically, the more habitual aspects of task performance, such as detecting salient stimuli and making distinct motor responses, are unaffected by ethanol. On the other hand, the cognitive aspects of performance tend to be sensitive to disruption by low doses of ethanol, a phenomenon that is also observed in humans. Steady and reproducible behavior tends to yield stable physiological correlates. The repetitive nature of operant tasks is an advantage for neurophysiological recording. Unit responses can be averaged over 150–200 trials in a typical behavioral session, thereby allowing the identification of precise and subtle relationships between unit activity and specific behavioral and cognitive components of the task.

As mentioned, it is essential to use valid animal models of specific cognitive processes to begin identifying the specific neural circuits and systems that are critical sites mediating the impairing effects of alcohol. We have been using an animal model of sustained visual attention that was developed and validated for rats and requires the identification of brief, unpredictable stimuli and continuous psychomotor performance (McGaughy and Sarter, 1995). Performance by the animals in this task closely resembles sustained attention performance in humans. In humans, alcohol appears to specifically target tasks in which accurate performance is dependent upon high levels of attention (Koelega, 1995). Indeed, tests of selective attention, sustained attention, and especially divided attention are sensitive to the disruptive effects of alcohol, even at low to moderate doses (0.25–1.00 g/kg; Rohrbaugh et al., 1988). The fundamental type of attentional impairment seen during alcohol intoxication is a decreased accuracy to correctly detect stimuli, particularly those that are brief and unpredictable, in combination with slowed reaction times in responding to those stimuli.

9.2.2.1 Behavioral Apparatus

Operant studies are carried out with the use of standard operant chambers (28 cm length × 21 cm width × 27 cm height) located inside light and sound attenuating shells (64 cm × 41 cm × 41 cm). Figure 9.1 illustrates that the front panel of each chamber contains one signal light (2.8 W) centered 6 cm above two response levers. Each response lever is located 7 cm above a grid floor. A tone generator is located on the back panel along with a water dispenser that delivers a single drop of water

(40 ml) into a recessed water port. The operant chamber is illuminated by a house light (2.8 W) located 5 cm above the central signal light on the front panel. A separate operant chamber is custom-built for behavioral testing concurrent with electrophysiological recording and is identical to the training chamber except that it has taller side walls (42 cm), a larger recessed water port to accommodate the head-mounted pre-amplifier, an infrared diode emitter and detector at the entrance to the water port, and is housed in a larger sound attenuating shell (50 cm ¥ 39 cm ¥ 54 cm). A CCD video camera is mounted through the top of the sound-attenuating shell and is connected to a monitor and video cassette recorder. Both types of operant chambers are controlled by personal computers interfaced with hardware and software developed by Med Associates Inc. (St. Albans, VT).

The timing of behavioral events is recorded simultaneously with the electrophysiological data. A 5 V (TTL) pulse is sent to a clock board (CTM05) at the onset of the signal light, tone, and houselight, at the time of right and left lever responses, at the time of water reinforcement delivery, and at the time of the water port entry. In addition, signals that code the start and end of each trial, length of signal light illumination, and "non-signal" events are also sent to the clock board. These behavioral events are inserted as flags into the electrophysiological data stream at a 0.1 ms resolution for off-line correlation with single unit activity (see section 9.3.4).

9.2.2.2 Animals and Behavioral Training

In the procedures described here, adult rats are used. Two important factors concerning the subjects should be considered. First, for rats to perform the operant procedures described in this article, they must be motivated to work for water (or food) rewards at the time of testing. By restricting the amount of water (or food) given in the home cage, the rats will be motivated to perform the task. For behavioral neurophysiology, we have found that the use of water as a reward is preferable to food because it leads to less electrical interference, due in part to muscle potentials associated with chewing. We make food continuously available, and water available for 5–30 minutes at a variable time delay after daily behavioral training or testing. The length of time the rats receive water is dependent on the amount of water they receive in the task, and their task performance. Rats are never restricted to the point that it impacts their health, or puts them below 90% of their *ad libitum* body weight.

Another important consideration in training animals to perform operant (or any behavioral) procedures is to handle them regularly outside of the training or testing context. Animals that are comfortable interacting with the experimenter yield more consistent behavior and show few signs of stress at the time of testing. We handle the rats daily when they first arrive in the facility, and approximately bi-weekly outside of the daily handling associated with training and testing. All rats start behavioral testing after 1–2 weeks of acclimation to the vivarium and daily handling.

9.2.2.3 Measuring Sustained Attention

For the sustained attention task, rats are initially trained on a two-lever barpress paradigm using a fixed ratio (FR-1) schedule of water reinforcement. The rats are shaped until they emit 50 responses on both the left and right response levers. After initial shaping, the rats are trained in three stages, each of which required the rats to discriminate the presence of brief, unpredictable signal events (brief illumination of the signal light) from non-signal events (non-illumination of signal light). Both signal and non-signal events are followed 1 sec later by a tone (200 msec) that initiates a 4-sec response window during which bar presses are scored as correct or incorrect based on the following rules:

1. Left lever responses on signal trials are correct, scored as "hits," and followed by water reinforcement.
2. Right lever responses on signal trials are incorrect, scored as "misses," and are not followed by reinforcement.
3. Right lever responses on non-signal trials are correct, scored as "correct rejections," and followed by water reinforcement.
4. Left lever responses on non-signal trials are incorrect, scored as "false alarms," and not followed by reinforcement.

Bar presses within the response window or failures to respond initiate a variable intertrial interval (ITI, 10 ± 3 sec). An equal number of signal and non-signal trials are randomly presented within a testing session. Each behavioral testing session includes a 36-min task period divided into three 12-min blocks of trials, that is both preceded and followed by 5-min task-free periods. The rats are trained 5–6 days per week and are required to reach criterion performance before advancing to the next stage of training. The criterion performance level is choice accuracy of at least 70% on both signal and non-signal trials with less than 30% total omissions for 3 consecutive days.

After reaching criterion performance in the first stage of training, which required the discrimination of only 500 msec signals from non-signals, the rats are transferred to the second stage, in which they are required to discriminate between the presence and absence of signals that vary in length (25, 50, and 500 ms). All three signal lengths are pseudo-randomly selected and equally presented throughout each testing session. After reaching criterion performance, the rats continue training on standard testing sessions but within the electrophysiology chamber. During the final stage of training, the rats are tested on a distractor version of the attentional task on every 4th day. The only difference between the two types of testing sessions is that, during distractor testing sessions, the house light flashes at 0.5 Hz during the second 12-min block of trials. Each rat receives approximately six distractor sessions before surgical implantation of recording electrodes.

For 1 week following surgery, the rats are left in their home cage with free access to food and water. Thereafter, access to water is gradually reduced and the rats subsequently are retrained to criterion performance. During postsurgical recovery and behavioral retraining, the microdrive is slowly advanced in 45-μm increments each day until the tips of the electrodes are positioned at the dorsal border of the recording region of interest. Electrophysiological recording begins after the rats return to the criterion level of performance. The rats are recorded and behaviorally tested 6–7 days per week and are given a minimum of three standard testing sessions between each distractor testing session.

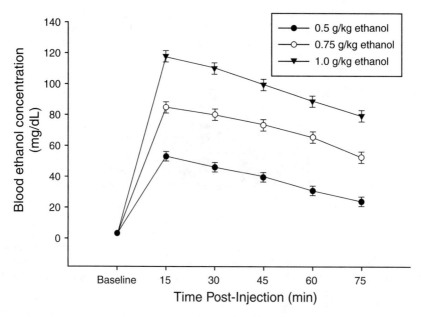

FIGURE 9.2 Changes in blood ethanol concentration (BEC) over time. Blood concentration was determined from the lateral tail vein blood using an alcohol dehydrogenase assay (Sigma Diagnostics, sigma-aldrich.com). Note that BEC remains relatively stable during behavioral and neurophysiological testing, which occurs between 15 and 60 min postinjection.

9.3 ASSESSING BEHAVIORAL CORRELATES OF UNIT ACTIVITY

To assess the acute effects of ethanol on single neurons as they relate to task performance, ethanol is given immediately before the rat is placed in the operant chamber, and 15 min before the start of testing. Ethanol is administered in a 10% (w/v) solution by intraperitoneal injection at doses of 0.0, 0.25, 0.5, 0.75 or 1.0 g/kg body weight. Figure 9.2 illustrates that blood ethanol concentrations, taken from rat-tail blood at 15-min intervals following ethanol injection, remain elevated throughout the period of time necessary to complete the 46-min behavioral testing sessions.

9.3.1 CHANGES IN ACTIVITY: RATE HISTOGRAMS

The ability to monitor activity in a group of single neurons from an awake, behaving animal is a powerful and unique window through which to appreciate the computations of the brain. A rate histogram that displays spontaneous neuronal activity over time is one of the simplest ways to assess the relationship between individual neurons and behavior. Figure 9.3 illustrates the activity of a neuron in the medial prefrontal cortex (mPFC) of a rat that is performing the sustained attention task. One can appreciate that the neuron "fired" at a base level of approximately 1.5 spikes/sec prior to the start of the task, and, when the task began, the activity of the neuron remained unchanged. During the second block of trials in the task, when the

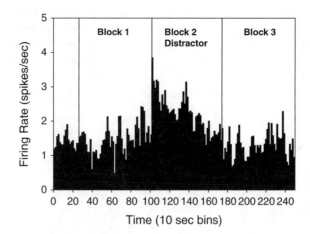

FIGURE 9.3 Ratemeter histogram plotting the mean firing rate of a single medial prefrontal cortex neuron over the course of sustained attention testing in which a visual distractor is presented at 0.5 Hz during the second block of trials. The neuron exhibited a distractor-induced increase in firing rate.

visual distractor was present, performance was impaired, and the unit increased activity to 2–3 spikes/sec and then returned to the previous level of activity when the distractor was removed.

Our data from 296 mPFC neurons (Gill et al., 2000) indicate during the 5-min pretask period and before the onset of behavioral performance, the mean firing rate (± SEM) of mPFC unit activity was 1.60 ± 0.11 spikes/sec. After the onset of behavioral performance during trial block 1, there was a slight increase in the mean firing rate of unit activity to 1.65 ± 0.11 spikes/sec. The presence of the visual distractor evoked alterations in the firing rate of a proportion of the mPFC units recorded during attentional testing. Twice as many mPFC units demonstrated distractor-induced increases in firing rate (as in Figure 9.3) when compared with the proportion of units that exhibited distractor-induced decreases in firing rate. Interestingly, the distractor-induced increase appears to involve a cholinergic modulation because the proportion of neurons that show these kind of rate increases in response to the distractor are significantly reduced when the cholinergic afferents to the mPFC are removed (Gill et al., 2000).

9.3.2 ACTIVITY PATTERNS: AUTO-CORRELATION

While rate histograms can reveal important information about changes in the activity of neurons, when viewed alone, they may obscure important alterations in the temporal patterning of activity. Auto-correlograms can provide a measure of spike discharge patterns by summing the activity of a neuron during a defined time interval around the occurrences of each spike in the neuron's own spike train. Any regularity in the timing between spikes becomes apparent as large numbers of spikes are accumulated. For example, auto-correlational analysis of mPFC single unit activity during attentional performance has identified four major patterns of spike discharge.

Single units are classified as either bursting (high frequency of repetitive spike discharge), non-bursting (low frequency of repetitive spike discharge), rhythmic (multiple bursts of synchronous spike discharges), or random (no distinct pattern of spike discharges). We have found (Gill et al., 1998) that the spike discharge pattern of mPFC units exhibited an equal distribution between the bursting, non-bursting classes, and random classes (appoximately 30% each) while the rhythmic class was the least frequent (10%). Individual neurons have a particular and stable "signature" that can be used for comparison, for example, across behavioral states or drug conditions, as in response to acute ethanol administration.

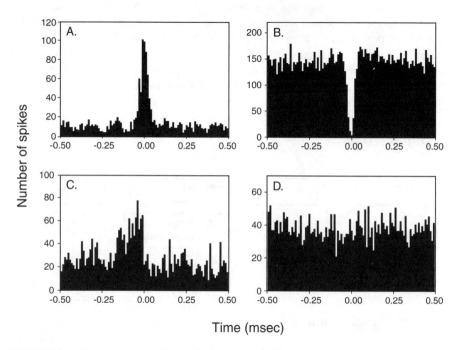

FIGURE 9.4 Cross-correlational plots for pairs of mPFC neurons. The four panels represent the four principle types of relationships observed among mPFC neurons: (A) Symmetric coactivation, (B) Inhibition, (C) Asymmetric coactivation, and (D) No relationship.

9.3.3 ENSEMBLE ACTIVITY: CROSS-CORRELAIONS

Cross-correlations are useful for analyzing the extent of cooperative activity and network properties within an ensemble of neurons. A cross-correllogram represents the probability of one (target) neuron firing with respect to the firing of another (reference) neuron, summed over each occurrence of the reference spike. Cross-correlational analysis conducted on all possible pairs of mPFC single units identified four distinct patterns of coordinated activity during sustained visual attention: symmetrical coactivation, inhibition, asymmetrical coactivation, and no relationship (see Figure 9.4). Note that when we distinguish between symmetrical and asymmetrical coactivations there is no intention to suggest that asymmetrical implies connectivity (see Hampson and Deadwyler, 1999). In our studies, a substantial proportion of all possible pairs of neurons (24%) exhibited a large degree of coactivation (top left panel), with approximately 42% of those pairs illustrating an asymmetrical pattern of coactivity (bottom left panel). A much smaller proportion of all possible pairs of units (13%) exhibited inhibitory patterns of coactivity (top right panel), while the largest proportion of pairs of units (63%) displayed no correlation in unit activity during sustained visual attention (bottom right panel). The established relations between pairs of neurons can change as a function of state, for example, in response to ethanol. A recent and fascinating application of cross-correlational analysis suggests that patterns of activity across ensembles of neurons change during behavioral exploration, and that these patterns are reproduced during the subsequent periods of rapid-eye-movement sleep (Louie and Wilson, 2001). With the application of cross-correlational analysis across neurons that are recorded simultaneously in different brain regions, it is possible to begin constructing directional circuit diagrams to undersand the flow of information that occurs during cognitive operations.

Signal Trials

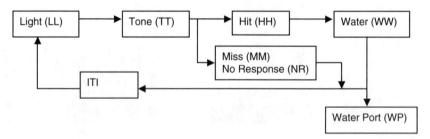

Non-Signal Trials

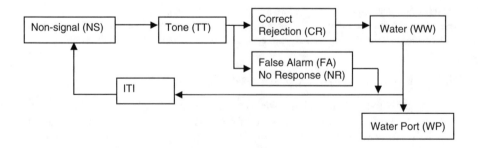

FIGURE 9.5 Behavioral events that occur during performance of the sustained-attention task. Each event generates a flag (indicated by two-letter abbreviation) that is time-stamped and entered into the physiological data stream. The flags form the basis for constructing peri-event time histograms.

9.3.4   BEHAVIORAL CORRELATES: PERI-EVENT TIME HISTOGRAMS

Peri-event time histograms (PETHs) are one of the most direct and intuitive ways to determine the temporal relationship between unit activity and behavioral performance. For each behavioral event that occurs during task performance, a "flag" is entered into the data stream with a unique timestamp. Figure 9.5 illustrates the behavioral events in the sustained attention task (coded as two-letter abbreviations). Individual single unit activity is typically summed in 10–50 ms bins within a 1–4-sec time frame around the time of individual behavioral events that occur during each testing session. For a given behavioral session, a multitude of PETHs can be generated. The amount of data used to generate each histogram varies as a function of the number of incidents of the behavioral event and the rate of neuronal firing. Figure 9.6 shows two simultaneously recorded neurons from the same electrode while the rat performed the sustained attention task. The firing rate of one neuron increases after the rat has entered the water port and begins consuming the water, while the other neuron increases activity *prior* to entry into the water port, and becomes completely quiescent during consumption. These kinds of variations among neurons with respect to their specific relation to a behavioral event are typical of neurons in the mPFC.

In our ongoing study of neurons recorded in the mPFC during performance of the sustained attention task, the majority of neurons show significant alterations in firing that are temporally correlated with specific behavioral events. Surprisingly, these behaviorally correlated alterations in unit firing are not elicited by the presentation of either signal lights or tones, but rather, are correlated with the preparation and production of operant responses and with the anticipation and consumption of subsequent water reward. Consequently, behaviorally correlated mPFC unit activity is classified into response-related and reward-related types. Response-related correlates are alterations in firing rate that were temporally correlated with either response preparation or response emission. Most units exhibit an increased rate of firing just prior to or during the emission of a response, although some units also exhibited decreased rates of firing. Some neurons have interesting selectivities such as response-type specificity in which the neurons fire only when the rat makes a correct (as opposed to an incorrect) response, irrespective of the side (left or right) of the response. Further, some neurons that fire selectively on correct responses also display trial-type specificity in which unit firing is selective to signal or non-signal trials. These firing selectivities impart special properties to the neurons that allow them to uniquely contribute to the behavioral and cognitive operations underlying task performance.

Most mPFC neurons exhibited significant alterations in firing rate that are temporally correlated with either the anticipation or consumption of water reward. The majority of these reward-related correlates consist of an increased rate of unit firing during the approach to the water port. Another sizable population of neurons exhibit increased rates of unit firing during the consumption of the water reward. It should be noted that the majority of mPFC units demonstrate more than one type of behavioral correlate during sustained visual attention performance.

In summary, in the sustained attention task, mPFC unit activity does not correlate with individual stimuli, signal lights or tone; however, there are significant event-

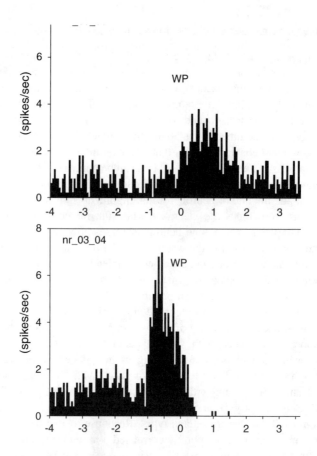

FIGURE 9.6 Peri-event time histograms for two different mPFC neurons recorded simultaneously during performance of the sustained-attention task. Both neurons increase their firing rate around the time of entering the water port (WP) after a correct response, but the neuron in the top panel fires after water port entry, whereas the neuron in the bottom panel increases activity prior to water port entry, and then falls silent.

related alterations in unit activity to both the response and reward, with the most prominent correlation consisting of increased activity just after the emission of correct responses and in anticipation of water port entry for reward consumption. These properties of mPFC neurons form a baseline for making comparison with neuronal characteristics observed following ethanol administration.

9.4 APPLYING THE BEHAVIORAL NEUROPHYSIOLOGICAL APPROACH TO COGNITIVE EFFECTS OF ETHANOL

To advance our understanding of the neural basis for alcohol-induced cognitive impairments, it is necessary to develop and use valid animal models of cognition.

Once established, neurophysiological studies can be designed to identify the specific neural circuits and systems supporting cognitive processing. These identified circuits are likely to be the critical sites underlying the deficits in cognition produced by ethanol.

9.4.1 ETHANOL-INDUCED ATTENTION DEFICIT

Using an operant animal model of sustained visual attention requiring the identification of brief, unpredictable stimuli in a continuous psychomotor performance, we have previously found that ethanol produces selective effects on the ability of rats to detect and respond to stimuli, and to sustain attention over time (Givens, 1997). The effects of ethanol are specific to the briefest (25 and 50 msec) signals, and interact with time such that there is a decrement in performance toward the end of the session. This type of *vigilance decrement* is well documented in humans (Koelega, 1995). The similarity of the effects of ethanol in rats to those seen in humans, both in terms of decreased accuracy and increased reaction times, further validate this animal model of attention and its extension into alcohol-related impairments of sustained attention.

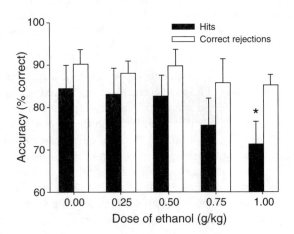

FIGURE 9.7 Dose-dependent effects of ethanol on accuracy in the sustained attention task. Ethanol impairs accuracy on signal trials at 1.0 g/kg i.p., but has no effect on non-signal trials. * p<0.05.

We report here an interim analysis of an ongoing study that ethanol produces dose-dependent effects on choice accuracy on the relative number of hits, whereas the relative number of correct rejections is not significantly altered by ethanol exposure (see Figure 9.7). The relative number of hits is attenuated after exposure to the two highest doses of ethanol, but only the 1.00 g/kg ethanol dose produces a statistically significant decrease when compared with saline treatment. In addition, the results demonstrate a trend for a greater decrease in the relative number of hits to the 25- and 50-msec signals within blocks 1 and 3 after acute exposure to the 0.75 and 1.00 g/kg doses of ethanol. Significant dose-dependent increases in the number of omitted trials occur after ethanol exposure, at the 0.75 and 1.00 g/kg doses. Further, errors of omission under saline and low dose ethanol conditions increase across trial blocks, and that effect is accelerated by the two highest doses of ethanol.

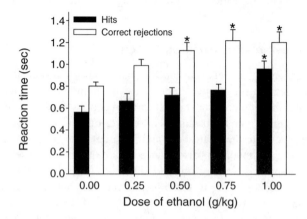

FIGURE 9.8 Dose-dependent effects of ethanol on reaction time in the sustained attention task. Ethanol increases reaction time at doses of 0.5, 0.75 and 1.0 g/kg i.p. on non-signal trials, and increases reaction time at the 1.0 g/kg i.p. dose on signal trials. *p<0.05

Preliminary results on response latencies demonstrate a significant dose-dependent increase in the latency to respond correctly on signal and non-signal trials after ethanol exposure. The response latency of hits increases only at the 1.00 g/kg dose of ethanol, while there are significant increases in the response latency of correct rejections at the 0.50, 0.75, and 1.00 g/kg doses of ethanol (see Figure 9.8). Similar to choice accuracy on signal trials, there is a trend for a greater increase in hit response latency to the shortest signal lengths and within block 3 after acute exposure to the two highest doses of ethanol. The choice accuracy on non-signal trials, or relative number of correct rejections, is not affected by the same concentrations of ethanol that significantly increase the latency to correctly respond on those same trials, suggesting a speed or accuracy trade-off in accurately detecting the absence of brief visual signals.

In accordance with previous work on sustained attention, the results from the current analysis demonstrate that acute ethanol exposure produces dose-dependent (1) decreases in the accuracy to detect brief signals, especially at the shortest signal lengths, (2) increases in the latency for correctly responding on both signal and non-signal trials, especially in the last block of trials, and (3) increases in errors of omission, especially late in the task.

The dose-dependent impairments of acute ethanol exposure on sustained attention are most notable on signal trials, especially at the shortest signal lengths, and during distractor testing sessions. These ethanol-related effects are comparable to the attentional effects noted after lesions of the basal forebrain cholinergic system (McGaughy et al., 1996), and after more limited cholinergic deafferentation of the cortex (McGaughy and Sarter, 1998). These parallel results further suggest that ethanol-related impairments in sustained attention may occur through a disruption at the level of the basal forebrain that results in a decreased level of cortical cholinergic tone critical for task performance, especially within the mPFC (see discussion below).

9.4.2 DISRUPTION OF MEDIAL PREFRONTAL CORTICAL (mPFC) ACTIVITY BY ETHANOL

9.4.2.1 Ethanol Decreases mPFC Unit Firing Rates

Our current data demonstrate varying effects of ethanol on the overall firing rates of mPFC single units during sustained visual attention. The majority of the single units recorded (approximately 75%) decreased their rate of firing (by at least 15%) after the 0.75 and 1.00 g/kg doses of ethanol relative to the percentage of units under saline treatment. A significantly smaller percentage of the single units (approximately 25%) increased their rate of firing (by at least 15%) after the 0.75 and 1.00 g/kg doses of ethanol relative to the percentage of units under saline treatment. Although a more complete understanding of the alteration in the absolute firing rates of mPFC single units after ethanol exposure is necessitated by this preliminary analysis, these initial findings do suggest that ethanol alters the firing rates of a majority of mPFC units at the same doses that impair sustained visual attention; however, as previously mentioned, simple firing rate changes do not address more subtle and behaviorally correlated changes in unit activity.

9.4.2.2 Ethanol Disrupts the Pattern of Auto-Correlated Spike Discharge

In addition to alterations in firing rate, our current data suggest that exposure to moderate doses of ethanol (0.5–1.00 g/kg) results in a variety of deviations from the normal pattern of spike discharge observed 24 h before and after ethanol treatment. Although 37% of the units recorded do not exhibit ethanol-related alterations in their normal pattern of spike discharge, the majority of units change their spike discharge pattern (e.g., bursting units becoming non-bursting, non-bursting units becoming bursting, and non-bursting units becoming rhythmic). Figure 9.9 illustrates the spike discharge patterns of two mPFC neurons simultaneously recorded from both cortical hemispheres. The middle row of panels illustrate a bursting unit that became rhythmic when exposed to a 1.00 g/kg dose of ethanol before returning to a bursting pattern 24 h later, while the bottom row of panels illustrate a non-bursting unit that also became rhythmic following ethanol exposure and returned to a non-bursting pattern 24 h later. These results indicate that ethanol significantly alters the normal pattern of spike discharge of mPFC single units at approximately the same doses that impair sustained visual attention.

9.4.2.3 Ethanol Alters Cross-Correlated Patterns of mPFC Unit Activity

Similar to the preliminary evidence on autocorrelated spike discharge patterns, ethanol dose-dependently affects the normal distribution of cross-correlated pat-

Cross-correlation across hemispheres

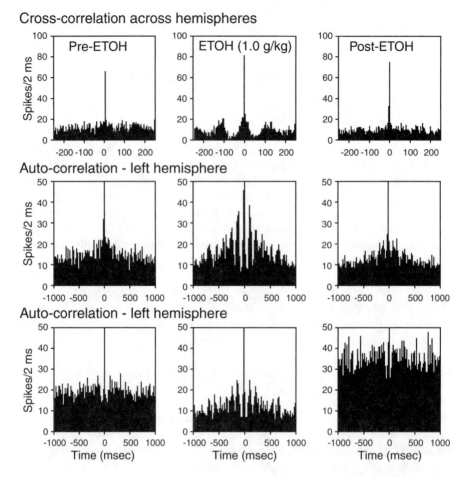

FIGURE 9.9 Effect of ethanol on patterns of single neuron activity and on the relationship of activities between two neurons in the mPFC during performance of the sustained attention task. The middle column of panels show the effect of 1.0 g/kg ethanol. Ethanol produced a shift to rhythmic activity in the auto-correlograms (bottom two rows of panels) and induced rhythmic coupling in the activity between the two neurons in the cross-correlogram (top row).

terns of spike discharge between pairs of mPFC units. Figure 9.9 also illustrates that the same two units that showed alterations in spike discharge patterns as revealed by auotocorrelograms also exhibited a rhythmic alteration in their pattern of coactivity after exposure to the 1.00 g/kg dose of ethanol, which also returned to the pre-ethanol pattern 24 h later. Thus, as these neurons become rhythmic following exposure to ethanol, they also become entrained to one another as mentioned above. Evidence from our studies assessing the cross-correlation between mPFC neurons following the loss of cortical cholinergic innervation suggests that under conditions of impaired attention, neurons become increasingly co-activated (Gill et al., 1998). Notably, these altered patterns of correlated activity occurred at the same doses of ethanol that impair sustained visual attention. These findings suggest that ethanol may bring about deficits in sustained attention by altering the synaptic relationships between mPFC neurons.

9.4.2.4 Ethanol Diminishes the Behavioral Correlates of mPFC Unit Activity

Examination of the data on the effects of acute ethanol exposure on mPFC unit activity indicate a significant dose-dependent attenuation in the expression of behaviorally correlated unit activity that coincides with impairments in overall behavioral performance. The top panel of Figure 9.10 illustrates a mPFC unit that exhibited a reward anticipatory correlate, as defined by the increase in unit activity that occurred just after correct rejections on non-signal trials and ceased as the rat entered the water port to consume the reward. Twenty-four hours later, when the rat was administered a 0.75 g/kg dose of ethanol, that same neuron no longer exhibited an increased rate of activity after correct responding on non-signal trials, even though a significant number of correct rejections were still emitted under ethanol exposure. When the rat was tested 24 h later under intact conditions, the reward anticipatory correlate following correct rejections returned to pre-ethanol levels. These data and results from other similar neurons provide support that acute ethanol exposure dose-dependently attenuates behaviorally correlated unit activity within the mPFC simultaneous with impairments in sustained visual attention; however, additional analysis is necessary to more thoroughly delineate the relationship between ethanol-induced alterations in behaviorally correlated mPFC activity and sustained visual attentional performance.

9.5 FUTURE DIRECTIONS

The behavioral neurophysiological approach as outlined above can be applied to many different behavioral and cognitive measures as well as a variety of different brain regions and circuits. Thus, by simply applying the available technology to specific brain structures and circuits and their known behavioral and cognitive functions, the future research projects using these techniques are virtually limitless. These applications are already under way in various laboratories, investigating phenomena from spatial memory (White et al., 2000; Ludvig, Chapter 11 this volume) to ethanol self administration (Janak et al., 1999; Chapter 10 this volume) .

However, beyond applying behavioral neurophysiological techniques to relevant brain structures and functions, advances in the ability to specifically combine behavioral neurophysiological techniques with other *in vivo* measures, such as microdialysis or voltametry (Ludvig et al., 1998; Rebek, 1998), allow for the powerful convergence of neurochemical and neurophysiological data. In addition, as the speed and capacity of computers continues to advance, so does the technology for collecting and analyzing increasingly larger numbers of simultaneously recorded single neurons (Nicolelis, 1999). The field of neural ensemble physiology holds great promise for allowing the simultaneous assessment of (and the interactions between) larger brain systems and networks.

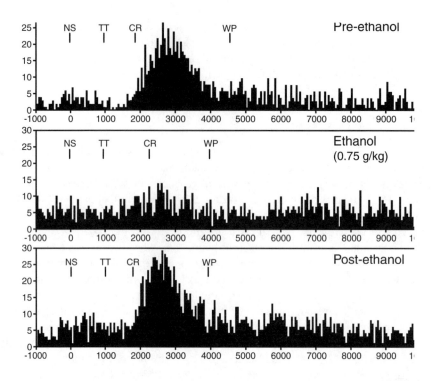

FIGURE 9.10 Effect of ethanol on peri-event time histogram surrounding the emission of a correct rejection. Following 0.75 g/kg i.p. ethanol, the neuron no longer showed a behavioral correlate to correct rejections. Behavioral flags as indicated in Figure 9.5.

We have begun to combine recording of mPFC neurons with local cholinergic deafferentation to assess the role of acetylcholine in the neural mediation of sustained attention. The results of bilateral cholinergic deafferentation of mPFC, along with those following unilateral deafferentation, suggest that attentional deficits occur only after the bilateral loss of cholinergic input and only under testing conditions requiring an increased level of attentional demand and flexibility. Interestingly, testing under conditions of increased attentional demand also markedly increase cortical acetylcholine efflux (Himmelheber et al., 2000) and the bilateral removal of that increased acetylcholine release within the mPFC results in decreased levels of attentional performance, suggesting that increases in cortical acetylcholine release modulate optimal levels of attentional performance, especially under conditions requiring effortful processing of brief visual stimuli. By applying advanced information on the role of acetylcholine as well as directed ethanol microinfusions into relevent brain regions, we can begin to unravel the mechanisms by which ethanol disrupts sustained attention.

9.6 CONCLUSIONS

This chapter gives an overview of the basic techniques used for assessing the role of single neurons in ethanol-induced changes in cognitive function. The application of these techniques to the study of sustained attention, and to the assessment of acute ethanol effects on the cognitive correlates of neural activity represent an initial attempt to describe the circuits involved in sustained attention that may be vulnerable to selective disruption by ethanol. By combining these techniques with other neurobiological techniques, as well as applying sophisticated neural ensemble analytic techniques, the future is bright for advancing our understanding of the specific neural circuits and systems responsible for ethanol-induced behavioral deficits.

ACKNOWLEDGMENTS

We thank Maria Yurrita for the data used in several figures, and Helen Sabolek, Leslie Rudzinski, and Suzanna Yan for assistance in data analysis. This work is supported by a grant from the National Institute of Alcohol Abuse and Alcoholism (RO1 AA12395).

REFERENCES

Buccafusco, JJ (2001). *Methods of Behavior Analysis in Neuroscience*. CRC Press.

Chapin, JK (1999) Population-level analysis of multi-single neuron recording data: Multivariate statistical methods. In *Methods for Neural Ensemble Recordings*. Nicolelis, AL (Ed) CRC Press.

Gill, TM et al. (1998). Correlated firing patterns of neighboring neurons within the rat medial prefrontal cortex during sustained visual attentional processing. *Soc Neurosci Absts*, 24, 169.

Gill TM, Sarter M and Givens, B (2000) Sustained visual attention performance-associated prefrontal neuronal activity: Evidence for cholinergic modulation. *J Neurosci* 20: 4745-4757.

Givens, B (1997). Effect of ethanol on sustained attention in rats. *Psychopharmacology* 129: 135-140.

Gray CM et al. (1995) Tetrodes markedly improve the reliability and yield of multiple single-unit isolation from multi-unit recordings in the cat striate cortex. *J. Neurosci Meth* 63: 43.

Hampson RE and Deadwyler SA (1999) Pitfalls and problems in the analysis of neuronal ensemble recordings during behavioral tasks. In *Methods for Neural Ensemble Recordings*. Nicolelis, A.L. (Ed) CRC Press.

Himmelheber AM, Sarter M, and Bruno JP (2000) Increases in cortical acetylcholine release during sustained attention performance in rats. *Cog Brain Res* 9: 313-25.

Janak PH, Chang JY and Woodward DJ (1999) Neuronal spike activity in the nucleus accumbens of behaving rats during ethanol self-administration. *Brain Res* 817: 172-84.

Koelega, H.S. (1995). Alcohol and vigilance performance: a review. *Psychopharmacology*, 118: 233-249.

Laubach M, Wessberg J and Nicolelis MA (2000) Cortical ensemble activity increasingly predicts behaviour outcomes during learning of a motor task. *Nature* 405: 567-71.

Louie K and Wilson MA (2001) Temporally structured replay of awake hippocampal ensemble activity during rapid eye movement sleep. *Neuron* 29: 145-56.

Ludvig N et al. (1998) Application of the combined single-cell recording intracerebral microdialysis method to alcohol research in freely behaving animals. *Alc Clin Exp Res* 22: 41-50.

McGaughy J and Sarter M (1995) Behavioral vigilance in rats: task validation and effects of age, amphetamine, and benzodiazepine receptor ligands. *Psychopharmacology* 117: 340-357.

McGaughy J, Kaiser T and Sarter M (1996) Behavioral vigilance following infusions of 192 IgG-saporin into the basal forebrain: selectivity of the behavioral impairment and relation to cortical AChE-positive fiber density. *Behav Neurosci* 110: 247-265.

McGaughy J and Sarter M. (1998) Sustained attention performance in rats with intracortical infusions of 192 IgG-Saporin-induced cortical cholinergic deafferentation: Effects of physostigmine and FG7142. *Behav Neurosci* 112: 1519-1525.

McNaughton BL, O'Keefe J and Barnes CA (1983) The stereotrode: A new technique for simultaneous isolation of several single units in the central nervous system from multiple unit records. *J Neurosci Meth* 8: 391-397.

Nicolelis AL, Ed. (1999). *Methods for Neural Ensemble Recordings*. CRC Press.

Pinault D (1996) A novel single-cell staining procedure performed *in vivo* under electrophysiological control: morpho-functional features of juxtacellularly labeled thalamic cells and other central neurons with biocytin or Neurobiotin. *J Neurosci Meth* 65: 113-36.

Rebec GV (1998) Real-time assessments of dopamine function during behavior: single-unit recording, iontophoresis, and fast-scan cyclic voltammetry in awake, unrestrained rats. *Appears Alc Clin Exp Res* 22: 32-40.

Rohrbaugh JW et al. (1988) Alcohol intoxication reduces visual sustained attention. *Psychopharmacology* 96: 442-446.

White AM and Best PJ (2000) Effects of ethanol on hippocampal place-cell and interneuron activity. *Brain Res* 876: 54-65.

10 Multichannel Neural Ensemble Recording During Alcohol Self-Administration

Patricia H. Janak

CONTENTS

10.1 INTRODUCTION

How is the physiology of the brain influenced by exposure to alcohol? What are the acute effects of alcohol on neuronal function, and how are these acute effects transformed into chronic effects both in direct response to a subsequent experience with the drug, such as tolerance, and in response to previous experience with the drug, such as alcohol-seeking behavior?

There are many ways to gather information on the immediate and long-lasting changes that alcohol induces in the nervous system and that are causal for the effects of alcohol on behavior. This chapter describes the application of an *in vivo* multichannel neural ensemble recording technique that allows for spike trains to be recorded from groups of individual neurons during behavior. Although extracellular single-unit recording has been applied to questions of alcohol and brain for some time (Hyvarinen et al., 1978; Chapin and Woodward, 1983, 1989; Schoener, 1984; Givens and Breese, 1990; Criado et al., 1995, 1997), there are both practical and conceptual advantages to the multichannel recording technique. Some of these advantages are discussed in Chapter 9. The microelectrode arrays used for the multichannel neural ensemble recording technique discussed in this chapter are fixed in place following implant; their position cannot be adjusted. The advantage of fixed placement is that it appears that the same unit can be followed for long periods of time (Nicolelis et al., 1998; Williams et al., 1999). A fixed electrode certainly would be a disadvantage were there only one electrode — the yield would be too low to justify the effort. Therefore, this technique relies on the use of one or more multiwire electrode arrays, typically allowing for 16 to 48 individual wires to be implanted in the rodent brain. The use of multielectrode arrays allows a high yield of data, offering obvious practical advantages.

The primary conceptual benefit of multichannel neural ensemble recording is that interactions among pairs and larger groups of neurons can be determined, and these interactions can be related to the ongoing behavior of the experimental subject. Interactions among neurons can be examined within and between brain regions. Hence, this technique offers the promise of new insights into dynamic principles of brain function.

10.2 DESCRIPTION OF THE TECHNIQUE

The principles underlying extracellular recording using metal microwires are discussed in Chapter 9. We use chronic implants of microwire arrays that can be repeatedly and easily attached to preamplifiers, digitizers, and computers for acquisition of neural data. The electrodes and their insertion into brain are described below, followed by an explanation of the collection of spike train data during ongoing behavior.

10.2.1 MULTIELECTRODE ARRAY

Historically, many neurophysiologists have made their own electrodes. Some of the earliest published examples of multichannel recordings collected from such a home-made microwire electrode array can be found in the work of Olds and colleagues (Komisaruk and Olds, 1968; Olds et al., 1972). The electrode arrays we use are commercially available from NBLabs (Dennison, TX). They are constructed from Teflon-insulated 50 µm stainless steel wire soldered to plastic female connectors. These electrode arrays are arranged in groups of 8 or 16 wires, with one stainless steel ground wire per eight recording electrodes. Details on these and related electrodes can be found in a recent review by Moxon (1999).

10.2.2 CHRONIC IMPLANTS OF MULTIELECTRODE ARRAYS

Depending upon the number of arrays one wishes to implant, the surgical procedure can be lengthy (3–5 h). Therefore, it is important to keep the subject warm using a heating pad and to pick an anesthetic that is appropriate for long surgeries. We use a ketamine (80 mg/kg) and xylazine (10 mg/kg) mixture. Hand-held or mounted drills are used to create openings in the skull large enough to encompass the wires of the electrode arrays. The eight-wire microwire arrays we use typically have a 1.5–2.0 mm total diameter. The electrode arrays are secured to the stereotaxic device using holders made of male connectors that fit into the female connectors of the electrode arrays. These fine-wire arrays lack the stiffness to penetrate brain tissue and to remain straight and true. Therefore, a stiffening agent is used. NBLabs coats the electrode array with an inert agent (polyethylene glycol) along the length of the array, leaving a customer-specified bare length of wire. Care must be taken to keep the insulation intact along the length of the wires. Because it is sometimes difficult to pierce the dura using the 50-μm wires, we gently make a small opening using fine forceps. The electrodes can be lowered to the desired depth over a period of minutes using micromanipulators. Once the electrodes are in place, very small fragments of gelfoam are gently placed around the electrodes to protect the surface of the brain; the electrodes are then secured to the skull using a small amount of dental acrylic (Jet dental acrylic, Lang Dental Mfg. Co., Inc., Wheeling, IL). All initial applications of dental acrylic are semi-liquid, which allows the acrylic to flow easily between the individual electrodes. After the acrylic hardens, the stiffening agent can be rinsed off the distal section of the array if desired. This process is repeated if multiple arrays are to be implanted. One can define skull quadrants by the four regions created by the intersection of the saggital suture and bregma; holes are drilled in each quadrant of the skull for placement of screws. At least one skull screw per quadrant should be used. More holes are drilled for ground wires, which are inserted 1–2 mm into cortex and are secured using a small amount of dental acrylic.

After all electrodes and ground wires are secured into place, and all screws have been placed, the connectors of the electrodes are arranged above the rat's skull in the desired final position using homemade holders secured to the stereotaxic device. Dental acrylic is carefully applied over the entire assemblage such that all metal and the bottom half of the connectors is covered. Care is taken to remove any dental acrylic that may have seeped over the subject's skin. If the original skin incision was longer than the final headset required, then sutures are used to close the tissue. Topical anesthetics (5% lidocaine ointment) are applied to the wound margin to reduce post-surgical discomfort. Topical antibiotics are used to combat infection.

The factors that contribute to good quality, long-lasting recordings are not entirely understood at this point (Williams et al., 1999). However, the longevity of the implants appears to improve if one takes great care in cleaning and drying the skull prior to the application of any dental acrylic. In addition, the use of prophylactic antibiotics and aseptic surgical technique also contributes to the success of the implants.

10.2.3 Spike Train Recording During Behavior

The primary objective of this recording technique is to acquire neural activity measures in the awake subject to better understand the neural bases for behaviors of interest. Therefore, in most cases, it is desirable to record neurophysiological data during task performance. Because subjects must be connected to a recording system, their range of motion is somewhat limited. This means that behaviors that can be measured in a relatively small environment work best. We typically examine operant behavior in our subjects because our primary interest in alcohol-seeking behavior is studied with the oft-used operant self-administration procedure. In addition, many sophisticated procedures that are based on operant conditioning and that can be applied to great effect in a relatively small chamber (1–1.5 feet square) have been developed over time.

Subjects are allowed to recover from the surgery for 1 week before recording commences. The exposed female connectors on the top of the rat's head are attached to a recording cable (headset cable) via a male connector that is inserted while holding the subject. Alternatively, the dental acrylic headstage may be gently grasped between the forefinger and thumb of one hand and the male connector inserted while the animal is unrestrained in a cage or experimental chamber.

The opposite end of the recording cable is then connected to a commutator to allow preservation of the signals from each separate channel during the subject's movement. The recording cable typically contains field effect transistors to boost the signal current. These cables can be homemade or purchased (NBLabs or Plexon Inc., Dallas, TX).

Several companies supply recording hardware and software (Sameshima and Baccala, 1999). A simplified view of the behavioral electrophysiological setup is depicted in Figure 10.1. The commercial products used to collect and analyze the data described here were produced by Plexon Inc. and by Biographics, Inc. (Winston-Salem, NC).

After passing through the commutator at the top of the behavioral chamber, the signals from each channel are amplified, filtered, and digitized, and the resulting data is stored on computer. Data acquisition software allows for spike sorting to be carried out by the user on-line. Typically, the spikes from each channel are sorted by separating those waveforms thought to represent firing from individual neurons from each other and from the noise, and then recording of the neural data can begin. The digitized waveforms can be saved to the hard drive. In addition, all times of occurrence of each individual neuron's extracellular action potential, or spike, are saved. These times are referred to as timestamps and the progression of these spikes through time is commonly referred to as the spike train. At this time, one also can collect behavioral data. The Plexon system can interface with any behavioral control system that can emit TTL pulses for each input and output, such as lever presses, tones, etc. Hence the times of each spike from each user-identified neuron and the times of each behavioral event are all saved together in the same data file as separate variables, greatly simplifying off-line analysis.

Data can be collected from the same subject across consecutive behavioral sessions. We typically record from a single subject for 5 d to 4 weeks, depending upon the stability and quality of the neural signals, and the experimental design. All the factors that contribute to long-lasting neural signals with good signal-to-noise ratios are not known but may depend, in part, on the quality of the surgery.

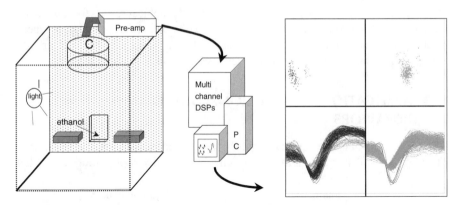

FIGURE 10.1 *In vivo* multichannel recording setup. Neural signals are acquired from rats during alcohol self-administration within a customized operant behavior chamber. Commutator (C) installed in center of chamber ceiling turns as the animal turns, and is designed to preserve the integrity of signals from each channel while allowing free movement of subject throughout experimental chamber. C is connected to a preamplifier that amplifies and filters signals. Neural signals are then sent through analog-to-digital boards, where additional amplification or filtering may take place, then to digital signal processing (DSP) boards under control of a personal computer (PC). Signals from each channel are sorted into waveforms that correspond to action potentials from individual neurons using specialized data acquisition software running on the PC. Examples of waveforms visible on two individual channels (wires) are shown on right. Two distinct waveforms are visible on each wire. Directly above the waveforms is principal component analysis for each wire, which aids in sorting and confirms separation of the waveforms into two separate units because two separate clusters are visible. Digitized and sorted waveforms and timestamps (time of occurrence of each waveform) are then saved to the PC along with times of occurrence of any behavioral events. (Figure also shown in color insert following page 202.)

We do not have the ability to track the same neuron from day to day with absolute certainty, as the nature of extracellular recording prevents us from ever identifying exactly from which cell the recorded potential arises. However, a combination of parameters is used to provide confidence that a particular recognized waveform does or does not represent the same neuron. We look for similar spontaneous firing rates, waveforms, and autocorrelograms as observed under baseline conditions. By these criteria, a majority of units are detected across more than 1 d. We believe we have been able to track the same unit for approximately 3–6 weeks.

Following the final recording session, the subject is deeply anesthetized, and the locations of the tips of the electrodes are marked by passing a 10–20 µA current for 10 sec through one or more wires per array to deposit iron ions into the surrounding tissue. Subjects are then transcardially perfused with PBS followed by a 4% paraformaldehyde/3% potassium ferrous cyanide solution. The potassium ferrous cyanide forms a blue reaction product with the deposited iron. Standard histological procedures are then used to slice and stain the tissue to visualize electrode locations.

The raw data files can be analyzed using homemade programs. However, software programs designed to import data files from various data acquisition systems are available, including Stranger (Biographics, Inc.) and NeuroExplorer (Plexon,

Inc.). These programs rapidly prepare numerical and graphical results for multichannel data including overall firing rates, autocorrelograms, cross-correlograms, peri-event histograms, and more. The numerical results can be exported into other software programs as needed for statistical analyses and other manipulations.

10.3 APPLICATION TO ALCOHOL-INDUCED BEHAVIORS

Because we are interested primarily in the neural mechanisms of the reinforcing effects of alcohol and alcohol-seeking behavior, our electrophysiological studies

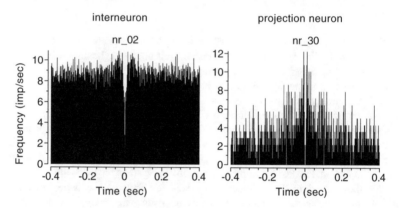

FIGURE 10.2 Major cell types within NAC are recognized by their autocorrelogram shape. Autocorrelograms depict the temporal pattern of the spike activity of a neuron relative to each individual spike of that same neuron; these histograms indicate the probability of firing of a given unit as described by all 1st-, 2nd-, and nth-order inter-spike intervals in the spike train. The autocorrelogram on the left is relatively flat except for a dip at time = 0 on the x-axis that corresponds to the refractory period for that cell. Hence this cell fires in a random, tonic fashion. The central peak of autocorrelogram on right indicates a greater likelihood for spikes to occur within about a 50-msec time window following a first spike. Hence this cell fires in a phasic, bursting pattern. The tonic pattern is thought to represent firing of an interneuron, while the lower rate phasic pattern is thought to represent firing of a medium spiny projection neuron (bin size = 2 msec).

have focused on the mesocorticolimbic system, especially the nucleus accumbens (NAC), and its excitatory inputs from the frontal cortex and the amygdala. About 95–97% of the rat NAC is composed of medium spiny neurons that are GABAergic projection neurons. The remaining proportion of neurons includes a few different classes of interneurons. Intracellular and extracellular recording studies in the striatum have characterized the firing rate and pattern differences between the projection neurons and the interneurons, and these can be used to separate the cell types (Wilson and Groves, 1981). For example, the medium spiny neurons tend to fire in a phasic manner, while the interneurons tend to fire in a tonic manner. These properties can be examined in the spike autocorrelograms, as depicted in Figure 10.2. From a practical perspective, however, the much greater number of medium spiny neurons relative to interneurons means that the great majority of the cells possibly recorded from the NAC will be medium spiny neurons.

In the following sections, results from recordings obtained from the NAC following the acute administration of alcohol and during alcohol-seeking behavior are described, along with analyses at both the single-neuron and multichannel levels.

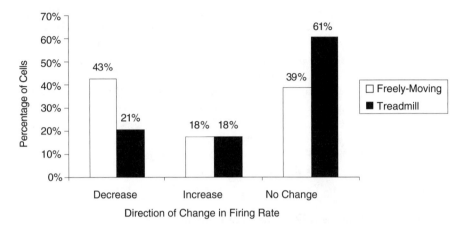

FIGURE 10.3 Effects of alcohol on firing rates vary with behavioral state. Proportion of NAC neurons that increased, decreased, or did not change mean firing rates following alcohol treatment. More NAC units are inhibited by a 1 g/kg IP injection of alcohol when subjects are free to move in open field than subjects whose locomotion is dictated by a treadmill. Fewer units changed firing rate following alcohol administration while walking on a treadmill, i.e., when the behavioral state was relatively similar before and after alcohol administration. Criterion for a decrease or increase was a mean rate change greater than 20% of baseline pre-alcohol firing rate (freely moving, n = 28 units; treadmill, n = 56 units).

10.3.1 PASSIVE ADMINISTRATION OF ALCOHOL AND IN VIVO RECORDING

Studies of the effects of alcohol on extracellular spike activity in the freely moving unanesthetized rat indicate that the behavioral state of the animal is a major determinant of the effects of alcohol on firing rates. For example, we injected rats with 1 g/kg alcohol and recorded neuronal activity within the NAC when animals either were free to move about a square behavioral chamber (40 × 40 × 50 cm) or were required to locomote on a treadmill (Janak et al., 1995). In both cases, increases, decreases, and no change in firing rates were observed within the populations of recorded neurons. However, subjects locomoting on the treadmill showed fewer decreases in firing rate, and more cells with no change (Figure 10.3). The behavior of subjects walking on the treadmill was greatly constrained by the experimental conditions such that it was relatively similar before and after alcohol administration. Because fewer alcohol-related changes were observed in these subjects, these data suggest that the behavior of the subject plays a major role in the effects of alcohol on neural activity within the NAC. This example emphasizes the importance of determining alcohol's effects on single-unit activity in the awake, behaving subject.

10.3.2 SELF-ADMINISTRATION OF ALCOHOL AND RECORDING

The neurophysiological basis for alcohol-seeking behavior cannot be determined completely without examining the neural activity that drives the behavior. Although lesion and drug studies have provided powerful evidence to support a role for the mesocorticolimbic brain circuits in drug and alcohol self-administration, the complementary studies of neural activity patterns allow for a detailed examination of precisely when these regions determine crucial behavioral components during a self-administration session. Multi-channel neural ensemble recording has made this endeavor much simpler than ever before. As discussed above, not only is yield increased, but also the interactions among the highly interconnected mesocorticolimbic regions, and the role of these interactions in self-administration behavior, can now be studied in detail. Below, the operant behavioral procedure used to study alcohol-seeking behavior is described followed by a description of electrophysiological results.

10.3.2.1 Self-Administration of Alcohol: Behavioral Procedures

The voluntary intake of alcohol by rats can be studied using operant procedures in which alcohol solutions are delivered following a response on a lever. Subjects typically are not water- or food-restricted, suggesting that their willingness to work for and consume significant amounts of alcohol reflects the reinforcing pharmacological properties of alcohol. However, many subjects will never consume enough alcohol to experience its pharmacological effects, because the taste of pure alcohol in water is unpalatable. The successful use of alcohol operant self-administration procedures was greatly enhanced by the development of a drinking initiation process that masks the unpalatable taste of alcohol with sweeteners such as sucrose or saccharin (Samson et al., 1988) during the early period of training. Operant self-administration of alcohol is studied using outbred rats, such as Long Evans or Wistar, and rat lines that have been bred based on alcohol intake or sensitivity (for example, the alcohol-preferring and alcohol non-preferring rats (Penn et al., 1978)). Recently, the operant self-administration of alcohol by mice has been reported (Middaugh and Kelley, 1999).

In our laboratory, subjects are first trained to lever-press for a solution of 10% sucrose in water (10S) by placing them in the operant chambers overnight. At this time, the following FR1 schedule is in effect: lever press → 0.5- to 1.5-sec variable delay → 0.5 sec tone → 0.5 sec fixed delay → delivery of 0.1 ml reinforcer solution. To facilitate acquisition of the response, subjects are water-restricted beginning 23 h before operant training. Most subjects acquire the lever-press response during 1–3 overnight sessions. Subjects that do not are then shaped by hand in additional sessions.

Following the overnight sessions, subjects begin daily limited access sessions, typically 30–60 min in length, for 5 to 7 d per week. During this first week, water restriction is gradually eliminated. At this time, the sucrose substitution procedure begins. The reinforcer for which subjects respond is altered every 2–3 d such that the concentration of alcohol in the solution is progressively increased, and then the concentration of sucrose in the solution is progressively decreased. The order of solutions we typically use is 10S, 10S/2E, 10S/5E, 10S/10E, 5S/10E, 2S/10E, and 10E (S = sucrose; E = ethanol). During the first week in which 10E is offered, a few sessions with 2S/10E are sometimes interspersed if responding, and subsequent intake, is particularly low. A minimum intake of 0.3 g/kg/30 min is desirable to ensure pharmacological relevance of the ingested alcohol.

After an additional 2 weeks of training, subjects typically have reached stable responding for 10E. At this time, increases in the response requirement (FR2 or FR3) may be made. When responding is again stable, experimental manipulations such as pharmacological treatment or surgery to implant microwire electrode arrays for *in vivo* electrophysiological recording can commence.

10.3.2.2 Self-Administration of Alcohol: Behavioral Neurophysiology

10.3.2.2.1 Analyzing Single Neurons

The first studies that applied the multichannel recording technique to addiction research measured neural activity within the NAC of rats self-administering intravenous cocaine (Carelli et al., 1993; Chang et al., 1994; Peoples and West, 1996). Since then, neural activity patterns within the NAC and related regions have been studied during heroin and nicotine self-administration as well (Chang et al., 1997, 1998; Lee et al., 1999). Our studies of neural activity within the NAC during alcohol self-administration found that over half of the recorded units show phasic changes in firing rate in relation to the behavioral events that occur during performance of this operant task (Janak et al., 1999). By using a behavioral schedule with programmed delays between important events (described above), we were able to track which behavioral event the observed neuronal changes were most closely associated with: the performance of the operant response, the onset of a tone cue that predicted alcohol delivery, or the delivery of the drops of alcohol, itself. Figure 10.4 depicts three typical examples of these types of behavioral correlates. No clear regional differences for the source of signals within the NAC have yet emerged.

Operant behavior can be divided into two phases, the initial appetitive or preparatory phase, which consists of the operant response (lever press or other), and the subsequent consummatory phase, during which the subject actively (oral reinforcers, sexual partners) or passively (intravenous reinforcers) experiences the reinforcer. Most drugs of abuse, including alcohol, induce increases in extracellular dopamine levels within the NAC (Di Chiara and Imperato, 1988). Also, rats will willingly lever press to receive injections of many drugs of abuse, such as cocaine (McKinzie et al., 1999), amphetamine (Hoebel et al., 1983) and PCP (Carlezon and Wise, 1996), directly into the NAC. Because of these findings, the contribution of the NAC to drug- and alcohol-seeking behavior mainly appeared to occur upon

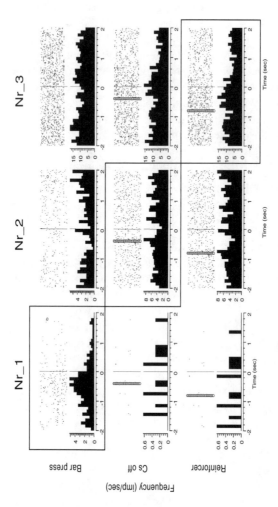

FIGURE 10.4 Three primary behavioral correlates observed in NAC during alcohol self-administration. Peri-event histograms and rasters show mean firing rate of individual neurons around time of occurrence of a behavioral event at time = 0. A different behavioral event is shown in each row, and three different simultaneously recorded neurons are shown, one in each column. First row shows activity of each of 3 units relative to bar press; Nr_1 significantly increased its firing rate prior to press. The second row shows activity of each unit relative to offset of 500 msec tone; Nr_2 significantly increased its firing rate upon tone onset, which is indicated by vertical line visible in raster at t = -0.5 sec. Third row shows activity of each unit relative to delivery of a drop of alcohol, 500 msec after tone offset; Nr_3 decreased its firing rate following alcohol receipt (bin size = 100 msec).

drug and alcohol receipt, the consummatory phase of the behavior. In fact, a prolonged inhibition upon reinforcer receipt appears in a large subset of NAC neurons when subjects are responding for oral alcohol solutions (47% of responsive cells; Janak et al., 1999), oral sucrose solutions (43% of responsive cells; Caulder, Chen, and Janak, unpublished observations), or intravenous cocaine or heroin (42% and 40% of responsive cells, respectively; Chang et al., 1998). In contrast, multi-channel recording studies have found that the majority of the excitatory responses that occur are observed during the appetitive phase of the trial, prior to the reinforcer period. These single-unit findings imply that the NAC may function in a qualitatively different way during the appetitive and consummatory phases of alcohol-seeking behavior. The further exploration of this idea provides an example of the application of multivariate statistics for ensemble analyses in a subsequent section of this chapter.

Variations of the basic operant self-administration procedure can be used to better understand the nature of the neural signals recorded within the NAC and elsewhere. We trained rats on a multiple schedule where the available reinforcer alternated every 5 min (Janak and Woodward, unpublished observations). Subjects were trained to insert their noses into a nosepoke operandum instead of lever-pressing. The liquid delivery ports for the two different reinforcers were located on either side of the centrally located nosepoke. These types of within-session comparisons between reinforcers allow us to compare the neural correlates for the two reinforcers. An example of an NAC unit that responded specifically to an alcohol + sucrose solution, but not to a sucrose-only solution, is shown in Figure 10.5. The finding that a proportion of NAC units respond differently to different reinforcers allows us to conclude that the signals reflect more than a general code for reward or reinforcement, and may represent something specific about a given behavioral goal such as preference (Chang et al., 1998).

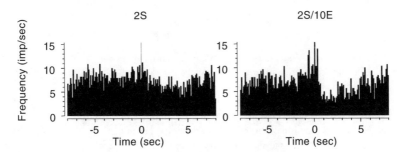

FIGURE 10.5 NAC units encode different reinforcers distinctly. Peri-event histograms representing mean spike activity of one neuron relative to delivery of a 2% sucrose solution and a 2% sucrose plus 10% alcohol solution. Both reinforcers were available on an alternating schedule every 5 min. Unit responds with a firing rate increase before and decrease after delivery of sucrose-alcohol solution, but not sucrose-only solution. (bin size = 100 msec).

The ability to record the same unit from session to session allows us to ask questions about the stability of the neuronal code over time. Figure 10.6 shows the lever press-related activity of one neuron recorded during two different sessions, 9 d apart. On both days, the neuron fired just prior to the lever press made to obtain alcohol. This type of finding indicates that the same neuron can be followed over days. This tracking ability should prove useful in studies examining changes in behavior, such as those that occur during response acquisition, extinction, or reinstatement.

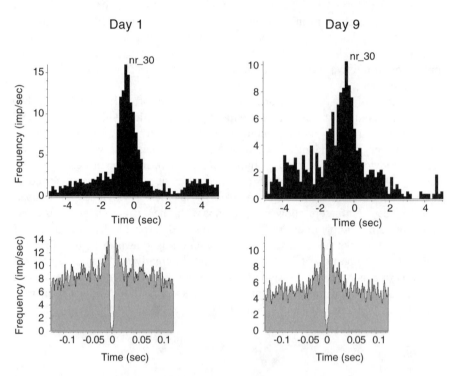

FIGURE 10.6 NAC unit's encoding of specific behavioral events is stable over time. Presence of an excitatory response prior to lever press during alcohol self-administration was apparent for this unit across two sessions conducted 9 d apart. Autocorrelograms are provided; similarity in shape of these histograms along with a similarity in waveform shape were used to identify the same cell across days. Average firing rate of this cell decreased over this recording period, either due to slow degradation of recording signal or changes in behavioral variables. (Subject was water-restricted for first but not second session.) For peri-event histograms, lever press occurred at time = 0; bin size = 150 msec. Autocorrelogram bin size = 0.8 msec.

10.3.2.2.2 Analyzing Data from Neural Ensembles

The major advantage of multichannel recording is the promise it holds for an increased understanding of the basic principles of brain function. This is because neurons within the same and across different brain regions are recorded simultaneously. We, therefore, can begin to understand how neurons work together, within and across brain regions, to accomplish complex behaviors. Our ability to understand the information contained within trains of simultaneously recorded neurons is in its infancy, although scientists have applied their energy to this problem for many years (Moore et al., 1966; Gerstein et al., 1985; Abeles et al., 1993; Shaw et al., 1993). We are guided in our attempts to apply neural ensemble analyses by the techniques that have proved successful for the fields of learning and memory (Wilson and McNaughton, 1994; Deadwyler et al., 1996), and somatosensory function and motor control (Nicolelis et al., 1993, 1998; Laubach et al., 2000). Recently, startling examples of the power of ensemble analyses have emerged; statistical models built from neuronal ensemble activity recorded from rats and non-human primates have been used to control mechanical devices (Chapin et al., 1999; Wessberg et al., 2000).

Our initial explorations of network function have centered on neural activity in the mesocorticolimbic system during alcohol self-administration. Linear discriminant analysis (LDA) can be applied in an attempt to identify brain states across groups of simultaneously recorded neurons that are related to specific aspects of self-administration behavior. The purpose is to try to understand better what information the neurons within a given region encode, and to what extent the encoding of that information is carried by neuronal ensembles, rather than individual neurons. LDA is a multivariate technique that attempts to find the best weighted linear combination of variables that will maximally distinguish among distinct classification groups. The application of this technique to multichannel neural ensemble recording is described in detail by Chapin (1999). Here we will discuss a brief example.

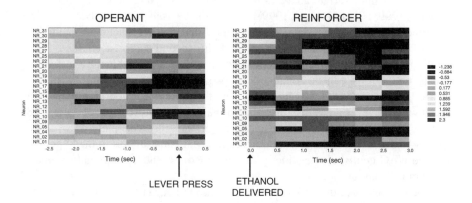

FIGURE 10.7 Patterns of neuronal activity within NAC are different during appetitive and consummatory phases of self-administration trial. Standardized firing rates of 21 simultaneously recorded NAC neurons are depicted relative to lever press response and alcohol receipt and consumption periods for a single alcohol self-administration session. Each neuron is represented in one row of contour plots, while time, in 500-msec blocks, is arranged across the abscissa. There are more excitatory responses during the appetitive phase of the trial. (Figure also shown in color insert following page 202.)

Recall that the single-neuron correlates recorded from the NAC suggest that the time period within a self-administration trial before reinforcement is qualitatively different from the time period when subjects receive reinforcement (see above). Consequently, it may be that the NAC contributes both to the appetitive and consummatory phases of operant behavior, but its contribution may be different for the two phases. Figure 10.7 is a contour plot that illustrates the neural activity of 23 simultaneously recorded neurons from one subject during these two phases. The plots show the standardized mean firing rates of each unit (arrayed along the y-axis) over a 3-sec time period (x-axis) around the operant response and around alcohol delivery. This type of figure illustrates the diversity of signals that exist at any one moment across a population of NAC neurons in the behaving subject. One can also see relatively more spike firing during the operant phase than during the reinforcement phase, as noted by the greater proportion of yellow and red colors during the operant phase. The two patterns of neural activity appear to be distinct. One way to examine this question quantitatively is to ask if the pattern of neural activity within the NAC during the appetitive phase is statistically distinct from that observed during the consummatory phase when the subject receives alcohol. LDA can be used to answer this question.

Many commercially available statistical packages contain LDA; SPSS (SPSS, Inc., Chicago, IL) was used for the example described here. The data variables consist of the spike activity of each neuron over 3 sec, divided into six 500-msec bins for both phases. The phase of the trial is indicated with a coding variable. In this case, as for many self-administration sessions, the number of cases is relatively few (74), while the number of variables (neurons, 23 units, six time bins each) is relatively many. A high ratio of cases (or trials) to variables is necessary for multivariate analyses to create models that do not over-fit the data (Chapin, 1999; Nicolelis et al., 1999). Therefore, we used factor analysis to reduce the dimensionality of the data set. We found that the first factor accounted for 15% of the variance in our data set, and the second factor accounted for an additional 6%. Both of these factors were then used as variables in the LDA. A linear combination can be found of the two factors that can correctly classify 98.6% of the time, on a trial-by-trial basis, the patterns of neural activity associated with the lever press response and with alcohol reinforcement (Figure 10.8). This objective measure of the relatedness of the observed patterns of neural activity to the phase of the self-administration trial adds support to the hypothesis that the NAC contributes distinctly to the appetitive and consummatory phases of operant behavior.

One can also use the spike activity of a single neuron to discriminate among brain states. When we determined the performance of each of our 23 neurons, we found that three units approached a classification as good as the ensemble, with percent correct classifications of 90.5, 91.9, and 94.6, respectively. We reconstructed our factors within an additional factor analysis, without the data from these three neurons, and conducted the LDA once again. Even without the very good single-neuron predictors, the spike activity of the remaining ensemble can be used to correctly classify the phase of the trial 97% of the time. These results suggest that there is redundant information contained within these networks, and that ensembles can encode behaviorally relevant information better than most individual neurons.

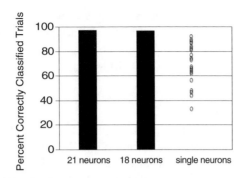

FIGURE 10.8 Classification of trial phase by linear discriminant analysis (LDA), used to analyze simultaneously recorded spike trains relative to trial phase (operant vs. consummatory). Simultaneously recorded spike trains were first subjected to factor analysis and LDA was conducted using the first two factors with a leave-one-out classification method. Spike activity of the ensemble could be used to correctly predict phase for each individual trial well above chance (21 neurons). Individual neurons also were used to classify phase of each trial; ensemble behaved similarly when three single best performing neurons were removed from ensemble (19 neurons). Results for individual neuron are shown as points (single neurons).

10.4 FUTURE DIRECTIONS

The application of multivariate statistical techniques such as LDA to data obtained from simultaneously recorded neurons in the behaving subject can shed light on distinct patterns of activity related to different phases of a behavioral trial, and perhaps to different phases of intoxication that may occur throughout an alcohol behavioral session. In addition, neural ensemble recording and multivariate analyses can be used to learn more about the neural circuits that mediate conditioned aspects of alcohol-seeking behavior. For example, what neural patterns are activated by conditioned stimuli that are associated with alcohol? How might these patterns of activity change during extinction or cue-induced relapse? Which brain regions contribute to these cue representations? Multi-channel neural ensemble recording coupled with appropriate statistical techniques can provide a powerful means to answer these and related questions regarding the neural basis of the behavioral effects of alcohol.

We can expect great advances in the analysis of ensemble data as we increase the accuracy of our picture of ongoing spike activity by increasing the number of neurons recorded. In addition, advances in appropriate data reduction techniques to simplify computation, and the development of new statistical methods to better characterize the neural code hidden within the spike activity of neural networks, will continue. The never-ending improvements in computing power will aid this endeavor. Principles of neural network function, and of the relation of neural network function to behavior, will emerge.

The technique of multichannel neural ensemble recording can be used in combination with other methods to gain further understanding of the neuronal processes underlying alcohol-related behaviors and to test the causality of observed neural patterns to behavior. Microinjection of pharmacological agents at afferent sites or at the recording site is one such method. Ensemble recording can also provide a functional measure of the effects of genetic alterations by recording neural activity within genetically manipulated animals. These types of studies that combine techniques will enhance our power to understand the mechanisms underlying alcohol-induced behavior.

10.5 CONCLUSIONS

The technique of *in vivo* multichannel neural ensemble recording in the awake, behaving rodent offers a means to further our understanding of the neurons and neural circuits that mediate the behavioral effects of alcohol. Great advances in computers and other commercially available electronic products now make this technique accessible to many researchers interested in behavioral neurophysiology. The ability to simultaneously record from dozens of individual neurons offers two advantages: 1) an increase in the total number of cells per animal, increasing the data collection rate; and, 2) the ability to examine interactions among groups of neurons, with the possibility of understanding how neural networks account for behavior.

ACKNOWLEDGMENTS

Supported by funds from the State of California for Medical Research on Alcohol and Substance Abuse through the University of California at San Francisco. I would like to give sincere thanks to Dr. Donald Woodward, for pioneering efforts in this field and for all his support.

REFERENCES

Abeles M et al. (1993) Spatiotemporal firing patterns in the frontal cortex of behaving monkeys. *J Neurophysiol* 70:1629-1638.

Carelli RM et al. (1993) Firing patterns of nucleus accumbens neurons during cocaine self-administration in rats. *Brain Res* 626:14-22.

Carlezon WA, Jr. and Wise RA (1996) Rewarding actions of phencyclidine and related drugs in nucleus accumbens shell and frontal cortex. *J Neurosci* 16:3112-3122.

Chang JY, Janak PH and Woodward DJ (1998) Comparison of mesocorticolimbic neuronal responses during cocaine and heroin self-administration in freely moving rats. *J Neurosci* 18:3098-3115.

Chang JY et al. (1994) Electrophysiological and pharmacological evidence for the role of the nucleus accumbens in cocaine self-administration in freely moving rats. *J Neurosci* 14:1224-1244.

Chang JY et al. (1997) Neuronal responses in prefrontal cortex and nucleus accumbens during heroin self-administration in freely moving rats. *Brain Res* 754:12-20.

Chapin JK (1999) Population-level analysis of multi-single neuron recording data: multivariate statistical methods. In: *Methods for Neural Ensemble Recordings* (Nicolelis MA, Ed), pp 193-228. Boca Raton: CRC Press.

Chapin JK andWoodward DJ (1983) Ethanol's effect on selective gating of somatic sensory inputs to single cortical neurons. *Pharmacol Biochem Behav* 18:489-493.

Chapin JK andWoodward DJ (1989) Ethanol withdrawal increases sensory responsiveness of single somatosensory cortical neurons in the awake, behaving rat. *Alcohol Clin Exp Res* 13:8-14.

Chapin JK et al. (1999) Real-time control of a robot arm using simultaneously recorded neurons in the motor cortex. *Nat Neurosci* 2:664-670.

Criado JR et al.(1995) Sensitivity of nucleus accumbens neurons in vivo to intoxicating doses of ethanol. *Alcohol Clin Exp Res* 19:164-169.

Criado JR et al. (1997) Ethanol inhibits single-unit responses in the nucleus accumbens evoked by stimulation of the basolateral nucleus of the amygdala. *Alcohol Clin Exp Res* 21:368-374.

Deadwyler SA, Bunn T and Hampson RE (1996) Hippocampal ensemble activity during spatial delayed-nonmatch-to-sample performance in rats. *J Neurosci* 16:354-372.

Di Chiara G and Imperato A (1988) Drugs abused by humans preferentially increase synaptic dopamine concentrations in the mesolimbic system of freely moving rats. *Proc Natl Acad Sci USA* 85:5274-5278.

Gerstein GL, Perkel DH and Dayhoff JE (1985) Cooperative firing activity in simultaneously recorded populations of neurons: detection and measurement. *J Neurosci* 5:881-889.

Givens BS and Breese GR (1990) Electrophysiological evidence that ethanol alters function of medial septal area without affecting lateral septal function. *J Pharmacol Exp Ther* 253:95-103.

Hoebel BG et al. (1983) Self-injection of amphetamine directly into the brain. *Psychopharmacology* 81:158-163.

Hyvarinen J et al. (1978) Effect of ethanol on neuronal activity in the parietal association cortex of alert monkeys. *Brain* 101:701-715.

Janak PH, Chang JY andWoodward DJ (1999) Neuronal spike activity in the nucleus accumbens of behaving rats during ethanol self-administration. *Brain Res* 817:172-184.

Janak PH, Laubach MG and Woodward DJ (1995) Ethanol alters the response properties of neurons in the striatum of the awake, behaving rat. Society for Neuroscience Abstracts, 21.

Komisaruk BR and Olds J (1968) Neuronal correlates of behavior in freely moving rats. *Science* 161:810-813.

Laubach M, Wessberg J and Nicolelis MA (2000) Cortical ensemble activity increasingly predicts behaviour outcomes during learning of a motor task. *Nature* 405:567-571.

Lee RS et al. (1999) Cellular responses of nucleus accumbens neurons to opiate-seeking behavior: I. Sustained responding during heroin self-administration. *Synapse* 33:49-58.

McKinzie DL et al. (1999) Cocaine is self-administered into the shell region of the nucleus accumbens in Wistar rats. *Ann N Y Acad Sci* 877:788-791.

Middaugh LD and Kelley BM (1999) Operant ethanol reward in C57BL/6 mice: influence of gender and procedural variables. *Alcohol* 17:185-194.

Moore GP, Perkel DH andSegundo JP (1966) Statistical analysis and functional interpretation of neuronal spike data. *Annu Rev Physiol* 28:493-522.

Moxon KA (1999) Multichannel electrode design: considerations for different applications. In: *Methods for Neural Ensemble Recordings* (Nicolelis MA, Ed), pp 25-45. Boca Raton: CRC Press.

Nicolelis MA et al. (1993) Dynamic and distributed properties of many-neuron ensembles in the ventral posterior medial thalamus of awake rats. *Proc Natl Acad Sci USA* 90:2212-2216.

Nicolelis MA et al. (1998) Simultaneous encoding of tactile information by three primate cortical areas. *Nat Neurosci* 1:621-630.

Nicolelis MA et al. (1999) Methods for simultaneous multisite neural ensemble recordings in behaving primates. In: *Methods for Neural Ensemble Recordings* (Nicolelis MA, Ed), pp 121-156. Boca Raton: CRC Press.

Olds J et al. (1972) Learning centers of rat brain mapped by measuring latencies of conditioned unit responses. *J Neurophysiol* 35:202-219.

Penn PE et al. (1978) Neurochemical and operant behavioral studies of a strain of alcohol-preferring rats. *Pharmacol Biochem Behav* 8:475-481.

Peoples LL, West MO (1996) Phasic firing of single neurons in the rat nucleus accumbens correlated with the timing of intravenous cocaine self-administration. *J Neurosci* 16:3459-3473.

Sameshima K and Baccala LA (1999) Trends in multichannel neural ensemble recording instrumentation. In: *Methods for Neural Ensemble Recordings* (Nicolelis MA, Ed), pp 47-60. Boca Raton: CRC Press.

Samson HH, Pfeffer AO and Tolliver GA (1988) Oral ethanol self-administration in rats: models of alcohol-seeking behavior. *Alcohol Clin Exp Res* 12:591-598.

Schoener EP (1984) Ethanol effects on striatal neuron activity. *Alcohol Clin Exp Res* 8:266-268.

Shaw GL et al. (1993) Rhythmic and patterned neuronal firing in visual cortex. *Neurol Res* 15:46-50.

Wessberg J et al. (2000) Real-time prediction of hand trajectory by ensembles of cortical neurons in primates. *Nature* 408:361-365.

Williams JC, Rennaker RL and Kipke DR (1999) Long-term neural recording characteristics of wire microelectrode arrays implanted in cerebral cortex. *Brain Res Brain Res Protoc* 4:303-313.

Wilson CJ and Groves PM (1981) Spontaneous firing patterns of identified spiny neurons in the rat neostriatum. *Brain Res* 220:67-80.

Wilson MA and McNaughton BL (1994) Reactivation of hippocampal ensemble memories during sleep. *Science* 265:676-679.

11 Combined Single-Neuron Recording and Microdialysis in the Brain of Freely Behaving Animals (Rodents and Non-Human Primates)

Nandor Ludvig

CONTENTS

11.1 INTRODUCTION

Alcoholism can be viewed as a complex behavioral disorder due to a specific, gradually developing, partly genetic and partly acquired intercellular communication error within the motivational, emotional and cognitive neural networks of the brain. The genetic component of this intercellular communication error has been suggested by clinical studies (Hesselbrock, 1995). Some pharmacogenetic animal models of alcoholism have been proposed (Li and McBride, 1995). The role of acquired factors is indicated by the necessity for repeated alcohol intoxications in the pathogenesis of the disease. The involvement of the motivational system is reflected in craving and alcohol-seeking behavior. The involvement of the emotional system is reflected in the positive reinforcing and anxiety- and stress-reducing effects of alcohol. The involvement of the cognitive system is reflected in the role of environmental cues and social and cultural influences in the development of alcoholism. The brain structures that compose these neural networks include, but are not limited to, the hypothalamus, ventral tegmental area, nucleus accumbens, amygdala, hippocampus and association cortex (Koob et al., 1998).

The intercellular communication errors that lead to the symptoms of alcoholism are ultimately due to abnormal molecular processes. The combined single-cell recording and microdialysis technique is an *in vivo* method that is suitable to obtain information on these molecular processes (Ludvig et al., 1995, 1998, 2000, 2001a). The concept of the method is shown in Figure 11.1. This technique, which my colleagues and I developed (Ludvig et al., 1994), has been used in several other laboratories to determine cellular drug actions in the prefrontal cortex (Dudkin et al., 1996), mesopontine nuclei (Thakkar et al., 1998) and basal forebrain (Alam et al., 1999). In fact, simultaneous electrophysiological recording and microdialysis were proposed for human studies in a neurosurgery setting (Engel, 1998). The method is useful for alcohol-related neuroscience research because it allows the monitoring of molecular and cellular effects of alcohol in brain, within the motivational, emotional and cognitive systems, during behavior and for relatively long periods.

Specifically, the following types of studies can be conducted with this procedure:

- Delivering ethanol, via microdialysis, into the environment of neurons in various brain structures. Detection of the ethanol-induced firing pattern changes allows one to map the acute cellular actions of ethanol in brain, *in vivo*, without the generalized behavioral effects of systemic ethanol administration or the confounding effects of anesthesia. The ethanol delivery can be repeated to examine the development of tolerance. So far, we have focused on the performance of these types of experiments.
- Delivering ethanol, via microdialysis, in combination with various neurotransmitter receptor agonists and antagonists, ion-channel blockers, second messenger modulators, and other drugs, into the environment of neurons in various brain structures. Detection of the developed firing pattern changes allows the examination of the molecular mechanisms that mediate the acute cellular actions of ethanol, *in vivo*.
- Delivering various drugs, via microdialysis, into the environment of neurons in various brain structures after systemic ethanol administration or

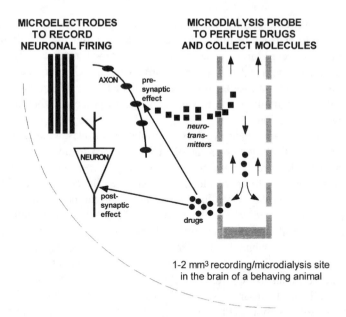

FIGURE 11.1 The concept of combined single-cell recording and intracerebral microdialysis technique. Drugs (e.g., ethanol), perfused via a microdialysis probe, diffuse to brain tissue and act on local neurons and axon terminals. The postsynaptic firing modulatory effects of delivered compounds are recorded with nearby microelectrodes. The presynaptic neurotransmitter release modulatory effects of the drugs can also be monitored by analyzing the microdialysates. The figure depicts only one neuron, but, in reality, the recording and microdialysis area includes a large number of cells. As indicated, the procedure can readily be performed in freely behaving animals.

> ethanol consumption. Detection of the drug-induced firing pattern changes can provide information on the cellular and molecular events in brain that accompany systemic ethanol exposure or natural ethanol intake.

This chapter will provide a brief theoretical background and a detailed description of the method. This will be followed by presenting some experimental results obtained with the technique. It will be pointed out how these results helped to develop new hypotheses on the mechanism of action of ethanol in brain. Possible future expansion of the method and its best utilization in alcohol-related neuroscience research will also be examined. Finally, the technique will be put in the context of other neuroscience methods and its advantages and disadvantages will be summarized.

11.2 THEORETICAL BACKGROUND

The combined single-cell recording and microdialysis method is based on the recognition that the microdialysis procedure causes such a minimal amount of tissue damage, and can provide such a constant solute exchange between the microdialysis probe and the surrounding interstitial space, that it is possible to record the electrical activity of single neurons in this area and to pharmacologically manipulate these cells by drugs delivered through the microdialysis probe ("reverse-dialysis"). From the drug-induced firing pattern changes, information can be obtained on the molecular machinery of the examined neurons. Thus, the method is not merely the addition of one technique to another, rather, it is an integrated procedure that allows one to collect novel information.

The recordings are made with extracellular electrodes. Chapters 9 and 10 describe this recording method in detail. It is sufficient to point out here that, in contrast to intracellular recordings, where membrane potential changes are measured with an "active and "reference" electrode positioned inside and outside the cell, respectively, extracellular recordings measure electrical potential field changes within the low resistance conducting medium of the extracellular fluid ("volume conductor") with both the active and reference electrodes positioned in this medium, outside the cell. When a bundle of microwires is used for recordings (Kubie, 1984; Ludvig et al., 1994; Kentros et al., 1998), the active electrodes are those that are closer to the cell than to the reference electrode. To capture action potentials, the active extracellular electrode must be close to the neuron, usually within about 50 μm from the cell membrane (Phillips, 1973). The shape and amplitude of extracellular action potentials depend on many factors, including the size of the cell, the position of the active electrode along the axon-soma-dendrite axis, the distance of the active electrode from the cell membrane, the conductivity of the extracellular medium, and the extent of action potential propagation within the neuron. In contrast, the frequency and temporal distribution (firing pattern) of extracellular action potentials depend primarily on the physiological function of the cell and the incoming synaptic inputs from local and remote neurons. Because these inputs act on ion channels, receptors, second messengers and genes, which are all sensitive to drugs, it is the neuronal firing pattern that is principally affected by reverse-dialyzed compounds (Ludvig et al., 1994; Dudkin et al., 1996; Thakkar et al., 1998; Alam et al., 1999). Therefore, this is the major cellular electrophysiological parameter that should be analyzed in recording and microdialysis studies. At the same time, the shape and amplitude of the waveforms should also be documented, as they provide valuable information on the stability of recording and the functional integrity of the recorded and dialyzed tissue.

Pharmacological manipulation of neuronal firing via intracerebrally implanted microdialysis probes utilizes the principle that drugs dissolved in the microdialysis fluid ("perfusate") diffuse through the probe-membrane and penetrate into the tissue deeply enough to reach, and act on, the recorded cells. The theory of microdialysis is thoroughly examined in Chapter 12. This chapter will discuss only some key considerations. It should be recognized that intracerebral drug administration via microdialysis involves two successive processes, each obeying a different set of laws. The first process is the movement of the drug molecules from the lumen of the microdialysis probe to the tissue interface ("delivery"). This is primarily determined by Fick's first law of diffusion. Thus, the rate of drug diffusion across the probe–membrane is proportional to the difference of drug concentration in the microdialysis fluid and the tissue interface. Also, this diffusion is directly related to the area of the microdialysis membrane and inversely related to thickness of this membrane. Therefore, increased membrane surface results in increased drug delivery. However, using larger microdialysis probes decreases the neuroanatomical site-specificity of the reverse-dialysis procedure.

The second process of drug administration via microdialysis is the movement of drug molecules away from the probe–tissue interface into the interstitial space ("penetration"). This is determined partly by the diffusion coefficient of the delivered drug and partly by biological factors. These biological factors include clearance of the drug molecules from the interstitial space by local blood flow, extracellular metabolism, and cellular uptake (Bungay et al., 1990). Drugs that are not taken up by the cells, metabolized or carried away by the local circulation to any great degree, such as sucrose, can accumulate in the interstitial space and penetrate deeply into the neuropil (Dykstra et al., 1992). Tissue penetration is also affected by the duration of drug delivery: longer duration leads to increased penetration distance. Nevertheless, after the establishment of steady-state reverse-dialysis, when the delivery rate and clearance rate are equal, the drugs do not penetrate farther into the tissue. Between the establishment of steady-state reverse-dialysis and the termination of drug delivery, the penetration distance remains unchanged. The recording electrodes should be located within this distance from the microdialysis probe to detect direct cellular drug effects. Tissue debris and blood clots around the electrode-array and the microdialysis probe, however minimal, can slow the movement of drug molecules toward the recording site. This factor is difficult to control, varies from experiment to experiment, and may well contribute to the variability of the data.

We have recently applied the mathematical model of quantitative microdialysis developed by Bungay et al. (1990) and advanced by Gonzales et al. (1998) to ethanol reverse-dialysis data obtained in the hippocampus of anesthetized rats (Ludvig et al., 2001a). Figure 11.2 summarizes the results. Within 2.4 min of initiating the perfusion of 1 M ethanol, a 45–70 mM extracellular ethanol concentration was established at a 400–500 μm distance from the microdialysis probe. In simultaneous single-cell recording and microdialysis experiments, clinically relevant ethanol concentrations can be achieved at the electrode-site about 400–500 μm from the probe.

A major advantage of using microwires for extracellular recordings and microdialysis probes for drug delivery is that neither of these tools is sensitive to head movements and intracranial pulsation. Therefore, these tools can readily be used in freely moving animals (Kubie, 1984; Ludvig et al., 1994, 1995, 1998, 2001a; Kentros et al., 1998). Insertion of the microdialysis probe into the brain causes blood–brain-barrier disruption and tissue damage in the vicinity of the probe (Benveniste and Huttemeier, 1990; Groothuis et al., 1998). However, this is irrelevant in combined single-cell recording or microdialysis studies as long as the neurons at the electrode site function normally. We have shown that, during intrahippocampal microdialysis, fine-tuned neuronal functions, such as the location-specific firing of place cells, remain completely intact at the recording site (Ludvig et al., 1998; Ludvig and Tang, 2000). Based on relevant neuroanatomical studies (Fox and Ranck, 1975) and on the consideration that, in the vicinity of the microdialysis probe, the neurons are damaged, it can be estimated that ethanol reverse-dialysis in the cell-packed CA1 region of the rat hippocampus affects the firing of about 15,000 physiologically

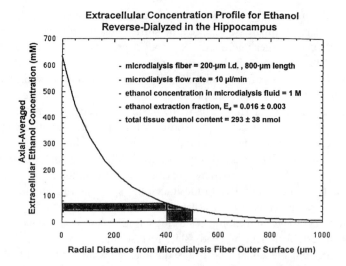

FIGURE 11.2 Extracellular concentration profile ethanol delivered via intrahippocampal microdialysis. The curve was constructed with the mathematical model of Gonzales et al. (1998), based on hippocampal ethanol content measurements in anesthetized rats. Microdialysis parameters, value for total tissue ethanol content, and *in vivo* extraction fraction, E_d, are as indicated. $E_d = 1 - (C_{out}/C_{in})$, where C_{out} is the ethanol concentration in the outlet tube of the microdialysis probe, and C_{in} is the ethanol concentration in the inlet tube of the microdialysis probe. The shaded areas indicate the concentration range of ethanol (45–70 mM) in the site of the recording electrodes positioned 400–500 μm from the probe. Note that, despite the extremely high ethanol concentration in the fluid that enters the microdialysis probe, the concentration of the drug at the recording site is actually within a clinically relevant range. (Reprinted from Ludvig N et al.; Evidence for the ability of hippocampal neurons to develop acute tolerance to ethanol in behaving rats. *Brain Research*, 2001, vol. 900, pp. 252-260. With permission from Elsevier Science.)

functioning neurons. (This is about 1% of the total hippocampal neuron population in this species.) Thus, in combined single-cell recording or microdialysis studies, the drug-induced neuronal firing pattern changes are due to both direct drug effects on the recorded cells and to indirect effects mediated by local synaptic connections within the recording or dialysis site. These effects can be distinguished by the determination of drug effects on all major neuron types within the examined area.

11.3 THE TECHNIQUE OF COMBINED SINGLE-NEURON RECORDING AND MICRODIALYSIS

11.3.1 ANIMALS

Most of our studies have been performed on freely moving rats (Ludvig et al., 1994, 1995, 1996, 1998, 2001a; Ludvig and Tang, 2000), the species used most frequently in alcohol-related neuroscience research. Therefore, the following methodological descriptions will focus on techniques tailored for rats. In collaboration with A. Rotenberg, my colleagues and I also performed a limited number of recording and dialysis sessions in freely moving mice. This study, employing the microelectrode design of Rotenberg et al. (1996), demonstrated that simultaneous single-cell recording and microdialysis can also be conducted in these animals. Because there is an increasing interest in using genetically altered mice for alcohol research (see Chapters 1 and 2), the few modifications in the recording and dialysis technique, which are necessary for applying it to mice, will also be mentioned. Finally, we adapted the recording and dialysis method for behaving squirrel monkeys (*Saimiri sciureus*) (Ludvig et al., 2000). In contrast to rats and mice, non-human primates have been used in alcohol research by only a few groups (Grant et al., 1997; Mandillo et al., 1998). Nevertheless, this may change in the future. Therefore, the method for simultaneous neuronal recording and microdialysis in monkeys will also be described.

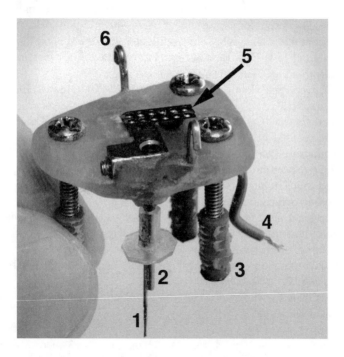

FIGURE 11.3 Microelectrode and microdialysis probe-guide assembly for rats. 1: microelectrode array; 2: microdialysis probe guide; 3: driving screw; 4: grounding wire; 5: Mill-Max socket; 6: stainless steel hook. Note the plastic ring around the electrode and probe-guide unit: it is pulled down during surgery to seal the craniotomy.

11.3.2 ELECTRODE AND MICRODIALYSIS PROBE ASSEMBLY

At present, this assembly is not available commercially. Therefore, investigators should fabricate it in their laboratory. Assembly samples and a detailed construction guide are available from the author. Figure 11.3 shows the design my colleagues and I use. The components of this device are listed below:

- Mill-Max sockets (distributed by Mouser Electronics, Randolph, NJ). We use 14-pin or 12-pin dual row profile sockets; for the latter the Mouser Electronics stock # is 575-853-93-012. Other types of sockets can also be used.
- Driving screws embedded in a nylon block. These are custom-made parts, designed by the author and manufactured by Small Parts (Miami Lakes, FL). Alternatively, similar driving screws can be constructed according to the description of Kubie (1984).
- Stainless steel hypodermic tube; part # R-HTX-26 (Small Parts). This is the tube for the microwire electrodes.
- Stainless steel microdialysis probe guide with fitting probe holders (Can-Do Services, Hopewell Junction, NY; contact: Laszlo Kando). The design for rat studies allows the removal and reinsertion of the microdialysis probe. The design for mouse studies allows the insertion of the probe only once, but this device is smaller and lighter than the version for rats.
- Stablohm 675 H-Formvar-coated nichrome wires, 0.001 inch in diameter (California Fine Wire, Grover Beach CA).
- GC Silver Print Conductive Paint, part # 22-201 (Newark Electronics, Newark, NJ).
- Dental cement (Cat # 51459, Stoelting, Wood Dale, IL). Other types of dental cements can also be used.
- Spectra and Por RC Hollow Fiber Bundles, MWCO: 18,000; part # 132226 (Spectrum Houston, TX).
- Fused silica tubing, part # TSP075150, TSP200300 and TSP320450 (Polymicro Technologies, Phoenix, AZ).
- There is no need to exactly copy this design, and the investigators can modify it as they wish. However, the following features are critical:
- The use of microwires, as they yield stable recordings
- The incorporation of three driving screws, as they provide stability for the assembly
- A less than 1 mm distance between the electrode-tube and the microdialysis probe guide.

The assembly is constructed as follows: A bundle of 11–13 microwires are placed in the hypodermic tube. It is glued to the microdialysis probe guide. This unit is glued to the Mill-Max socket, with the wires connected to the pins of the socket with the conductive paint. The driving screws are glued to this assembly, and the entire apparatus is embedded in dental cement. To one of the pins of the Mill-Max socket, a larger grounding wire is connected. As shown in Figure 11.3, two stainless steel hooks are embedded in the dental cement. These are used during the experiments to secure the device to the recording cable. The length of the microelectrode tube varies according to the depth coordinate of the examined brain site.

We fabricate the microdialysis probes in our laboratory, allowing strict control over each fabrication step. Our 200-μm-diameter and 800-μm-long concentric probes are modified versions of the design of Sved and Curtis (1993). These modifications are described in a previous publication (Ludvig et al., 2000). Commercially available

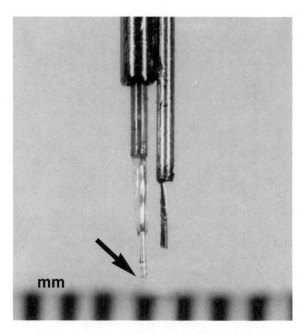

FIGURE 11.4 Microelectrode array and the microdialysis probe, as adjusted under a dissection microscope prior to implantation. Arrow points to the tip of the concentric probe. After adjustment, the probe is removed from the probe-guide and replaced with a stylet. The microdialysis probe is reinserted into the probe guide at the beginning of the recording and dialysis session.

microdialysis probes can also be used. Before each experiment, we place the microdialysis probe into the probe-holder and adjust the distance between the bottom of this holder and the tip of the probe. This critical step should be performed under a dissection microscope. As Figure 11.4 shows, the optimal horizontal distance between the microelectrode array and the microdialysis probe is 400–600 μm, and the tip of the probe should also be about 400–600 μm below the tip of the array. This is important to ensure drug diffusion to both the soma and the dendrites of the recorded cells during the experiments.

The assembly is essentially the same for rats and monkeys. For mice, smaller driving screws and microdialysis probe guides can be used, with sockets containing only 8–10 pins; hooks are not necessary.

11.3.3 ELECTRODE AND MICRODIALYSIS PROBE IMPLANTATION

This procedure is usually preceded by behavioral training. For example, if place cells are recorded in rats, the animals are trained for 1–2 weeks to chase food-pellets scattered on the floor of the test chamber (Ludvig, 1999). The monkeys are either trained for 2–4 weeks to sit in a primate chair, or they are trained to retrieve food pellets from food ports placed on the walls of the test chamber (Ludvig et al., 2001b).

The implantation technique for rats and mice is essentially similar. Here, we describe our protocol for rats. The animal is anesthetized with 60 mg/kg pentobarbital, i.p., and placed in the stereotaxic apparatus. The skull is exposed, and four anchoring screws are placed in the frontal and occipital bones. A 2-mm-diameter craniotomy is drilled above the targeted brain site, according to the atlas of Paxinos and Watson (1998). The dura and pia mater are excised, and the microelectrodes are introduced into the brain so that the tip of the electrode array is about 0.5 mm above the targeted area. The tip of the microdialysis probe guide should be below the brain surface but 1–3 mm above the tip of the electrode array. This arrangement causes minimal tissue damage and assures the safe insertion of the microdialysis probe during the experiments. The lumen of the probe guide is protected with a removable stylet. After the assembly is positioned, it is anchored to the skull with dental cement applied around the anchoring screws and the cuffs of the driving screws. One of the anchoring screws also serves to ground the rat: the grounding wire is connected to this screw with conductive paint. A plastic ring with sterile bone wax and Panalog ointment on its surface seals the craniotomy. Finally, the assembly is covered with a protective tape, and the skin is approximated. The animals are ready for the recording and dialysis sessions within 4–6 d. Figure 11.5 shows a rat wearing the implanted electrode and microdialysis probe assembly before an experiment.

For squirrel monkeys, the surgical tools and the stereotaxic apparatus are autoclaved. After the injection of Atropine (0.05 mg/kg, i.m.) and Bicillin (100,000 units /kg, i.m.), anesthesia is induced with a mixture of Ketamine (11 mg/kg, i.m.) and Xylazine (0.5 mg/kg, i.m.). The animal is weighed, the head and extremities are shaved, and the non-invasive blood pressure, SpO_2 and ECG monitors are placed. The animal is preoxygenated with 100% O_2 through a facemask, and the anesthesia is deepened with 2.5% Halothane for 5 min. Laryngoscopy is performed, and via endotracheal intubation the anesthesia is maintained with 1.2%–2% Isoflurane in oxygen using a Narkomed Compact system. The tail vein is cannulated for fluid (lactated Ringer) administration. A Propaq 106 can be served to monitor blood pressure, ECG, SpO_2, respiratory rate, end tidal CO_2, as well as rectal temperature. After positioning the head, the skull is exposed, and a 3-mm-diameter craniotomy is made with a dental drill. The coordinates of the center of craniotomy depend on the targeted brain structure. An excellent stereotaxic atlas of squirrel monkey brain is available (Gergen and MacLean, 1962). Two anchoring screws are placed in the bone around the craniotomy. The dura mater and the pia mater are excised from the brain surface, and the microelectrode unit of the electrode and probe assembly is introduced into the brain. The grounding wire of the assembly is connected to one of the screws with conductive paint. The assembly is secured to the skull and the screws with dental cement, and the craniotomy is sealed with a plastic ring, as in rats. Next, the base of a protective cap is positioned around the assembly and secured to the skull with interior and exterior screws and dental cement (Ludvig et al., 2000). The removable cover of the protective cap is then attached to the base. The skin is approximated, the anesthesia is discontinued, and the animal is extubated before being returned to the home cage. The neurological status of the animal is monitored and documented twice a day for a full week, before the recording and dialysis sessions start.

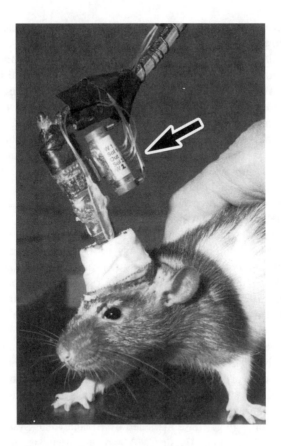

FIGURE 11.5 Experimental rat with chronically implanted electrode and microdialysis probe assembly. The microelectrodes are connected to the recording cable through operational amplifiers to eliminate movement artifacts. The microdialysis probe, already inserted into the brain, is connected to the microdialysis tubing. Note the minivalve (arrow), secured to the operational amplifiers. This novel, remote-controlled device allows switching between artificial cerebrospinal fluid (ACSF) and the drug solution on the head of the animal, close to the recorded and dialyzed brain site. As a result, drugs can be delivered to the neurons rapidly, within 1–2 min.

11.3.4 Recording and Microdialysis Setup

The apparatus for combined single-cell recording and microdialysis studies can be built with either traditional or advanced devices. A traditional setup comprises the following pieces of equipment:

- Recording and microdialysis cable, equipped with impedance-lowering operational amplifiers. This can be obtained from G-tech (Cortland Manor, NY) or Brooklyn Biomedical Engineering Services (Brooklyn, NY). The cable provides connection between the intracerebrally implanted electrode and microdialysis probe assembly and the commutator and swivel unit (see below).

- Electrical commutator coupled with a liquid swivel. Such a unit can be purchased from Dragonfly R & D, (Ridgeley, WV). Alternatively, the liquid swivel of Bioanalytical Systems (West Lafayette, IN) can be integrated into the commutator of Crist Instruments (Hagerstown, MD). The commutator should have at least 10 channels to transmit the electrical signals from the microelectrodes and to connect the operational amplifiers to their power supply. The liquid swivel should have two channels to allow the flow of microdialysis fluid to and from the implanted microdialysis probe.
- A patch panel, made in the investigator's laboratory. It is this panel that distributes the electrical signals from the commutator to the amplifiers, oscilloscopes and the computer interface.
- Differential AC amplifiers (e.g., from A-M Systems, Everett, WA). These filter and amplify the electrical signals transmitted through the commutator.
- Oscilloscopes and computer to acquire, display, store and analyze the collected extracellular signals. My co-workers and I use the data acquisition and analysis hardware and software by DataWave Technologies (Longmont, CO). This software runs on DOS computers.
- A four-way valve (Valco Instruments, Houston, TX), connected to the liquid swivel. This valve, placed on the ceiling of the test chamber, directs either the control or the drug solution to the implanted microdialysis probe (via the liquid swivel). In our apparatus, it usually takes 20–30 min for the drugs to travel from the valve to the brain.
- Syringe pumps (e.g., those marketed by Bioanalytical Systems; Harvard Apparatus, Holliston, MA; or CMA and Microdialysis, Solna, Sweden) drive the control and drug solutions to the valve.

These basic components can be complemented with an automatic fraction collector to collect the dialysates for subsequent neurochemical analyses. The setup may also contain a video system to monitor the behavior of the animal during the recording and dialysis sessions. The animal moves in an electrically shielded test chamber, with the commutator and swivel unit, the valve, and the camera placed on the ceiling. We have used this type of setup in many experiments (Ludvig et al., 1994, 1995, 1996, 1998; Ludvig and Tang, 2000).

The advanced setup contains the same type of recording and microdialysis cable, amplifiers, oscilloscopes, computer software and hardware, and syringe pumps. However, the control and drug solutions are alternated with a remotely controlled, lightweight (1.4 g) minivalve (U.S. patent pending; application # 09/579,199). This minivalve is secured to the operational amplifiers of the cable (Figure 11.5). Thus, in contrast to the large, heavy traditional valves placed on the ceiling of the test chamber, the minivalve is placed directly on the head of the animal. As a consequence, the drugs reach the brain within 1–2 min, allowing the detection of drug-induced neuronal firing changes almost instantly. In turn, this enables the investigator to intracerebrally perfuse drugs and examine their cellular effects during transient behavioral states (e.g., drinking or grooming). With the use of the minivalve, the onset of drug diffusion into the tissue can be clearly defined. Finally, the minivalve makes it possible to examine the effects of many consecutively perfused drugs without unnecessarily lengthening the experiments. The minivalve is connected, via a special cable, to a commutator and swivel unit rotated by a servo system. These can be obtained as one package, complemented with a patch panel ("Signal Distributor") from G-Tech. The minivalve is regulated by a microprocessor-based controller (G-Tech).

11.3.5 Conduct of a Recording and Microdialysis Session

When using rats, the session starts with transferring the animal to the test chamber and wrapping it in a soft cloth. This simple procedure makes the use of sedatives unnecessary for probe insertion and cable connection. While the animal is restrained, the microdialysis probe is perfused with artificial cerebrospinal fluid [ACSF; pH = 7.4; total osmolarity = 311.2 mOsm; composition (in mM) 150 Na, 155 Cl, 1.4 Ca, 3.0 K, 0.8 Mg, 1.0 P] filtered through Nylon Acrodisc syringe filters. Flow rates ranging from 1.5 µl/min to 10 µl/min are used. For 5 min, the flow of ACSF in the probe is observed. If no leakage occurs and the fluid flow is constant, the microdialysis probe is inserted through the probe guide into the brain. This is followed by plugging the recording cable into the electrode connector. The cable must be equipped with impedance-lowering devices (e.g., operational amplifiers) to eliminate movement artifacts from the recordings. After connecting the rat to the experimental setup, the animal is released into the test chamber. For about 12–18 h, the brain tissue is allowed to recover from the microtrauma caused by probe insertion (Ludvig et al., 1996; Thakkar et al, 1998; Alam et al., 1999). After this period, the electrode /probe assembly is advanced to detect action potentials in the examined brain site ("searching for cells"). To recognize neuronal action potentials, the extracellular signals are amplified (10,000×), filtered (300–10,000 Hz), displayed on oscilloscopes, and digitized at a sampling rate of 40,000 Hz. The digitized signals are acquired and displayed with a computer. We use the Discovery Data Acquisition program of DataWave Technologies for this purpose. When action potentials with amplitudes 2–3× higher than the background noise are consistently recorded over at least 30 min, data collection can start. The importance of this observation period cannot be overemphasized, for studying drug effects on unstable or low signal-to-noise ratio action potentials cannot produce meaningful information.

The data collection period consists of a minimum of three phases. In the first, the control data are stored on the hard disk. During this 10–30-min phase, control solution (e.g., ACSF) is perfused in the microdialysis probe. In the second phase, the control solution is replaced with a drug solution in the microdialysis probe. This is accomplished with the use of a valve. The duration of drug reverse-dialysis usually lasts 5–45 min (Ludvig et al., 1994; Thakkar et al., 1998; Alam et al., 1999). In the third phase, the valve is switched back to its original position to again direct the control solution to the brain. This washout and recovery period should last until the drug-exposed cells regain, at least partially, their predrug firing pattern. The described design can be extended to involve multiple drug deliveries separated by sufficiently long washout and recovery periods. For example, my colleagues and I examined the effects of two consecutive ethanol perfusions followed by a single N-Methyl-D-Aspartate (NMDA) perfusion in the hippocampus (Ludvig et al., 2001a). This study uncovered the development of acute cellular tolerance to ethanol.

The rat can stay in the test chamber with continuous microdialysis and recording for several days, with food and water provided. Multiple drug reverse-dialysis studies can be run for as long as 4 d in the same animal. If ethanol fails to change the firing of local neurons even at sufficiently high concentrations, it can be due to technical problems. For example, the neurons may be situated too far from the microdialysis probe and are not reached by the ethanol molecules. It is also possible that tissue reactions around the microdialysis probe decreased the movement of ethanol molecules through the interstitial space. These can be tested by subsequently perfusing a "reference" drug (e.g., NMDA) with reliable and known effects on the examined cells. Inactivity of this reference compound suggests technical problems. However, if the reference drug does induce firing pattern changes, it suggests diminished sensitivity of the cells to ethanol.

During the recording and dialysis procedure, the rat behaves freely in the test chamber, explores the environment, eats, drinks, rests, sleeps, and manifests various other behaviors (e.g., grooming). This allows the studying of cellular ethanol effects in the intact, physiologically functioning brain. However, it should be recognized that most neurons spontaneously change their firing pattern as the behavior changes. Therefore, it is absolutely necessary to collect the control and drug-exposure data during constant behavior. This is why it is important to train the rats to perform a behavioral task, such as chasing food pellets on the floor of the test chamber. This keeps them moving continuously for long periods, so the data can be collected during fairly similar behaviors. To avoid exhaustion, resting intervals should separate data collections. Because of the profound influence of behavior on the firing of many cell types in brain, videotaping the activity of the animals in the test chamber is important. We use a videotracker system from Ebtronics (Elmont, NY) to follow the head positions of the animals during the recording and dialysis sessions (Ludvig, 1999). This system includes a light-emitting diode (LED) secured to the operational amplifiers, a video camera, and a tracker that generates X and Y coordinates for head positions at 60 Hz. Furthermore, the Discovery Data Acquisition Program allows manual entering of "flags" into the data file to indicate various behaviors. This provides an additional tool to document behavioral changes during data collections.

Recording electroencephalogram (EEG) activity may be part of the experiments. Such recordings are recommended, as they can help to distinguish between slow-wave and paradox sleep states, and can help to recognize abnormal electrophysiological events, such as electrographic seizures or spreading depression (Ludvig and Tang, 2000). The dialysates can be collected for neurochemical analysis. Optimally, determination of various neurotransmitters and other extracellular compounds should be part of the recording and dialysis experiments. Such measurements can provide information on the presynaptic effects of the reverse-dialyzed drugs. Importantly, even if the investigator does not have the means to analyze the dialysates, the flow of microdialysis fluid in the tubing system must be monitored. This is achieved by regularly measuring the flow rate of the fluid that leaves the brain. If this flow rate is significantly lower than the preadjusted flow rate of the perfusate, it indicates either rupture or clogging of the microdialysis probe. In this case, the experiment should be terminated, as it will yield false results. After completing the data collection, the rats are sacrificed for histological studies. These studies are necessary to identify the localization of the tip of the microelectrode array and that of the microdialysis probe (Ludvig et al., 1996).

For studies in mice, the recording and dialysis sessions can be run in the same manner. For studies in monkeys, the investigator has two alternatives. Namely, the monkeys are either placed in a traditional primate chair for the duration of the recording and dialysis sessions (Ludvig et al., 2000), or move freely in three dimensions in a large test chamber (Ludvig et al., 2001b). If the monkeys are seated in a chair, the experiments are simpler, because the commutator and swivel unit is not needed. The recording cable can be directly connected to the patch-panel, and the tubing of the microdialysis probe can be directly connected to the syringe pumps. Furthermore, New World monkeys (Superfamily Ceboidea) and Old World monkeys (Superfamily Cercopithecoidea) both can be studied with this arrangement. However, long-term (> 5 h) experimental sessions of chaired-in monkeys are difficult to conduct, neuronal firing can be examined only during a limited set of behaviors, and restraint can stress the animals. These problems are eliminated in recording and microdialysis studies in freely moving monkeys (Ludvig et al., 2001b). However, such studies require a special test chamber. We have tested this new methodology only in squirrel monkeys.

It should be pointed out that, after completing the recording and microdialysis sessions, the microdialysis probe can be removed, and the effects of systemic drug administrations on the firing of the cells can be examined. This allows, for example, comparing the effects of local and systemic ethanol administrations on neuronal firing in brain.

11.3.6 DATA ANALYSIS

While the results of every experiment should be documented, data from experiments that yielded unstable action potentials and inconsistent microdialysis fluid flow should be excluded from data analysis. Extracellular recordings derived from flawless experiments can be analyzed as described in a previous paper (Ludvig, 1999). The raw data, usually stored in 5–120 min files, are subjected to the critical step of action potential discrimination. For this purpose, we use the CP Analysis Program of DataWave Technologies. Using the "cluster-cutting" method, action potentials with similar waveform parameters are discriminated. The discriminated action potentials are overlaid and their waveforms are examined. The interspike intervals of the discriminated waveforms are also determined. Action potentials that have similar waveform parameters and shape, and follow each other with interspike intervals longer than the refractory period (>2.5 ms in the case of hippocampal pyramidal cells), are very likely generated by the same single neuron.

Using the discriminated action potentials, several analyses can be made. The rate of action potentials can be plotted as a function of time. These firing-rate histograms provide the basic information on the average and maximum frequency of the discharges of the neurons before, during and after drug delivery. To obtain a finer picture of cellular firing pattern, the stream of action potentials can be played back and examined segment by segment. For example, examination of 1-min segments of place cell recordings reveals bursts of action potentials that last for 2–10 sec and are initiated by the highest-amplitude spikes (Ludvig, 1999). The spatial properties of the firing of the neurons can also be analyzed. This is especially important when hippocampal complex-spike (pyramidal) cells are studied, because these neurons characteristically display location-specific discharges. The spatial properties of these neurons can be analyzed with the MapMaker software of ESCO (Mt. Kisco, NY), as described (Ludvig, 1999). To use this program, the extracellular electrical signals and the head-position data should be simultaneously collected. The program divides the area of the test chamber into pixels, determines the average firing rate of a discriminated neuron in the area of each pixel, and generates a color-coded map to illustrate the spatial distribution of the firing rates.

In the experiments, each neuron is exposed to a minimum of three treatments (pre-drug ACSF, drug, and post-drug ACSF), and the average firing rates during these treatments are measured. Thus, these studies follow the "subjects as their own control" design, with the average firing rates being the dependent variables. The appropriate statistical analysis for comparing the means of these data is repeated measures ANOVA. When the recording and dialysis sessions yield data from only one cell type, there is one within-subjects factor with a minimum of three levels. When the recording and dialysis sessions yield data from several cell types, repeated measures ANOVA with between-subject factors can be used. For *post-hoc* analysis, Tukey or Newman-Keuls tests are appropriate. Multifactorial repeated measures ANOVA can be used for analyzing firing rate data in cellular ethanol tolerance studies. In these studies, the neurons are exposed to two consecutive ethanol perfusions and firing rate data are collected before and after each drug exposure (Ludvig et al., 2001). The main effects of ethanol exposure and ethanol application time, and the interaction of these two factors, are calculated.

The experimental sessions yield not only cellular electrophysiological data, but also EEG and behavioral data. These should also be examined. For example, we showed the EEG seizure-inducing effect of intrahippocampal NMDA perfusions (Ludvig and Tang, 2000) and demonstrated the lack of effect of intrahippocampal ethanol perfusions on the movement pattern of rats (Ludvig et al., 1998).

11.4 EXPERIMENTAL RESULTS

11.4.1 DIVERSE NEURONAL ACTIONS OF ETHANOL IN THE HIPPOCAMPUS OF RAT AND MONKEY

Intrahippocampal microdialysis with 0.1 M and 0.5 M ethanol was found to be ineffective to alter the firing of hippocampal neurons in freely moving rats (Ludvig et al., 1995).

Increasing the ethanol concentration in the microdialysis fluid to 1 M can produce clear firing pattern changes (Ludvig et al., 1998, 2001). Statistical analysis of the average firing rates of CA1/CA3 complex-spike (pyramidal) cells (n = 12) before, during and after a 30-min perfusion of 1 M ethanol yielded 3.4 ± 0.9 Hz, 1.4 ± 0.5 and 2.9 ± 0.8 Hz (mean ± S.E.M.) values, respectively.

Repeated measures ANOVA and Newman-Keuls multiple comparison test revealed that during ethanol perfusion the neurons fired at a significantly lower rate ($p < 0.01$) than before the drug exposure. On the other hand, the firing rates before and after ethanol exposure did not differ significantly ($p > 0.05$), indicating that the neurons regained their pre-ethanol firing pattern after the drug had been washed out of their extracellular space. My co-workers and I also studied the effects of ethanol reverse-dialysis on the firing of CA1/CA3 interneurons (theta cells). Again, ethanol predominantly caused a firing rate decrease. Although this firing rate suppressant action was the predominant effect of ethanol on both pyramidal cells and interneurons (Figure 11.6), not all neurons responded to ethanol in the same manner. While in most neurons only firing rate suppression was observed (Figure 11.6A and C), in some, this effect was preceded by a brief firing rate increase (Figure 11.6B and D).

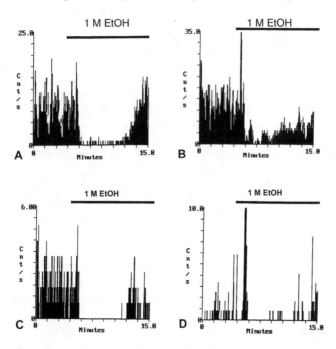

FIGURE 11.6 The effects of ethanol, reverse-dialyzed into the hippocampus proper of freely behaving rats, on the firing of interneurons (Panel A and B) and pyramidal cells (Panels C and D) in the microdialysis area. Firing rate histograms are shown; Y axis: firing rate (spike counts per sec), X axis: data collection time (15 min for each case). The data for each histogram were collected in different animals. Note that the baseline firing rate of the interneurons is much higher than that of the pyramidal cells. As Panels A and C show, delivering ethanol into the microenvironment of the cells via microdialysis causes a rapidly developing and reversible firing rate suppression. However, in other cells (Panels B and D) such firing-rate suppression is preceded by a brief, initial firing rate increase. Note the recovery of the pre-ethanol firing pattern while the cells are still exposed to the drug.

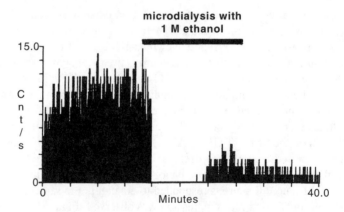

FIGURE 11.7 The effect of ethanol, reverse-dialyzed into the hippocampus proper of a chaired but awake squirrel monkey. No head restraint was used during the recording and dialysis procedure. Firing rate histogram is shown, as in Figure 11.6. Note the constant basal firing rate of the neuron before drug exposure, and the abrupt suppressant effect of ethanol on this firing pattern.As in rats (Figure 11.6), the neuron starts to regain its pre-ethanol electrical activity while the drug is still in its extracellular environment. However, during the time frame of data collection, the recovery of this cell's firing pattern was incomplete. (Reprinted from Ludvig N et al.; Delivering drugs, via microdialysis, into the environment of extracellularly recorded hippocampal neurons in behaving primates. *Brain Research Protocols*, 2000, vol. 5, pp. 75-84. With permission from Elsevier Science.)

Furthermore, neurons that did not respond to ethanol and cells that responded with firing rate increase were also detected (Ludvig et al., 2001a). Thus, as in other areas of the rat brain (Wayner et al., 1975), ethanol can exert diverse cellular electrophysiological effects in the rat hippocampus. We extended our ethanol-reverse dialysis studies to the primate hippocampus (Ludvig et al., 2000). Intrahippocampal microdialysis with 1 M ethanol in chaired but awake squirrel monkeys resulted in robust firing-rate suppression in the local neurons (Figure 11.7).

These studies provided the following new information:

- Ethanol can suppress the firing of hippocampal neurons via direct action on the neurons themselves. This conclusion could not be drawn from previous studies (Grupp and Perlansky, 1979) where ethanol was administered systemically, affecting all brain areas that project to the hippocampus.
- Ethanol can suppress the firing of both pyramidal cells and inhibitory interneurons in the hippocampus. Thus, the firing rate suppressant action of ethanol on pyramidal cells is not due to a simultaneous increase in the firing of inhibitory local circuit neurons.
- Ethanol can exert its predominant firing suppressant effect with or without a preceding brief firing-rate increase. This further demonstrates the different responsiveness of hippocampal neurons to ethanol exposure.
- Ethanol can also suppress the firing of hippocampal neurons in primates.

This raises the possibility that ethanol-induced hippocampal firing-rate suppression may be a contributing factor in the mediation of the memory-impairing effect of alcohol in humans.

11.4.2 EFFECTS OF ETHANOL ON PLACE CELLS

During free movement in a relatively large space, hippocampal pyramidal cells in rats can produce a unique firing pattern. Namely, they can markedly increase their firing rate at specific spatial locations. These locations are called "firing fields" and the cells that produce these discharges are called "place cells" (O'Keefe and Nadel, 1978; Ludvig, 1999). It is thought that these neurons play a central role in forming an internal representation of space, called the "cognitive map." Because acute alcohol intoxication impairs spatial information processing and memory formation, it was recently hypothesized that place cells might mediate this specific ethanol effect. Two groups set out to test this hypothesis in rats, using different experimental paradigms. Matthews et al. (1996) injected 2 g/kg ethanol intraperitoneally, and tested the effect of this treatment on the firing of place cells. This paradigm has the advantage that the whole body is exposed to ethanol, similar to what occurs in life during alcohol intake. However, the disadvantage of this approach is that the administration of 2 g/kg ethanol in rats induces locomotor depression, making it difficult to study the spatial properties of the firing of place cells. Our group administered 1 M ethanol via microdialysis into the hippocampus and tested the effects of this treatment on the electrical activity of local place cells (Ludvig et al., 1998). The advantage of this approach is that it does not affect locomotion, therefore, the spatial properties of the firing of place cells can be readily studied. The disadvantage of this approach is that it creates an artificial condition where ethanol is present only in the microenvironment of a limited number of cells. Nevertheless, both groups independently found that, while ethanol neither terminates the firing of place cells nor alters the coordinates of firing fields, the drug markedly decreases the location-specific firing of these neurons. Figure 11.8 represents the results of one of our experiments. The significance of these studies is that they revealed the deteriorating effect of ethanol on the function of place cells. We propose that one of the cellular mechanisms, through which alcohol intoxication impairs spatial memory, is the suppression of the location-specific firing of hippocampal place cells.

11.4.3 EVIDENCE FOR ALCOHOL TOLERANCE AT THE CELLULAR LEVEL IN THE HIPPOCAMPUS

While examining the effects of localized ethanol perfusions on hippocampal neurons, my colleagues and I observed that once the drug induced a clear firing pattern change it was often difficult to repeat this action within the next hours. Therefore, we set out to systematically study this phenomenon in freely moving rats. In our experimental design, intrahippocampal microdialysis with 1 M ethanol was performed for 30 min, and this procedure was repeated after a 1-h washout period. The recording and dialysis session was completed with a subsequent NMDA perfusion (500 μM for 10–20 min) to test the functional integrity of the microdialysis probe. Previously,

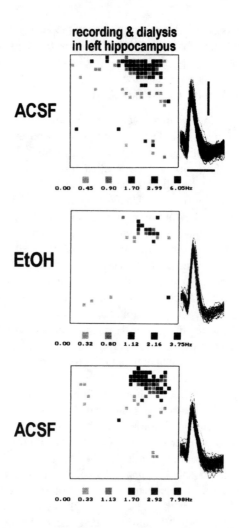

FIGURE 11.8 The effect of ethanol, reverse-dialyzed in the hippocampus, on the location-specific firing of a place cell in the microdialysis area. Firing rate distribution maps are shown, generated with the MapMaker software. The maps indicate the average firing rate of the place cell within the areas of pixels during 15-min data collection periods, as the rat moved around chasing food pellets on the floor of a rectangular test chamber. Black pixels represent areas where the cell fired with the highest firing rates; yellow pixels represent areas where the cell did not fire at all, and white areas represent areas not visited by the animal. During microdialysis with ACSF (upper panel), the cell displayed a clear location-specific firing (6.05 Hz) in the northern part of the chamber. When ethanol (1 M for 30 min) was perfused (middle panel), the location-specific firing rate of the cell dropped to 3.75 Hz. After washing out ethanol from the recording site, full recovery of the original location-specific firing (7.90 Hz) was observed (lower panel). The similarity of the corresponding, overlaid action potential waveforms on the right of each panel indicate that the data were collected from the same place cell throughout the recording and dialysis session. Calibrations: horizontal bar: 0.5 msec; vertical bar: 0.1 mV. (Figure also shown in color insert following page 202.)

we had characterized the cellular electrophysiological effects of such NMDA per-fusions (Ludvig and Tang, 2000). For statistical analysis, the average firing rates of eight stable ethanol-responding and subsequently recovering neurons within the dialysis area before and during the first and second ethanol perfusions were used. Before the first ethanol perfusion, the control firing rate was 2.66 ± 1.05 Hz (mean $\pm$ S.E.M.). This significantly decreased to 0.40 ± 0.20 Hz during the first ethanol exposure ($p < 0.05$; repeated measures ANOVA and Newman-Keuls test). Further-more, the neurons started to regain their pre-ethanol firing rate while the drug was still present in their microenvironment. In contrast, when the ethanol perfusion was repeated, the firing rates of the same cells before (1.65 ± 0.86 Hz) and during (1.93 ± 0.60 Hz) the ethanol exposure were not different ($p > 0.05$; repeated measures ANOVA and Newman-Keuls test). In an additional statistical analysis, factorial repeated measures ANOVA revealed significant interaction between ethanol expo-sure and ethanol application time ($F_{1,7} = 10,352$; $p = 0.015$). This suggested that the firing suppressant effect of ethanol depended on the time of application. Subsequent NMDA perfusions still caused robust firing pattern changes, indicating that the lack of effect of the second ethanol perfusion was not due to decreased microdialysis probe permeability or gross cellular dysfunction. These experiments were conducted with a traditional recording and microdialysis setup. When ethanol was delivered repeatedly into the hippocampus with the use of the minivalve, a similar phenomenon was observed. Namely, the second ethanol perfusion was ineffective (Figure 11.9). Those neurons that did not respond to the first ethanol perfusion did not respond to the second either.

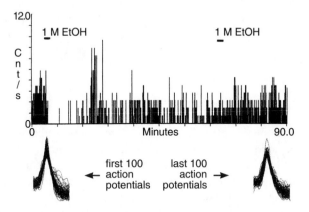

FIGURE 11.9 Phenomenon of acute cellular tolerance to ethanol in the hippocampus of a freely moving rat, as revealed with the use of the minivalve shown in Figure 11.5. The firing rate histogram demonstrates the effects of two consecutive 2-min, intrahippocampal microdialyses with 1 M ethanol on the electrical activity of a local pyramidal cell. The first ethanol perfusion suppressed the firing of the cell (asterisk). The cellular ethanol action developed instantly after the activation of the minivalve, because the device was placed on the head of the animal allowing rapid ethanol delivery into the recording site. Repetition of the same ethanol delivery 1 h later caused no firing-rate alteration, indicating the development of tolerance.

The significance of these studies is that they provided evidence for the ability of hippocampal neurons to develop acute tolerance to a single exposure of clinically relevant concentration of ethanol, *in vivo* (Ludvig et al., 2001a). Such cellular tolerance to ethanol was previously shown in the cerebellum of anesthetized rats (Palmer et al., 1992) and in the medial septum during systemic ethanol administration (Givens and Breese, 1990). Our recording and microdialysis experiments demonstrated that this phenomenon occurs in awake and behaving animals and that it develops in brain without influences from the rest of the body.

The role of tolerance in the pathogenesis of alcoholism has long been suspected (Tabakoff and Hoffman, 1988). Indeed, clinical studies showed that subjects with a positive family history of alcoholism express acute alcohol tolerance at a greater level than those with negative family history (Martin and Moss, 1993). Demonstration of the existence of acute neuronal tolerance to ethanol in brain may shed new light on the relationship between tolerance and alcoholism. As we proposed, the diminished responsiveness of a tolerant neuron to alcohol may well be merely one manifestation of a diverse and long-lasting plastic change in its molecular machinery (Ludvig et al., 2001a). While this may not affect the basal firing rate of the cell, the responsiveness of the neuron to specific inputs can certainly change. In turn, this can slightly but significantly modulate the signaling of the ethanol-tolerant cell to its target neurons. This altered signaling can be temporary if no repeated alcohol exposure occurs within a critical time window. However, it is possible that, after repetitive alcohol exposures, these signaling modifications become permanent, leading to severe intercellular communication errors. Consistent with our general theory on alcoholism (Ludvig et al., 1998), the gradual establishment of such an abnormally communicating, alcohol-specific, "domineering hypernetwork" across the motivational, emotional and cognitive systems can ultimately produce the behavioral symptoms of alcoholism.

11.5 FUTURE DIRECTIONS

As the combined single-cell recording–intracerebral microdialysis method integrates electrophysiological, neuropharmacological, and neurochemical techniques and can be applied to both freely behaving rodents and monkeys, it offers a virtually unlimited number of novel experimental designs and studies. At the same time, the advantage of this technique — its integrative nature — makes it more difficult to use than other, less complex methods. Therefore, it seems to be the most prudent way, at least for the near future, to apply combined single-cell recording and microdialysis to only a limited number of carefully selected projects. These projects must address fundamentally important problems in alcohol research, which, at the same time, can appropriately be studied with recording and dialysis experiments. Here I would like to point out only three of such potentially important projects, while readily admitting that the reader may have better ideas. Indeed, stimulating new ideas with this chapter is as much my goal as to provide help for those who would like to utilize integrative neurobiological techniques for their alcohol-related research.

First, recording and microdialysis studies can be useful in mapping the electrophysiological and neuropharmacological characteristics of neurons in the brain of genetically modified mice relevant to alcohol research. For example, transgenic mice that lack neuropeptide Y as a result of targeted gene disruption show increased alcohol consumption compared with wild-type mice (Thiele et al., 1998). Many other transgenic mouse models, as well as various inbred mouse strains, possess molecular abnormalities leading, directly or indirectly, to altered alcohol sensitivity or alcohol consumption. However, it is unknown how these molecular abnormalities actually alter the firing pattern, ethanol responsiveness and pharmacological properties of neurons within the motivational, emotional and cognitive systems. The recording and microdialysis method can extract precisely this information, bridging the gap between genetics and behavior in these precious mouse models.

Second, our described method can determine whether the neurons of the amygdala and nucleus accumbens respond with different firing patterns to ethanol in alcohol-preferring and non-preferring rats. Furthermore, this method can elaborate the molecular and neurochemical mechanisms that mediate ethanol actions in these critical brain areas. It is known that microinjection of various drugs into these nuclei can profoundly influence ethanol intake (Koob et al., 1998). However, the underlying neuronal firing pattern alterations have not been identified. Furthermore, much is known about the plastic molecular changes that take place in the nucleus accumbens after chronic ethanol consumption (Fitzgerald and Nestler, 1995). But it is completely unclear which of these molecular changes are actually responsible for the firing pattern alterations that ultimately affect behavior and alcohol consumption. These problems can be adequately addressed with the combined single-cell recording and microdialysis method, especially if it is conducted parallel on alcohol-preferring and non-preferring rats.

Third, the adaptation of this method to freely moving monkeys (Ludvig et al., 2001b) can open up new possibilities in studying the cellular and molecular mechanisms of craving for alcohol. In my opinion, it is important to examine this problem at the appropriate phylogenetic level, in anthropoid non-human primates. The cytoarchitecture and cellular electrophysiology of the emotional, cognitive and motivational systems are different in primates and rodents. Therefore, it is reasonable to assume that the cellular and molecular mechanisms that mediate alcohol effects in these systems are different, as well. Controlled cellular and molecular studies in the human brain cannot be done. But such studies, including simultaneous recording and microdialysis experiments during free movement, can be performed in non-human primates, at least in squirrel monkeys (*Saimiri sciureus*). Unfortunately, attempts to induce excessive ethanol drinking in squirrel monkeys, thus to study alcohol craving in this species, have been unsuccessful (Mandillo et al., 1998). This is probably because these animals have a strong taste aversion to ethanol. However, this taste aversion may not be an unbreakable barrier, and consistently associating ethanol drinking with pleasant events (e.g., sweet food, larger space, interaction with another monkey, etc.) for a long period may well translate into increased alcohol consumption even in these animals. Recording and microdialysis studies in the prefrontal cortex and amygdala and nucleus accumbens circuitry in freely moving, socially interacting and alcohol-seeking squirrel monkeys would be a very exciting addition to the existing animal models for alcoholism.

Finally, it should be stressed that the neurochemical analysis of the collected dialysates is absolutely necessary to exploit the full potential of the method and to obtain information not only on postsynaptic cellular electrophysiological events but also on presynaptic neurotransmitter release modifications. We have yet to add an HPLC or other analysis device to our experimental setup, and our future projects should incorporate such tools for analyzing the dialysates.

11.6 CONCLUSIONS

Combined single-cell recording and microdialysis in the brain of freely moving animals has been demonstrated to be a useful integrative neurobiological method in various laboratories (Ludvig et al., 1994; Dudkin et al., 1996; Thakkar et al., 1998; Alam et al., 1999). Its application to alcohol research resulted in the following new information:

- The predominant firing suppressant effect of ethanol on hippocampal neurons is largely due to its direct action on these cells, in both rodents and monkeys (Ludvig et al., 1995, 1998, 2000, 2001a).
- Ethanol decreases the frequency of the location-specific firing of hippocampal place cells. This may be one of the cellular mechanisms through which alcohol disrupts spatial memory formation (Ludvig et al., 1998).
- Hippocampal neurons are able to develop acute tolerance to a single exposure of clinically relevant concentration of ethanol, without influence from the rest of the brain and the body (Ludvig et al., 2001a).

The main advantages of the method are that it is able to:

- Extracellularly deliver ethanol, alone or in combination with other drugs, into specific brain sites in freely moving mice, rats and monkeys
- Record the ethanol- and drug-induced neuronal firing pattern changes for many hours and even days
- Collect molecules from the extracellular space of the recording and microdialysis site, allowing the researcher to distinguish presynaptic and postsynaptic ethanol and drug effects.

The main disadvantage of the method is that its establishment is time-consuming and requires expertise in cellular electrophysiology, microdialysis and behavioral procedures. Also, the delivered drugs affect each neuron in the recording and microdialysis site both directly (by acting on its cell body and dendrites) and indirectly (by acting on synaptically connected local neurons). The direct and indirect cellular effects are often difficult to separate. Furthermore, the extracellular concentration of the delivered ethanol and other drugs in the recording and microdialysis site is hard to accurately determine. However, as this chapter shows, these problems can be addressed and successfully surmounted.

Comparing it with *in vitro* methods, the combined single-cell recording and microdialysis method offers the advantage of collecting cellular and molecular information in intact neural networks, during behavior. However, *in vitro* methods using slice or tissue culture preparations have a much better control over the inputs and extracellular environment of the recorded and pharmacologically manipulated cells. Furthermore, *in vitro* techniques have the ability to examine membrane potential changes and visualize the very cells that are being recorded. Our recording and dialysis method does not have these abilities. In summary, the combined single-cell recording–intracerebral microdialysis method is an integrative neurobiological technique that allows information on how molecular processes are translated into cellular electrophysiological changes to produce behavior. But, as the method examines the complex continuously changing dynamic system of the brain of freely behaving animals, the experimental results are much more difficult to interpret than the data obtained in simpler, *in vitro* systems. Thus, paying utmost attention to every single detail of the recording and microdialysis procedure and analyzing the data with powerful statistics are essential parts of this enterprise.

ACKNOWLEDGMENTS

I am grateful to Drs. Peter M. Bungay and Steven E. Fox for reviewing the theoretical section of the manuscript. I also would like to express my gratitude to Drs. Bella T. Altura and Burton M. Altura for their advice and encouragement throughout the presented studies. This work was supported by a grant from the National Institute on Alcohol Abuse and Alcoholism (RO1 AA 10814).

REFERENCES

Alam MN et al.(1999) Adenosinergic modulation of rat basal forebrain neurons during sleep and waking: neuronal recording with microdialysis. *J Physiol* 521:679-690.

Benveniste H and Huttemeier PC (1990) Microdialysis — theory and application. *Prog Neurobiol* 35:195-215.

Bungay PM, Morrison PF and Dedrick RL (1990) Steady-state theory for quantitative microdialysis of solutes and water *in vivo* and *in vitro*. *Life Sci* 46:105-119.

Dudkin KN, Kruchinin VK and Chueva IV (1996) Neurophysiological correlates of improvements in cognitive characteristics in monkeys during modification of NMDA-ergic structures of the prefrontal cortex. *Neurosci Behav Physiol* 26:545-551.

Dykstra KH et al. (1992) Quantitative examination of tissue concentration profiles associated with microdialysis. *J Neurochem* 58:931-940.

Engel J Jr (1998) Research on the human brain in an epilepsy setting. *Epilepsy Res* 32: 1-11.

Fitzgerald LW and Nestler EJ (1995) Molecular and cellular adaptations in signal transduction pathways following ethanol exposure. *Clin Neurosci* 3:165-173.

Fox SE and Ranck JB Jr (1975) Localization and anatomical identification of theta and complex spike cells in dorsal hippocampal formation of rats. *Exp Neurol* 49:299-313.

Gergen JA and MacLean PD (1962) A Stereotaxic Atlas of the Squirrel Monkey's Brain (*Saimiri sciureus*). Bethesda: Public Health Service Publication No 933.

Givens BS and Breese GR (1990) Electrophysiological evidence that ethanol alters the function of medial septal area without affecting lateral septal function. *J Pharmacol Exp Ther* 253:95-103.

Gonzales RA et al. (1998) Quantitative microdialysis of ethanol in rat striatum. *Alcohol Clin Exp Res* 22: 858-867.

Grant KA et al. (1997) Discriminative stimulus effects of ethanol and 3α-hydroxy-5α-pregnan-20-one in relation to menstrual cycle phase in cynomolgus monkeys (*Macaca fascicularis*). *Psychopharmacology* 130:59-68.

Groothuis DR et al. (1998) Changes in blood-brain barrier permeability associated with insertion of brain cannulas and microdialysis probes. *Brain Res.* 803:218-230.

Grupp LA and Perlanski E (1979) Ethanol-induced changes in the spontaneous activity of single units in the hippocampus of the awake rat: a dose-response study. *Neuropharmacology* 18:63-70.

Hesselbrock VM (1995) The genetic epidemiology of alcoholism. In: *The Genetics of Alcoholism* (Begleiter H and Kissin B, Eds.), pp 17-39. Oxford: Oxford University Press.

Kentros C et al. (1998) Abolition of long-term stability of new hippocampal place cell maps by NMDA receptor blockade. *Science* 280:2121-2126 .

Koob GF et al. (1998) Neurocircuitry targets in ethanol reward and dependence. *Alcohol Clin Exp Res* 22:3-9.

Kubie JL (1984) A driveable bundle of microwires for collecting single-unit data from freely moving rats. *Physiol Behav* 32:115-118.

Li T-K and McBride WJ (1995) Pharmacogenetic models of alcoholism. *Clin Neurosci* 3:182-188.

Ludvig N, Fox SE and Potter PE (1994) Simultaneous single-cell recording and microdialysis within the same brain site in freely behaving rats: a novel neurobiological method. *J Neurosci Meth* 55:31-40.

Ludvig N et al. (1995) The suppressant effect of ethanol, delivered via intrahippocampal microdialysis, on the firing of local pyramidal cells in freely behaving rats. *Alcohol,* 12:417-421.

Ludvig N et al. (1996) Manipulation of pyramidal cell firing in the hippocampus of freely behaving rats by local application of K$^+$ via microdialysis. *Hippocampus* 6:97-108.

Ludvig N et al. (1998) Application of the combined single-cell recording and intracerebral microdialysis method to alcohol research in freely behaving animals. *Alcohol Clin Exp Res* 22:41-50.

Ludvig N (1999) Place cells can flexibly terminate and develop their spatial firing. a new theory for their function. *Phys Behav* 67:57-67.

Ludvig N and Tang HM (2000) Cellular electrophysiological changes in the hippocampus of freely behaving rats during local microdialysis with epileptogenic concentration of N-methyl-D-aspartate. *Brain Res Bull.* 51:233-240.

Ludvig N et al. (2000) Delivering drugs, via microdialysis, into the environment of extracellularly recorded hippocampal neurons in behaving primates. *Brain Res Protocols* 5:75-84.

Ludvig N et al. (2001a) Evidence for the ability of hippocampal neurons to develop acute tolerance to ethanol in behaving rats. *Brain Res.* 900:252-260.

Ludvig N et al. (2001b) Single-cell recording in the brain of freely moving monkeys. *J Neurosci Meth* 106:179-187.

Mandillo S, Titchen K and, Miczek KA (1998) Ethanol drinking in socially housed squirrel monkeys. *Behav Pharmacol* 9:363-367.

Martin CS and Moss HB (1993) Measurement of acute tolerance to alcohol in human subjects. *Alcohol Clin Exp Res* 17:211-216.

Matthews DB, Simson PE and Best PJ (1996) Ethanol alters spatial processing of hippocampal place cells: a mechanism for impaired navigation when intoxicated. *Alcohol Clin Exp Res* 20:404-407.

O'Keefe J and Nadel L (1978) *The Hippocampus as a Cognitive Map*. Oxford: Clarendon Press.

Palmer MR, Harlan TJ and Spuhler K (1998) Genetic covariation in low alcohol-sensitive and high alcohol-sensitive selected lines of rats: behavioral and electrophysiological sensitivities to the depressant effects of ethanol and the development of acute neuronal tolerance to ethanol *in situ* in generation eight. *J Pharmacol Exp Ther* 260:879-886.

Paxinos G and Watson C (1998) *The Rat Brain in Stereotaxic Coordinates*. 4th ed., New York: Academic Press.

Phillips MI (1973) Unit activity recording in freely moving animals: some principles and theory. In: *Brain Unit Activity During Behavior* (Phillips MI, Ed) pp 5-40. Springfield IL: Charles C. Thomas.

Rotenberg A et al. (1996) Mice expressing activated CaMKII lack low frequency LTP and do not form stable place cells in the CA1 region of the hippocampus. *Cell* 87:1351-1361.

Sved AF and, Curtis JT (1993) Amino acid neurotransmitters in nucleus tractus solitarius: an *in vivo* microdialysis study. *J Neurochem* 61:2089-2098.

Tabakoff B, Hoffman PL (1988) Tolerance and the etiology of alcoholism: a hypothesis and mechanism. *Alcohol Clin Exp Res* 12:184-186.

Thakkar MM et al. (1998) Behavioral state control through differential serotonergic inhibition in the mesopontine cholinergic nuclei: A simultaneous unit recording and microdialysis study. *J Neurosci* 18:5490-5497.

Thiele TE et al. (1975) Ethanol consumption and resistance are inversely related to neuropeptide Y levels. *Nature* 396:366-369.

Wayner MJ, Ono T and Nolley D (1975) Effects of ethyl alcohol on central neurons *Pharm Biochem Behav* 3:499-506.

12 Quantitative Microdialysis for *in vivo* Studies of Pharmacodynamics

Rueben A. Gonzales, Amanda Tang and Donita L. Robinson

CONTENTS

12.1 INTRODUCTION

Major advances in biomedical research often accompany technical innovations, and this has certainly been the case with alcohol research. Some of the more fruitful advances in alcohol research have been driven by the interdisciplinary nature of the approaches used to tackle the critical neuroscience questions in the field. Numerous examples of this synergy between seemingly unrelated approaches exist, but the driving force behind the techniques of this chapter is the convergence between neurochemical and behavioral techniques. Although incredible advances have been made in understanding the molecular and cellular mechanisms of ethanol, the study of how these molecular and cellular changes affect behavior is not as advanced. A firm understanding of the mechanisms that underlie ethanol-related behavior requires a combination of behavioral analysis along with *in vivo* measurements of cellular or chemical activity in the brain. To this end, microdialysis has been extensively used in the past 15 years to monitor neurochemical activity in the conscious, freely moving animal.

Conceptually, microdialysis is a relatively simple technique in which a semi-permeable membrane is implanted into a specified brain area to allow sampling of the local extracellular environment. The diffusion of small molecules across the membrane, including neurotransmitters, occurs through a concentration gradient that is maintained across the dialysis membrane. The gradient is set up through the release of neuroactive molecules by the tissue on one side of the membrane, and, on the other side of the membrane, the removal of the molecules by a flow of fluid driven by a perfusion pump. The constant flow of fluid through the probe allows serial samples to be taken, and the power inherent in the technique is in the design of time course experiments in which experimental manipulations can be changed over the period of continuous sampling. The concentration of analyte in dialysate (fluid sample collected at the outlet of the probe) is presumed to follow changes in the concentration of the analyte in the extracellular fluid. However, under the most commonly used experimental conditions, the dialysate concentration of the analyte represents an unknown fraction of the concentration that exists in the extracellular fluid.

Microdialysis is useful in alcohol research in several ways. Changes in selected neurochemicals due to behavioral or pharmacological manipulations can be monitored in the dialysate. In this way, *in vivo* ethanol administration has been shown to increase dialysate levels of dopamine in the mesolimbic system and the corpus striatum (Imperato and Di Chiara, 1986; Di Chiara and Imperato, 1988; Yoshimoto et al., 1991; Blanchard et al., 1993; Kiianmaa et al., 1995; Yim et al., 1998; Yim et al., 2000). Another interesting application of microdialysis to pharmacological studies is the use of the microdialysis probe as a drug delivery device. Thus, ethanol can be added to the perfusate, and it will then diffuse out of the probe into the tissue to perfuse a local area. Such studies have been carried out to determine effects of local application of ethanol on cellular and neurochemical activity surrounding the probe (Wozniak et al., 1991; Benjamin et al., 1993; Yan et al., 1996; Yim et al., 1998). A third valuable application of the microdialysis method to alcohol research is to monitor the time course of ethanol appearing in the brain (Ferraro et al., 1990; Nurmi et al., 1994). However, until recently, most alcohol researchers who have used the microdialysis technique have ignored the possible pharmacodynamics information that can be gained with this method.

Conventional microdialysis procedures monitor *qualitative* changes (either positive or negative) in the concentration of neurochemicals (or drugs) that appear in the dialysate. But neither the mechanism of the change, if detected, nor the absolute magnitude of the change can be determined. To gain insight into potential mechanisms of changes in dialysate concentration of an analyte or to provide information about true extracellular tissue concentrations, specific *quantitative* microdialysis techniques must be used. One might argue that quantitative microdialysis techniques are not necessary for most applications in neuroscience, particularly in view of the increase in effort required for the already labor-intensive method. However, the interpretability of quantitative microdialysis data is clearly much higher than with the conventional qualitative techniques that are most often used. Furthermore, biomedical sciences are moving as a whole to more quantitative approaches to allow development of rigorous predictive models. Quantitative microdialysis methods are sure to play a role in establishing some of these new theoretical approaches. The purpose of this chapter is to review the theory and application of quantitative microdialysis, emphasizing its use in the pharmacodynamic analysis of actions of ethanol in the brain.

12.2 MICRODIALYSIS THEORY

Soon after the introduction of the microdialysis technique for neuroscience applications, the theoretical basis of microdialysis began to be explored. A major theoretical issue for which no consensus has yet been reached is how to calibrate probes to allow prediction of brain concentrations of an analyte from dialysate concentrations. Although the field has not agreed on the best method for determining the true extracellular concentration of neurotransmitters during microdialysis, it is clear that using *in vitro* calibrations for *in vivo* measurements, one of the first proposed methods for calibration, is not valid. Furthermore, the complexity of the dynamic process of diffusion and transport processes during microdialysis is now beginning to be appreciated, and the difficulties inherent in placing the microdialysis method on a firm quantitative basis are clear. In spite of these difficulties, several quantitative methods have been and continue to be employed to gain some insight into mechanisms of actions of ethanol in brain.

A preliminary examination of the microdialysis process will show why the initial proposed method of using an *in vitro* calibration will not work for quantitating results from an *in vivo* experiment. For illustration, consider a typical microdialysis situation *in vivo*: a probe placed in the nucleus accumbens to monitor dopamine. If the probe is perfused with a dopamine-free solution, the direction of dopamine flux will be from the tissue into the probe lumen. Dopamine produced by the tissue has to diffuse through the tissue and the dialysis membrane before it reaches the probe lumen where it can be carried away in the perfusate flow. Thus, there are two major barriers to diffusion: the tissue and the membrane. Under steady-state conditions there will be a constant fraction of dopamine originating in the tissue extracellular space that diffuses into the probe. It follows that the characteristics of the membrane as well as the tissue will affect the fraction of dopamine reaching the inside of the probe. However, in the *in vitro* situation the major barrier to diffusion is the membrane only. Therefore, the fraction of dopamine diffusing into the probe *in vivo* must be less than that *in vitro* because of the added resistance to diffusion contributed by the tissue. This brief description illustrates why *in vitro* calibration of microdialysis probes is not a valid method for quantitation of *in vivo* results. The argument given here is not dependent on the analyte in question, and is therefore applicable to any diffusible substance including ethanol.

A few selected quantitative microdialysis methods will be highlighted below. More detailed mathematical descriptions of the theory behind each method are available. The emphasis within each method selected here will be on understanding of the underlying concepts by descriptive illustrations rather than a development of the mathematical framework that governs the microdialysis processes. In this section the background of the technique will be given, and some of the technical details involved in carrying out the method will be described in the next section.

12.2.1 LÖNNROTH METHOD (NO NET FLUX)

The method of no net flux was introduced in 1987 by Lönnroth et al. This method is a general one that, in principle, will allow determination of the true extracellular concentration of any diffusible analyte in any tissue in which the probe can be stabilized. The Lönnroth method is elegant in its simplicity. As with all of the methods to be discussed in this chapter, there is a dramatic increase in the experimental effort required to obtain the final results. The technique works by manipulating the concentration of the desired analyte in the perfusate. The concentration of the analyte is varied so that concentrations above and below that expected in the tissue are perfused through the probe. Analysis of the amount of the analyte recovered in the dialysate relative to that perfused through the probe allows the determination of two key parameters that characterize the microdialysis system: the true extracellular concentration and an index of the fraction of the analyte that can cross the membrane.

For illustrative purposes, consider the case of dopamine. In conventional intracerebral microdialysis, the perfusate solution contains selected ions and other small molecules to mimic the cerebrospinal fluid. Dopamine is not included in the perfusate. This maximizes the amount of dopamine that diffuses into the probe and can be detected in the dialysate. The net flux of dopamine is from the tissue toward the probe or inward with respect to the probe (Figure 12.1, left panel). In the Lönnroth method, additional perfusions are carried out. For example, dopamine could be added to the perfusate in a very high concentration so it is higher than the basal level of dopamine in the extracellular fluid. Under these conditions, the dopamine will diffuse out of the probe down its concentration gradient, and there will be a net loss of dopamine from the perfusate. This will cause the dialysate concentration to be less than the original perfusate concentration. The net flux of dopamine in this situation will be outward with respect to the probe (Figure 12.1, middle panel). One can now imagine that an intermediate concentration of dopamine can be added to the perfusate (greater than zero but less than the maximal concentration used above) so that it will exactly match that found on the other side of the dialysis membrane in the extracellular fluid. This is the "no net flux" condition in which the perfusate concentration of dopamine will be equal to the dialysate concentration of dopamine (Figure 12.1, right panel). In practice, the experimenter does not know the exact

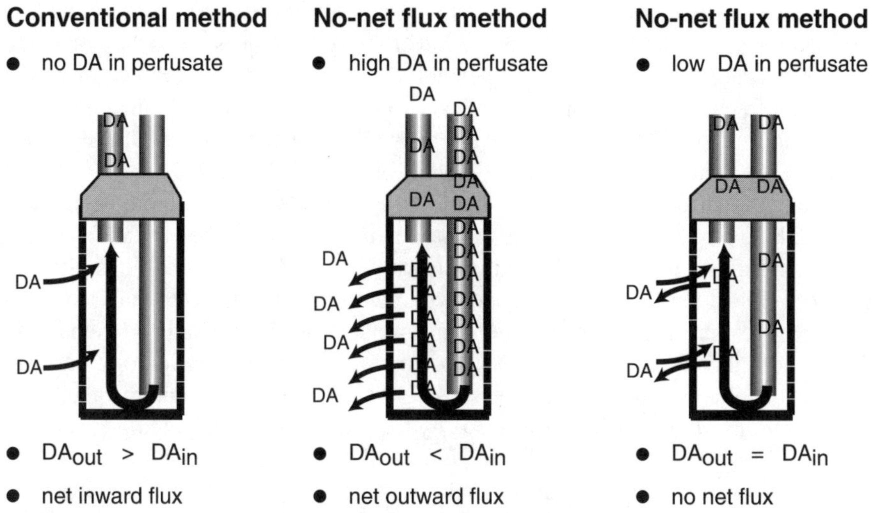

Conventional method **No-net flux method** **No-net flux method**

- no DA in perfusate
- high DA in perfusate
- low DA in perfusate

- $DA_{out} > DA_{in}$
- net inward flux

- $DA_{out} < DA_{in}$
- net outward flux

- $DA_{out} = DA_{in}$
- no net flux

FIGURE 12.1 Mass and fluid flows during no net flux microdialysis conditions. Panels illustrate a microdialysis probe placed in a dopamine rich region of brain. Arrows indicate movement of perfusate through a microdialysis probe along with flux of dopamine (DA) across probe under varying experimental conditions during a Lönnroth experiment. For each panel, movement of perfusate fluid (thick arrow) is from inlet tubing into probe lumen, and then fluid exits probe through outlet tubing. Dashed lines represent dialysis membrane. Density of DAs represents dopamine concentration. Left panel illustrates diffusion of dopamine into a probe from extracellular environment. Middle panel shows net diffusion of dopamine from probe to extracellular space. Right panel illustrates condition of no net flux in which concentration of dopamine in perfusate exactly matches that of extracellular fluid.

concentration of dopamine to add to the perfusate to match the extracellular concentration, but rather wishes to determine this value. Therefore, a series of concentrations that bracket the unknown concentration are used. This may require some trial and error if there is no indication from the literature or theory as to what the extracellular concentration should be. Simple regression methods can then be used to determine the point of no net flux that reflects the true extracellular concentration.

The other major parameter to be obtained from the Lönnroth method is an index of efficiency of diffusion of an analyte that crosses the probe membrane. This measure is often referred to as recovery or extraction fraction (E_d). The Lönnroth method assumes that the flux of an analyte is independent of the direction of flux, i.e., the mass transfer of analyte across the microdialysis probe membrane depends only on the magnitude of the concentration difference across the membrane, not its direction. Using the illustration of dopamine microdialysis, the dopamine flux into the probe is proportional to the concentration difference whether the flux is toward the probe or away from it. The net flux of dopamine will increase as the concentration difference increases. If the system is well behaved, there will be a smooth transition around the point of no net flux, and a plot of the flux against the perfusate concentration will be linear. The slope of this plot is a measure of the extraction fraction for the analyte under the specified conditions of probe geometry, membrane composition and flow rate. However, the assumption that the flux of an analyte is identical whether it is directed toward or away from the probe has recently been questioned for dopamine and similar molecules that undergo metabolic inactivation. The interpretation of the extraction fraction in terms of biochemical mechanisms of action of ethanol and the implications of the non-equivalence of direction of flux of an analyte will be discussed below in the section on mathematical models of microdialysis.

From the discussion above it should be clear that the Lönnroth method is applicable to any analyte, including ethanol. However, one of the limitations of the original Lönnroth method is that it is valid only at steady-state conditions. Steady-state conditions can be achieved under well-controlled experimental environments for *in vivo* experiments to study neurotransmitters like dopamine or glutamate. In the case of ethanol administration *in vivo*, however, the concentration of ethanol in brain extracellular fluid is rarely at steady state. The transient nature of ethanol concentrations *in vivo* makes its analysis by the original Lönnroth method problematic. Similarly, if a pharmacological treatment like ethanol administration produces a transient change in the extracellular level of a neurotransmitter like dopamine, then the original Lönnroth method is not valid. However, in these cases a modification of the Lönnroth method to enable analysis of transient conditions has been proposed (Justice, 1993; Olson and Justice, 1993).

The general concept for analyzing transient conditions with the Lönnroth method is the same as with the original method. Solutions are perfused through the probe using concentrations of the analyte above and below the extracellular level. The major difference is that in the original method, the different solutions are pumped sequentially through a probe placed within a single subject. In contrast, the Lönnroth method for transient analysis uses different solutions in different subjects. For the steady-state method the E_d and point of no net flux can be determined for a single animal (within subject analysis), whereas, for the transient method, the E_d and point of no net flux are determined for a group of animals (between subject analysis).

12.2.2 Mathematical Models

Interpretation of quantitative microdialysis results has been aided by the development of mathematical models of microdialysis that are based on fundamental chemical and biological processes. Early models were proposed and based on the well-established Fick's law of diffusion (Jacobson et al., 1985; Amberg and Lindefors, 1989; Lindefors et al., 1989; Kehr, 1993). Although some of the early models were relatively simple, it soon became clear that more complex models were needed that could account for the contribution of probe factors as well as the surrounding tissue to the overall behavior of a microdialysis system. Bungay and colleagues made a major advance in 1990 with the publication of a steady-state theory of microdialysis (Bungay et al., 1990). The model was also extended for analysis of transient conditions (Morrison et al., 1991). More recently, Michael and colleagues have suggested significant new additions to the theoretical basis of microdialysis (Lu et al., 1998; Peters and Michael, 1998). These two models will be briefly discussed below with an emphasis on the implications of each model for interpretation of quantitative microdialysis experiments.

12.2.2.1   Steady-State Microdialysis Theory

Bungay et al. (1990) were the first to explicitly incorporate biochemical processes in the tissue surrounding a microdialysis probe in a mathematical model for microdialysis. By doing so, they separated the factors that influence diffusion of an analyte into probe-based factors and tissue-based factors. The originally published theory was based on the idea that the mass transport of an analyte is driven by the concentration gradient across the membrane, but is also affected by the resistance to diffusion in the form of the probe and the tissue. The probe and tissue resistances are in series because an analyte has to cross both. An assumption of this approach is that there is no difference in the mass transport of an analyte whether it is inward or outward with respect to the probe.

One of the major features of the model of Bungay et al. (1990) is that it allows the prediction of steady-state concentration gradients that are achieved within the tissue. This information is difficult to obtain by direct experimental observation, but is critical for a rigorous interpretation of potential mechanisms and pharmacodynamic effects of drugs like ethanol. Consider again our example of dopamine microdialysis in which there is no dopamine present in the perfusate. Under steady-state conditions, there will be a certain distance from the probe that will not be influenced by the continuous draining of dopamine from the extracellular space toward the probe. At this distance the extracellular concentration of dopamine will be constant. As one moves closer to the probe, the extracellular concentration of dopamine will begin to drop due to the mass transport of dopamine into the probe. As the distance from the outer surface of the probe membrane becomes smaller, the drop in extracellular dopamine concentration will continue to decline until it reaches a lower limit at the probe-tissue interface. The volume of tissue, in which there is a decrease in extracellular dopamine concentration relative to a distant site undisturbed by the probe, represents the volume of tissue from which dopamine is sampled by the probe. This volume of tissue defines the spatial extent of the region surrounding the probe for which any interpretation of results may be made, and therefore, it is extremely useful to know this information for interpreting pharmacological studies. The drop in extracellular concentration surrounding the probe will also occur for ethanol after it is given systemically, and this has important implications for pharmacological analysis of the effects of ethanol (Yoshimoto et al., 1991; Yim et al., 1998). The converse situation also applies in which ethanol is perfused through the probe so that it bathes the area surrounding the probe membrane. The volume of tissue that is being exposed to ethanol in this situation contains a gradient in which the concentration of ethanol close to the probe is higher than that far away from the probe. Application of the steady-state theory of Bungay et al. (1990) allows the prediction of these important values.

In addition to the above important contribution, application of Bungay's model helps to interpret data obtained from Lönnroth plots in terms of neurotransmitter dynamics, such as release and uptake. Specifically, the steady-state theory predicts that the slope of a Lönnroth plot should be sensitive to changes in the clearance of the analyte. The slope of the plot, which is equivalent to the extraction fraction, will decrease as the clearance decreases. On the other hand, the slope will increase if the clearance is enhanced. For dopamine, uptake is the primary clearance mechanism, and therefore, changes in the uptake system can possibly be detected by monitoring changes in the *in vivo* extraction fraction. This has been shown by measuring the effect of cocaine, a known inhibitor of dopamine transport, on the *in vivo* extraction fraction for dopamine in rat brain (Justice, 1993; Parsons and Justice, 1994; Smith and Justice, 1994). Similar to this, the effect of an inhibitor of acetylcholinesterase, the primary clearance mechanism for acetylcholine, is to decrease the *in vivo* extraction fraction for acetylcholine (Vinson and Justice, 1997). There is one major caveat concerning this type of interpretation of the *in vivo* extraction fraction that arises from the steady-state theory. That is, as the value for clearance *in vivo* increases, *in vivo* extraction fraction will be closer and closer to the *in vitro* extraction fraction obtained in a well-stirred condition. Because dopamine has an avid uptake system, it is unclear how easy it will be to detect potential increases in dopamine uptake. The implications of the impact of changes in clearance on *in vivo* extraction fraction for ethanol are likely to be minimal. This is because there will probably not be major changes in clearance of ethanol in the brain since it is primarily removed through blood flow.

In certain cases, it may be possible to interpret changes in the x-intercept of a Lönnroth plot in terms of a change in the release of a neurotransmitter, rather than a change in uptake. As discussed above, a change in the slope of a Lönnroth plot can be interpreted in terms of a change in uptake of a neurotransmitter. One must keep in mind, however, that this does not rule out a potential change in release that may occur also. There may be simultaneous changes in both. It is likely that a change in uptake will influence the x-intercept of a Lönnroth plot (point of no net flux). Therefore, it is difficult to interpret directly the potential mechanism for a change in the x-intercept; it may be due to a change in release or uptake. On the other hand, if an experimental treatment does not alter the slope of a Lönnroth plot, and there is a corresponding change in the x-intercept (point of no net flux), this can be interpreted in terms of a change in the release of a neurotransmitter. The caveat here is that there may be small changes in the clearance of the neurotransmitter that may contribute to a change in the x-intercept, but may be too small to be detected by examination of the slope.

12.2.2.2 Recent Modifications of Microdialysis Theory

Michael and colleagues have recently challenged one of the major assumptions of Bungay's steady-state theory, at least as it applies to a neurotransmitter like dopamine (Peters and Michael, 1998). This assumption is that the flux of dopamine across the membrane is independent of the direction of the flux. In Michael's formulation, the flux of analyte is separable depending on whether the analyte originates from the lumen of the probe or the tissue. This separation has important implications for the magnitude of the flux across the membrane when the tissue contains an active clearance mechanism, as is the case for dopamine via the dopamine transporter. Thus, Michael's model predicts that the molecular flux for dopamine is greater when it originates from the probe compared with the flux in the opposite direction when dopamine originates in the tissue. These two processes based on the direction of flux predict that recovery of dopamine from the tissue is much less than extraction of dopamine by the tissue from the probe. Michael and colleagues have interpreted this differential flux in terms of a probe recovery factor and a tissue extraction factor.

The idea of a difference between recovery and extraction has several important implications for interpretation of quantitative microdialysis data. For example, there is no direct method to determine recovery because one must know the true extracellular level of dopamine *a priori*. Therefore, Michael's theory has called into question the validity of using the Lönnroth method for determining the value of the extracellular concentration. In practice, however, it is not possible to separate recovery from extraction. During the performance of a Lönnroth experiment, the perfusate concentration of dopamine is altered, but the total flux of dopamine is a combination of the influences of recovery and extraction. It is the total flux that is measured with this method, and the origin of dopamine, whether exogenous or endogenous, is irrelevant. Another important implication of Michael's analysis is the possibility that the functional nature of the tissue may be changed surrounding the probe. If uptake of dopamine is significantly decreased in the peri-probe region, then the recovery will not match that which is found in areas of the tissue far from the probe. This again calls into question the accuracy of the Lönnroth method for predicting the true extracellular concentration. In this scenario, the Lönnroth method will underestimate the true extracellular concentration.

12.2.3 SYNTHESIS

Although Michael's theoretical analysis is novel, the model does not suggest a practical way to determine either extraction or recovery. In view of this, the Lönnroth method remains a valid way to obtain an estimate of the true extracellular concentration as well as an index of the probe efficiency. However, one must continue to exercise caution regarding the accuracy of the estimates. Bungay has recently incorporated into his model the potential changes in release and uptake processes surrounding the probe that may be caused by tissue trauma during probe insertion (Bungay et al., 1999). This new analysis supports Michael's suggestion that the estimate of extracellular concentration of a neurotransmitter like dopamine from the Lönnroth method will be lower than that of undisturbed tissue (Lu et al., 1998). Unfortunately, the magnitude of the error that might be introduced in the estimate of extracellular concentration due to these factors will remain unknown until an independent and reliable measure of extracellular concentration can be obtained.

Despite these theoretical problems with interpretation of quantitative microdialysis data, the utility of the method remains important especially for gaining insight into the dynamics of the tissue factors that influence dialysate concentrations. Thus, both Bungay's and Michael's models support the idea that the slope of a Lönnroth plot (extraction fraction) is sensitive to changes in clearance (or neurotransmitter uptake) in a predictable way. A decrease in clearance will decrease the slope and an increase in clearance will increase the slope. Although this is an indirect method to monitor potential changes in neurotransmitter clearance, it is one of the few methods that will provide this information while an animal is conscious and freely moving.

12.3 METHODOLOGICAL ASPECTS

Several key steps are necessary to carry out a quantitative microdialysis experiment. In general, the following components are needed:

- A microdialysis probe
- A syringe pump for performing slow perfusions
- A subject
- A histological method to verify probe placement in the brain
- An analytical method to detect and quantify the analyte of interest

Detailed descriptions of various aspects of performing a microdialysis experiment have been previously published (see Chapter 11). Rather than repeating these details, the purpose of this section is to emphasize critical points that could influence the results of a quantitative microdialysis experiment. The focus of the discussion will be on two types of quantitative microdialysis experiments: the Lönnroth method and estimating the concentration profile for ethanol. The experimenter has a variety of choices of conditions to use, and an analysis of the particular experimental constraints is important for optimizing the chances of success.

12.3.1 MICRODIALYSIS PROBES

Probes can be constructed in the laboratory or obtained commercially. The four basic designs are concentric, side-by-side, U-shaped, and transverse (Figure 12.2). Our

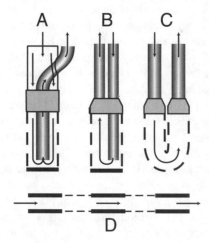

FIGURE 12.2 Basic designs of microdialysis probes. For each of diagrams, arrows show direction of perfusate flow, and dashed line designates active dialysis area. Thick lines in each diagram represent epoxy coating or other barriers to diffusion. Concentric design (A) has single tube extending into lumen of dialysis fiber that is plugged at distal end. Side-by-side design (B) has both inlet and outlet tubing inserted into dialysis fiber, and fiber is sealed at both ends. U-shaped probe (C) usually requires a thin wire (not shown) inserted into lumen of probe to provide rigidity. Transverse probe (D) has to be coated with epoxy or other impervious substance to define active dialysis area for bilateral sampling.

lab uses homemade probes of the side-by-side design (Pettit and Justice, 1991). All of these designs have been successfully used in conventional dialysis experiments, but the microdialysis theories discussed earlier are based on a concentric geometry. Much of the experimental work with quantitative microdialysis has been performed with the side-by-side design. The magnitude of deviations of the predictions based on the concentric geometry compared with other probe designs is not known. It seems likely that, for many of the concepts described above, there will be qualitative agreement among the various probe designs. For example, the basic interpretation of the parameters from a Lönnroth plot (slope and x-intercept) will be independent of the design.

However, other critical factors must be considered when choosing the probe design for quantitative microdialysis experiments. One factor is the overall size of the probe. This is important because a larger probe will produce greater tissue trauma, and, therefore, it may be more difficult to interpret changes in quantitative dialysis parameters in terms of changes in neurotransmitter dynamics. It is clear that there is tissue damage surrounding the probe, and it is likely that neurotransmitter dynamics have been altered relative to undisturbed tissue. Therefore, using the smallest probe available would be prudent. (The smallest probes generally available are in the range of 250 μm in diameter.) However, this must be balanced against the sensitivity of the analytical system available for chemical analysis of the dialysates. The probe recovery or extraction fraction is positively related to the probe surface area, so a small probe will have a decreased extraction fraction relative to a larger one. A U-shaped probe will generally have a larger extraction fraction than the other designs, but will cause more tissue trauma because of its larger diameter. If experiments are designed to deliver alcohol to a local tissue area, such as with combined microdialysis and single-cell recording, a smaller probe will be advantageous for both limiting the tissue damage and limiting the area of tissue that is perfused by alcohol (see Chapter 11).

Another consideration is the volume of tissue from which the probe is sampling or that is being perfused by a drug, such as ethanol. Smaller brain areas require small probes, but if the target brain region is large, then it is to the experimenter's advantage to use a longer probe. The converse situation is that in which ethanol is delivered to the tissue. Here, it may be better to limit the area that is being perfused by the ethanol, and a shorter probe will help in this regard. Concentric probes are the only ones modeled with regard to the volume of tissue being exposed to ethanol. A concentric design will produce a cylindrical volume of tissue with relatively high local concentration of ethanol. The spread of the ethanol away from the probe will vary in a radial manner, i.e., it is symmetrical around the probe in all directions. The transverse probe will also produce a similar type of symmetrical spread of ethanol. However, the side-by-side and U-shaped probes may produce an asymmetric distribution of ethanol to the local area surrounding the probe. It is advisable to confirm the distribution of ethanol in the tissue when performing local perfusions through the probe if a U-shaped or side-by-side design is used.

Given the growing realization of the potential role of tissue trauma in the interpretation of quantitative microdialysis results, some consideration should also be given to probe implantation procedures. Although there is no consensus in the field regarding the best time to begin sampling after placement of the probe into the brain, one should minimize the potential damage to the tissue that is inevitable. This issue is particularly important if neurotransmitter dynamics will be analyzed. In general, the longer the probe is in place, the more damage will be present surrounding the probe. For dopamine it is advisable to begin sampling at least 8 hours after placing the probe in the tissue to allow some recovery of the tissue from the initial trauma. On the other hand, we have recently shown that for sampling of ethanol, the time after probe implantation (from 2 to 48 h) does not affect the *in vivo* delivery of ethanol (Robinson et al., 2000). This could be due to the high diffusivity of ethanol and the fact that its clearance is not dependent on the functional state of surrounding neurons or glia. As with the implantation of any mechanical probe into living brain tissue, lowering of microdialysis probes should be done in a smooth, slow motion to minimize shearing forces. Use of a microdrive or stereotaxic device for precise control of the rate of lowering of the probe may be advantageous or necessary in some cases.

12.3.2 PERFUSION FLOW RATE

The choice of perfusion flow rate to use for a quantitative microdialysis experiment should be carefully considered. Flow rate is a major factor that influences the extraction fraction (Bungay et al., 1990). The extent of diffusion across the membrane increases as perfusion flow rate decreases until the limit of zero flow rate is approached. Therefore, decreasing the flow rate increases the concentration of analyte recovered or delivered by the probe. However, this might present a problem because the volume of sample collected will be much lower, which can lead to difficulty with sample handling. Unless the analytical system is optimized to handle these small sample volumes (such as with microbore HPLC or capillary electrophoresis), decreasing the perfusion flow rate may not be helpful. This difficulty can be partially overcome by increasing the sampling time, although this presents additional potential problems due to changes in the basal physiological state during the sampling period. Furthermore, as sampling times become longer, there is a greater need to control potential changes in the dialysate concentration of the analyte due to evaporation.

Perfusion flow rate also has a major influence on the practical performance of the probe in terms of the pressures generated during the probe perfusion. The major factors that determine the intraprobe pressure during perfusion are the length and inner diameter of the probe outlet tubing. Commercially available tubing for microdialysis generally has a large enough inner diameter (approximately 100 μm) to prevent the generation of significant backpressures. Laboratory constructed probes may have inlet and outlet tubing of 40 μm inner diameter, which will generate significant backpressure (around 1 atm) even at relatively low flow rates (1–2 μl/min). It is advisable to keep the pressure that builds up within the probe and inlet and outlet lines as low as possible to reduce stress on the various junctions between the various pieces of tubing that make up the fluid pathway. Leaks are a common problem in any microdialysis experiment, and reduced pressure within the lines will lessen the chances for leaks or catastrophic failure of the membrane.

In addition, increased pressure within the probe may influence the fluid flow across the membrane, and, therefore, will influence the mass transfer of an analyte (Bungay and Gonzales, 1996). This will obviously influence any measurement of extraction fraction. The major theories of microdialysis discussed earlier assume that diffusion is the basic process that drives mass transfer (although Bungay has modified his theory to take into consideration bulk fluid loss through the membrane). However, there is another factor related to the intraprobe pressure that is usually not considered in microdialysis experiments. That is, the elasticity of the probe membrane itself. The probe membrane has a high degree of elasticity, and is balloon-like in behavior. If the pressure within the probe increases, the diameter of the probe will correspondingly increase, and this will increase the effective size of the probe. One consequence of this could be further compression of the cells surrounding the probe followed by additional trauma and/or inhibition of recovery from trauma. This is a relatively unexplored phenomenon in microdialysis, but this suggests that probe pressure changes should be avoided as much as possible once the probe is implanted within the tissue.

One other factor to consider when choosing perfusion flow rate is the length and volume of the transfer lines that carry the perfusate from the syringe to the probe. Lönnroth experiments involve several changes in perfusion solutions, and the more volume in the transfer lines the longer it takes for new perfusion solution to reach the probe and then to exit the probe outlet line. If an experiment requires long transfer lines, choosing a higher flow rate will reduce the time to achieve steady state with the new solution.

12.3.3 PERFUSATE CHANGES MAY INFLUENCE THE ANALYSIS

The independent variable for a Lönnroth experiment is the concentration of the perfusate, and, for the within-subject design, each animal receives several defined perfusion solutions. Potential problems that might arise from the mechanics of solution changes during a microdialysis experiment are worth mentioning. The major issue here relates to possible changes in perfusion flow rate that could be caused by the solution change. Changing perfusion solutions during a microdialysis experiment involves using a liquid switch or manually stopping the flow to change solutions. A liquid switch (several designs are available commercially) is ostensibly a better way to perform the switch because flow rate changes and associated pressure changes are minimized during the switching procedure (see Chapter 11). However, liquid switches are not foolproof, and changes in flow rate precipitated by activation of the switch may occur. It is advisable to thoroughly test a liquid switch in an *in vitro* experiment before trusting it for an *in vivo* experiment.

The manual method of switching solutions is inherently problematic because the flow must be stopped in the middle of the experiment, and this will produce major changes in pressure and possibly in probe diameter, with the attendant problems described above. Changes in probe diameter could then lead to additional or further trauma of the tissue surrounding the probe. Another issue surrounding the manual method for changing perfusion solutions is the time it takes for the flow rate to come back to normal. This is critical, because extraction fraction is sensitive to flow rate, and it can take several minutes for the flow rate to proceed from zero to the desired steady-state level.

In addition, the time it takes for the new solution to reach the probe and then exit the outlet tubing needs to be considered. In some cases, the volume of the transfer lines (including a liquid swivel, if present) from the syringe containing the perfusate to the probe can be considerable. The experimenter should confirm the amount and account for this "dead" volume when planning the timing of sample collections during a Lönnroth experiment. The possible consequence of underestimating the dead volume is that the concentration of the analyte in the solution reaching the probe will not be exactly what is in the syringe, but rather, it will reflect a mixture of the previous solution with the desired solution. Because of the variability in flow rate, as well as possible mixing of the syringe solutions that can occur as a consequence of solution changes, it is strongly recommended that samples should be used for analysis only after confirmation that the extraction fraction has reached steady state. This can be done by plotting each sample taken after a perfusate solution change to monitor for any trends in the concentration of the analyte over time that do not stabilize. If this is the case, then additional samples should be taken to account for the equilibration time.

12.3.4 STABILITY OF ANALYTE

Stability of the analyte of interest is usually considered with regard to the method used for chemical analysis, but this issue may play a critical role in the accuracy of quantitative analysis and the interpretation of quantitative microdialysis experiments. The relevance of stability clearly depends on the analyte. For example, catecholamines are notoriously unstable because they easily undergo autooxidation. However, adherence to good analytical technique minimizes the influence of this issue. In direct contrast to this, analysis of amino acid neurotransmitters may be complicated by contamination from bacterial sources, which could lead to overestimation of actual values that originate from tissue being dialyzed. In this case, including a series of blank determinations to monitor the potential sources of external contamination is critical. Appropriately maintained water purification systems are essential to minimize this problem.

A third example of a stability problem is specific to the measurement of ethanol — its volatility. The high volatility of ethanol is easy to underestimate because the amounts of ethanol to be analyzed in a quantitative microdialysis experiment are usually high relative to other analytes, and, therefore, ethanol is easy to detect. Ethanol is vulnerable to evaporation from the sample at many points in the experiment, such as during dialysate or tissue sample collection and during storage before sample analysis. The chance for potential loss of ethanol due to its evaporation from a dialysis sample while it is being collected will increase as the sample collection time increases. In our laboratory, we have not observed major problems with ethanol volatility during sample collection times up to 15 min (Crippens et al., 1999; Yim et al., 1998; Yim et al., 2000). However, it is critical that samples be transferred to a sealed vial as soon as possible after collection of the complete sample. Furthermore, handling of the vial during this transfer step could enhance the artifactual loss of ethanol due to heat transfer from hands. Keep in mind that dialysate sample volumes are in the range of 5–30 microliters, and significant heat can be transferred from two fingers that are holding the vial to the ethanol solution, and thereby enhance evaporative loss of the ethanol. Analyzing ethanol in tissue samples presents additional problems with possible loss of ethanol. For example, estimation of the ethanol concentration profile in brain tissue during steady-state microdialysis requires quantitation of the amount of ethanol in the tissue. Handling of the tissue during preparation for analysis may involve exposure of the tissue to the atmosphere and loss of ethanol from the surface of the tissue. Some of the loss during this period may be unavoidable, but it should be minimized as much as possible. Finally, ethanol volatility during storage of the sample must be considered. It is imperative to ensure that a good seal on the sample vial is maintained during the storage period. In general, analyzing the samples within a day or so of collection is advisable. If storage is required, the samples should be stored in a refrigerator in an appropriately sealed vial to prevent loss of ethanol from the sample or gain of ethanol from the environment. Vials with caps containing rubber septa or Teflon backed rubber septa have proven to be reliable in our laboratory (Crippens et al., 1999; Yim et al., 1998; Yim et al., 2000). If long term storage is necessary, standards and blanks should be included along with the samples to monitor possible changes in ethanol concentration during the storage period.

12.3.5 STATISTICAL ANALYSIS OF A TRANSIENT LÖNNROTH EXPERIMENT

Analysis of a steady-state Lönnroth experiment involves determination of the slope and intercept of a plot generated from a single subject. The parameters can then be grouped and analyzed by traditional analysis of variance techniques. However, the analysis of a transient Lönnroth experiment is more complex. Most experiments using the transient Lönnroth method will be designed to determine whether a particular experimental treatment changes either the slope or x-intercept. We have adopted the method of extra sum of squares to carry out this type of analysis (Kenakin, 1997). Essentially, the basis of the method is to use an F test to determine whether sharing parameters across regressions will significantly increase the sum of squares associated with the entire experiment (Robinson et al., 2000; Yim et al., 2000).

The first step in the analysis is to define a set of data to which a linear regression will be applied at each time point in the experiment. The x-values are the perfusate solutions used. The y-values are the average of the difference between perfusate and dialysate concentration across all subjects in the group for each perfusate solution. Next, the data are grouped according to the experimental treatments. For example, a transient Lönnroth experiment usually has a set of basal samples that are replicates over a few time points along with a set of data corresponding to the sampling times after an experimental treatment. The set of basal data can be collapsed and analyzed as a single regression equation. On the other hand, each data set obtained after the experimental treatment (for example, an injection) should be kept as a separate regression for each time point.

The next step is to fit all of the regressions simultaneously and let all slopes and x-intercepts vary independently. This is denoted the unconstrained fit, which is associated with a residual sum of squares (SS_u) and degrees of freedom (df_u). Note that one can fit the x-intercept directly by using the alternative form of the slope-intercept equation, $y = m(x-a)$, where y is the net flux of the analyte, m is the slope, x is the perfusate concentration, and a is the x-intercept. The use of this form of the equation allows a direct statistical analysis of whether the x-intercepts are different. A mean sum of squares (MS_u) can then be defined for the unconstrained fit by dividing the SS_u by the df_u. This MS_u for the unconstrained fit will be the denominator for the computation of the F ratio.

The numerator for the F ratio comes from the extra sum of squares introduced into the overall regression analysis by forcing the sharing of parameters across regressions. For example, if there are two (or more) overall groups in the experiment, such as a saline group and an ethanol-injected group, the regressions are fit to the data so that all slopes are shared and all x-intercepts are shared. In other words, the entire data set will be fit using one slope and one x-intercept. This will yield a constrained fit. The difference in the sum of squares between the constrained and unconstrained fits is the extra sum of squares (SS_{diff}). Furthermore, sharing parameters across regressions will reduce the overall degrees of freedom (df) for the analysis, and the difference in the df between the constrained and unconstrained fits (df_{diff}) is the df for the SS_{diff}. Dividing the SS_{diff} by the df_{diff} yields the mean sum of squares for the difference between the fits, the MS_{diff}. The MS_{diff} is the numerator for the F ratio.

Now an F value can be calculated by dividing the MS_{diff} by the MS_u. If $p <$ 0.05 for this F value, there is an overall statistically significant difference between the fits, which could be due to a difference in slopes or x-intercepts. If there is a significant difference, the slope and x-intercept can then be tested individually in a similar manner. That is, to test whether there is a difference in slopes, a constrained fit is done by letting the x-intercepts vary independently for each regression, while forcing the slopes to be shared. An F value is generated as above, using the MS_u as the denominator. An independent F test can then be done to see whether there is a difference in x-intercepts. This general procedure can be repeated in a stepwise manner until all significant differences have been found. Our laboratory uses the NLR routine in the SPSS software package to set up the regressions and to obtain the SS and df values for each fit that is required. The SS and df values are then exported to a computer spreadsheet to calculate the final MS and F statistics.

In general, the initial tests will be done to compare regressions between experimental groups, for example, an ethanol-treated group and a control group. If there is a difference between the groups, the next level of tests will be calculated to determine whether there is any effect of the treatment within each group. At this level of analysis, the basal regression is compared with the posttreatment regressions. If a difference is detected at this step, additional tests can be carried out to determine which of the post-treatment regressions is different from the basal regression. We use a Bonferroni correction for the multiple comparisons at this level to maintain an experiment-wide alpha of 0.05.

12.4   APPLICATION OF QUANTITATIVE MICRODIALYSIS TO ALCOHOL RESEARCH

Examples of the application of three types of quantitative microdialysis experiments to alcohol research will be given in this section. The first will be the use of a between-subject Lönnroth experiment to gain information regarding the mechanism of action of ethanol on the mesolimbic dopamine system in rat brain. The second example will illustrate a similar method for determination of ethanol concentrations in rat brain after administration of ethanol by the intravenous route. Finally, an example of the estimation of the tissue concentration profile for ethanol will be provided during steady-state perfusion of ethanol through a microdialysis probe in the rat striatum.

12.4.1   Does Ethanol Alter Dopamine Release or Uptake in the Nucleus Accumbens?

The mechanism of action of ethanol on various neurotransmitter systems is an area of active research. The mesolimbic dopamine system has been a focus of alcohol research for many years. It is clear that systemic ethanol administration increases dialysate dopamine levels when monitored from the nucleus accumbens (Imperato and Di Chiara, 1986; Di Chiara and Imperato, 1988; Yoshimoto et al., 1991; Blanchard et al., 1993; Kiianmaa et al., 1995; Yim et al., 1998; Yim et al., 2000). Recently, we applied the Lönnroth method to gain insight into whether ethanol was increasing the release of dopamine or inhibiting its uptake (Yim and Gonzales, 2000). Our initial study compared the effects of saline and ethanol (1 g/kg) injections given by the intraperitoneal route. The between-subject Lönnroth method was chosen because we wanted to monitor the time course of any potential effect of ethanol, and, therefore, this was a transient experiment rather than one carried out at steady state. The dialysate conditions chosen were those used routinely in our laboratory to show that i.p. ethanol stimulated the dialysate level of dopamine using conventional microdialysis. The perfusate flow rate was 2.0 μl/min and the probe length was 2 mm. The dopamine concentrations used were 0, 4, 8, and 12 nM dopamine. Previous work had suggested that the basal level of dopamine using similar methods should be around 5 nM (Parsons and Justice, 1992). We expected ethanol to increase dialysate dopamine to approximately 30–40% above the basal levels. Four groups of rats were run in the experiment. Each rat in a group was perfused with one concentration of dopamine during the entire experiment. Dialysate samples were taken every 10 min before and after the injection of either saline or ethanol. After analysis of the dopamine concentrations in each sample by HPLC-EC, the data were analyzed by plotting the difference between perfusate and dialysate dopamine concentration (net flux) as a function of the perfused concentration. This plot was generated for each time point before and after the injection. A linear regression on each set of data was then performed to determine the slope and x-intercept. The slopes (extraction fractions) and x-intercepts (points of no net flux or true extracellular concentration) were then plotted over time to facilitate comparison of basal values to the values obtained after the injection. Regression analysis was used to determine the statistical significance of the effects of ethanol and saline on the slope and x-intercepts of the Lönnroth plots.

The results we obtained showed that the extraction fraction did not change over the time course of sampling for either the saline group or the ethanol group (the mean extraction fraction was 0.15 for the saline group and 0.14 for the ethanol group). Similar to this result, the saline group did not show any significant changes in the true extracellular concentration during the experiment (8.8 ± 0.4 nM). In contrast, the ethanol injection produced a transient, but significant, increase in the true extracellular concentration of dopamine from 9.4 ± 0.4 nM for the basal samples to 13.2 ± 1.8 nM after the i.p. injection of ethanol. The results were interpreted to suggest that the major effect of the ethanol injection was to increase dopamine release rather than inhibit dopamine uptake. However, there are two caveats with this interpretation. First, it may have been difficult to detect a decrease in the extraction fraction caused by an inhibition of dopamine uptake by ethanol because the extraction fraction was already low; that is, there may have been a floor effect. Second, we did not perform an *in vitro* characterization of the probes before the *in vivo* experiment. This would have enabled us to attribute any effects of ethanol to tissue changes rather than variations in the probe itself.

A second experiment was carried out using a lower perfusate flow rate to replicate and extend the findings of the first experiment. The flow rate in the second experiment was 1.0 µl/min. The experimental conditions were generally the same as described above except that the sampling interval was 15 min and the dopamine concentrations were 0, 5, 10, and 15 nM. Figures 12.3 and 12.4 illustrate the results of an i.p. injection of 1 g/kg ethanol on the slope and x-intercept of Lönnroth plots generated before and after the injection. The *in vivo* extraction fraction for dopamine under basal conditions was 0.24. This value matches well with what would be predicted based on the relationship between E_d and flow rate. Regression analysis indicated no significant change in the E_d for dopamine during the course of the experiment. In agreement with the previous experiment (Yim and Gonzales, 2000), the true extracellular concentration (x-intercept of the Lönnroth plots) was significantly increased after the injection of ethanol. Overall, the results agree with the previous experiment and strengthen the conclusion that the major action of acute ethanol administration is to selectively increase dopamine release rather than inhibit dopamine uptake.

A final note of caution should be mentioned regarding this interpretation. The influence of flow rate on the E_d partly depends on the clearance. According to the steady-state theory of Bungay et al. (1990) the E_d will increase as the clearance increases. Large decreases in clearance can then be detected as a decrease in E_d. It is not clear how large the inhibition of uptake must be to influence the E_d. This will partly be dependent on the precision of the measurement of E_d. Therefore, the lack of significant change in E_d for dopamine in our experiments produced by an ethanol injection cannot be interpreted as evidence that there is no effect of ethanol on dopamine uptake *in vivo*. There could be small changes in uptake that are beyond the limit of detection of this method. Nevertheless, it is safe to conclude that the major effect is an increase in dopamine release compared with an inhibition of uptake.

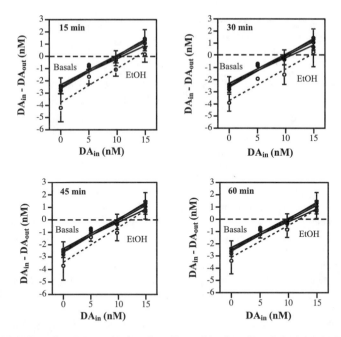

FIGURE 12.3 Transient Lönnroth plots for effect of 1 g/kg ethanol (i.p.) on dopamine from rat nucleus accumbens. Effect of ethanol on slope and x-intercept of Lönnroth plots is shown for various times after injection. Four basal samples were taken before injection, and their regressions are shown (closed circles, solid lines) in each of panels to facilitate comparison with samples taken after injection. A single regression corresponding to time after injection is also shown in each panel (open circles, dashed line). Basal regressions clearly overlap and are not significantly different from each other. However, regressions for 15, 30, and 45 min after injection are significantly shifted. Regression analysis indicated that x-intercept was significantly changed after injection, but slope was not. Regression for data collected 60 min after injection was not significantly different from basal samples.

12.4.2 Determination of Brain Concentrations of Ethanol

Blood ethanol concentrations are routinely monitored in behavioral experiments with ethanol. Few studies have looked at brain ethanol concentrations after ethanol administration because the analysis required removal of the brain, and therefore, it required a large number of experimental subjects to describe the time course. Intracerebral microdialysis is now being used to monitor ethanol in brain tissue in a way that allows serial samples to be taken so that a time course can be followed for a single animal (Ferraro et al., 1990; Yoshimoto and Komura, 1993; Nurmi et al., 1994; Nurmi et al., 1999). The ability to follow changes in dialysate ethanol concentration makes possible the study of pharmacodynamics of neurochemical responses to ethanol. However, calibration of the probes for *in vivo* use is necessary to determine quantitatively the concentrations of ethanol in brain extracellular fluid.

We recently performed a series of tests to validate a method to calibrate microdialysis probes for *in vivo* studies of ethanol (Robinson et al., 2000). We compared three methods for determining brain concentrations of ethanol: transient Lönnroth

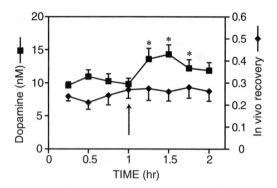

FIGURE 12.4 Time course of changes in *in vivo* E_d for dopamine and true extracellular concentration before and after 1 g/kg ethanol (i.p.). Slopes and x-intercepts of Lönnroth plots in Figure 12.3 are shown as a function of time. Slopes correspond to *in vivo* E_d for dopamine, and x-intercepts correspond to true extracellular concentration of dopamine. * indicates significantly different (P < 0.05) from basal samples by regression analysis. Error bars represent asymptotic standard errors obtained from SPSS output. ■ = dopamine concentrations taken from regressions. ◆ = *in vivo* E_d for dopamine taken from regressions.

method, *in vivo* delivery, and *in vivo* recovery. As discussed previously in this chapter, the transient Lönnroth method will yield a value for the true extracellular concentration of an analyte. *In vivo* delivery refers to delivering ethanol to the tissue by perfusing various concentrations of ethanol through a probe while the animal has no ethanol in its system, and an *in vivo* E_d for ethanol can be determined as a calibration factor. *In vivo* recovery refers to recovering ethanol from the tissue by varying the concentration of ethanol *in vivo* while perfusing a solution with no ethanol through the probe. The *in vivo* delivery method is the easiest and most straightforward method, and it can be done for each probe. Unfortunately, the *in vivo* delivery method systematically overestimated the E_d, which led to an underestimation of the brain ethanol concentrations. Similar to this, the transient Lönnroth method also underestimated the brain extracellular levels of ethanol. The *in vivo* recovery method is too unwieldy to perform on a routine basis, but this method turned out to be the most accurate and reliable. We propose using a combination of the *in vivo* recovery and *in vivo* delivery methods. Thus, a group of animals is run through the *in vivo* recovery method, and the results from this experiment are used to adjust the values obtained from the *in vivo* delivery method, which can be done on a day-to-day basis.

The rationale for our suggestion is that we found evidence for a directional bias in the flux of ethanol across the probe membrane. This occurred both *in vitro* and *in vivo*, and is illustrated in Figure 12.5. *In vitro* the E_d was 0.31 ± 0.01 (mean ± SEM, n = 6) when ethanol was perfused through the probe using our standard 2 mm probes and 2.0 µl/min flow rate (the beaker solution was well stirred and kept at 37°C). However, the E_d was 0.24 ± 0.01 when ethanol was in the beaker while perfusing the probe with a solution containing no ethanol. This is a surprising finding because most tested compounds show no difference in E_d that depends on the direction of the flux of the analyte (Zhao et al., 1995). Moreover, there is no readily

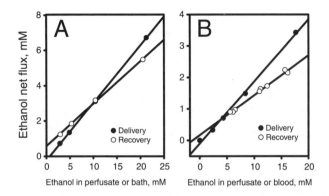

FIGURE 12.5 Effect of diffusion direction on ethanol E_d in representative microdialysis probe, shown *in vitro* (A) and *in vivo* (B). Data are plotted as diffusion of ethanol across probe membrane as a function of ethanol concentration. Ethanol delivery refers to diffusion of ethanol from perfusate into bath or tissue. Ethanol recovery refers to diffusion of ethanol into probe from bath or tissue. The slope of the linear regression indicates ethanol E_d, which was lower for ethanol recovery than delivery both *in vitro* and *in vivo*. Lines shown are linear regressions, and r^2 values were at least 0.99 in all cases.

apparent theoretical reason that explains the origin of this difference. However, the fact that it occurs *in vitro* suggests that the composition of the membrane may be partly responsible. At present, we do not know whether this directional bias in E_d for ethanol occurs with probes other than the ones we construct in our laboratory. Additional data with other probe membrane materials is needed before the generality of this effect can be ascertained.

The present example illustrates our proposed probe calibration procedure for ethanol and then shows how it can be used to determine the brain levels of ethanol after intravenous injection. The calibration procedure is based on determining the discrepancy between the inward and outward flux of ethanol. Therefore, the *in vivo* E_d for ethanol must be measured under conditions in which the flux of ethanol is out of the probe into the tissue as well as the opposite direction. In principle the experiment involves two phases. One phase, the *in vivo* delivery phase, consists of perfusing several concentrations of ethanol through the probe and measuring the loss of ethanol from the perfusate at each concentration. The second phase, the *in vivo* recovery phase, involves administration of several different concentrations of ethanol *in vivo*, and for each concentration, a blood sample and a dialysate sample are taken simultaneously. A problem with the *in vivo* recovery phase is that ethanol pharmacokinetics are not straightforward, and it is difficult to maintain a steady-state concentration of ethanol in the body. It is critical to try to maintain a steady-state concentration of ethanol in the blood to make sure that the concentrations of ethanol in blood and in the brain extracellular fluid are at steady state with respect to each other. It is assumed that at steady state, the extracellular concentration of ethanol in the blood will be equal to the concentration of ethanol in the brain. This is a reasonable assumption because ethanol is known to be a highly diffusible molecule with no blood brain barrier.

To perform the experiment, rats were fitted with a jugular catheter and prepared for microdialysis by implanting a guide cannula. The catheter was used to deliver ethanol intravenously and also to remove blood samples at the appropriate time. The procedure for microdialysis from the nucleus accumbens was begun using our standard conditions (probe length 2 mm, perfusate flow rate 2.0 μl/min). For the *in vivo* delivery phase of the experiment, we routinely use concentrations of ethanol of 2.5, 5, 10, and 20 mM. The perfusate solutions were switched in random order, and flow rates were monitored and confirmed after every switch. Also, samples of the perfusate were taken for ethanol analysis along with the dialysates. Dialysate samples were taken in duplicate at least 6 min after each switch to ensure that steady state had been achieved, as determined by the volume of our system. After all samples at all concentrations were collected, the perfusate was switched back to normal artificial cerebral spinal fluid for at least 30 min before starting the *in vivo* recovery phase of the experiment. Ethanol in the dialysate and perfusate samples was quantitated by headspace gas chromatography. The data from the *in vivo* delivery phase were analyzed by plotting the difference in ethanol concentration between the perfusate and the dialysate as a function of the perfusate concentration. This is essentially a Lönnroth plot. The slope of the regression through the points represents the *in vivo* delivery E_d. A representative plot for one of our probes is shown in Figure 12.5B.

During the *in vivo* recovery phase of the procedure, it is necessary to maintain a steady state of blood ethanol concentrations, as described above. To overcome the difficulty of maintaining such blood ethanol levels, we used an inhibitor of alcohol dehydrogenase, 4-methylypyrazole, to slow the metabolism of ethanol and its clearance from the blood. The perfusate was normal artificial cerebral spinal fluid throughout the procedure. An injection of 4-methylypyrazole was given (0.8 mg/kg) through the jugular catheter. An intravenous infusion of ethanol (0.25 g/kg over a 10 min period) was given 30 min later. The catheter was flushed with saline after each infusion. Blood and dialysate samples were taken 60, 90, and 120 min after the ethanol infusion. Dialysate samples were collected for 2 min in duplicate at each time point. During the collection of the dialysate samples, a blood draw was taken for ethanol analysis. Following the last sample, the ethanol infusion and sampling procedure was repeated twice. Ethanol was quantitated as described above. The dialysate ethanol concentration is then plotted as a function of the blood level, and the slope of the regression is the *in vivo* E_d. This is also shown in Figure 12.5B for the same representative probe. Examination of the blood ethanol concentrations showed that the 4-methylpyrazole effectively slowed the metabolism of ethanol so that the concentration decreased by about 10% during the 60 min of sampling after each infusion of ethanol.

When we repeated the procedure for a group of probes, the E_d for ethanol diffusion out of the probe was 0.20 ± 0.01, and the E_d for ethanol diffusion in the opposite direction was 0.13 ± 0.01 (mean ± SEM for both values, n = 6). We then defined an *in vivo* E_d ratio (0.65 ± 0.03) for the overall experiment by dividing the *in vivo* recovery E_d by the *in vivo* delivery E_d. The *in vivo* E_d ratio can now be used to correct the directional bias in E_d that occurs with *in vivo* delivery. For example, subsequent probes can be calibrated using the *in vivo* delivery method. However,

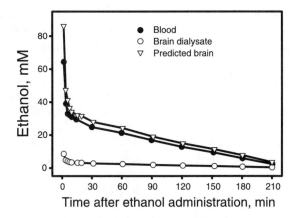

FIGURE 12.6 Whole blood, dialysate, and predicted extracellular brain ethanol concentrations after 1 g/kg ethanol administration (intravenous) in a representative rat. Extracellular brain ethanol concentrations were predicted from dialysate using *in vivo* delivery E_d measured for probe then corrected for diffusion direction bias. Resulting brain ethanol concentrations are very similar to those in blood, as expected from previous tissue studies. Note the high concentrations of ethanol achieved in brain and blood during first few minutes after intravenous bolus injection. These concentrations are not seen with the more commonly used technique of intraperitoneal injection at the dose given.

the E_d obtained with this method will overestimate the E_d by an average of 1.5 fold. To correct for this, the *in vivo* delivery E_d can be multiplied by the *in vivo* E_d ratio (0.65 for the present example) to predict the E_d that would be obtained from an *in vivo* recovery procedure. To calculate the predicted brain extracellular concentration of ethanol, the dialysate concentration is divided by the corrected E_d. Figure 12.6 illustrates the use of this probe calibration technique that combines both *in vivo* recovery and *in vivo* delivery. The method predicts that the extracellular concentration of ethanol in the brain matches well with the blood ethanol concentrations measured at various points after a bolus intravenous infusion of ethanol (1 g/kg).

Bungay et al. (2001) have recently analyzed the validity of using an E_d value obtained from a steady-state experiment to determine extracellular concentrations under transient conditions. The analysis suggests that there will be a systematic error in the predicted extracellular concentrations due to the lag period for transient conditions to approach steady state. The error will occur for both the ascending and descending phases of blood levels of a drug such as ethanol. However, our empirical data has validated the method we propose here. It is possible that the systematic error predicted by the analysis of Bungay et al. (2001) is too small to significantly affect a highly diffusible drug such as ethanol.

In summary, the data we obtained using probes constructed in our laboratory demonstrated a directional bias with respect to the E_d for ethanol. The E_d obtained with the *in vivo* delivery technique was significantly higher than the E_d obtained with *in vivo* recovery. Determination of brain extracellular concentrations during an experimental procedure, such as self-administration of ethanol, will occur under conditions that are similar to the *in vivo* recovery situation. Therefore, it is more appropriate to use the E_d associated with *in vivo* recovery of ethanol. However, performing an *in vivo* recovery calibration procedure for every probe and animal is not practical, and we suggest that individual probes be characterized using the *in vivo* delivery method to obtain E_d. A separate group of animals can then be used to perform a more rigorous comparison of *in vivo* delivery E_d and *in vivo* recovery E_d. Any differences can then be adjusted by using the ratio between the two. Overall, this proposed method accounts for individual differences in probe efficiency, while also correcting for the directional bias in ethanol flux. It is not clear at present whether the directional bias for ethanol flux will be present for other probes. Our data suggest that each laboratory should determine this on an individual basis.

12.4.3 TISSUE CONCENTRATION PROFILE FOR ETHANOL

One of the advantages of the microdialysis technique for pharmacological studies is that the probe can be used to deliver drugs as well as sample the extracellular fluid. In this way, several studies have been published in which ethanol solutions have been perfused through a microdialysis probe to expose the surrounding tissue to ethanol (Wozniak et al., 1991; Yan et al., 1996; Yim et al., 1998). Although the concentration of ethanol in the perfusate can be well controlled, the concentration of ethanol that reaches the tissue is more difficult to determine. However, interpretation of any potential neurochemical effect of ethanol delivered in this manner requires knowledge of the concentrations that may be producing the effect. In fact, when ethanol is delivered through a microdialysis probe, a concentration gradient exists in which the concentration of ethanol in the tissue varies as a function of the radial distance from the probe and axial distance along the probe. Therefore, any effects of ethanol observed when delivered through a probe are summed across a series of concentrations that are attained within the tissue. In any case, it is important to have a reliable estimate of the range of tissue concentrations of ethanol that result from diffusion across a probe membrane. For some analytes, the concentration gradient can be measured by perfusing the radiolabeled compound and determining the concentrations in the tissue by autoradiography (Dykstra et al., 1992; Dykstra et al., 1993). However, this is not feasible for ethanol due to its high volatility. At present, the best approach is to apply Bungay's steady-state model to estimate the tissue concentration profile for ethanol. We have now used this approach to estimate tissue concentrations of ethanol in rat striatum and hippocampus. Below we describe some of the experimental issues that are critical in carrying out this type of analysis. Modifications of the original steady-state theory and computational details for generating the predicted extracellular concentration profiles are not given here, but these can be found in Gonzales et al. (1998) and Ludvig et al. (2001).

Several key experimental parameters need to be determined to produce the estimate of the tissue concentration profile for ethanol. The extraction fraction for ethanol *in vitro* and *in vivo* is required. This can be determined by perfusing an ethanol-containing solution through the probe and measuring the loss of ethanol through the probe. The *in vitro* determination should be done in a beaker that is maintained at 37°C and is well stirred. In either case, the steady-state level for E_d is reached rapidly (within 10 min). For perfusion of a solution of 1% ethanol at 2.0

μl/min through a 3 mm probe we obtained an *in vitro* E_d of 0.39 ± 0.02 (n = 6). The E_d value we obtained for microdialysis in rat striatum *in vivo* was 0.21 ± 0.01 (n = 8).

Another important parameter needed is the total ethanol content in the tissue surrounding the probe during a steady-state perfusion with ethanol. Ideally, this determination will be carried out using the same probes for which the *in vivo* and *in vitro* E_d has been determined. Having all of this experimental information allows the calculation of clearance of ethanol that in turn allows an estimate for the tissue diffusivity of ethanol, D_t. The determination of tissue ethanol content is a difficult procedure because ethanol is so highly diffusible and volatile. A consequence of the high diffusivity is that the spatial variation in ethanol concentration in the tissue will not be readily maintained during the handling of the tissue prior to analysis. Maintaining the concentration gradients that exist in the tissue during microdialysis is not absolutely critical for producing the eventual estimate of the concentration profile. However, if ethanol diffuses away from the probe, the ability to detect it in the tissue will decrease because it will eventually be diluted to a value below the detection limit of the analytical instrument. This could contribute to an underestimate of the tissue content that will lead to an overestimate of the clearance. Similarly, the potential loss of ethanol during handling of the tissue prior to analysis of the ethanol content will result in an overestimate of the clearance. We freeze the tissue as rapidly as possible to stop the diffusion of ethanol before analysis of the tissue ethanol content by gas chromatography. For our striatal study, the rat was sedated with pentobarbital, and after steady-state perfusion was achieved, the head was removed by decapitation followed by dropping the whole head into liquid nitrogen. The frozen brain was then excised from the frozen head taking care to prevent thawing. The brain was then mounted on a cryostat chuck, the brain was sliced, and the tissue fragments were collected and analyzed for ethanol. This procedure resulted in a reproducible profile of ethanol content in brain slices, which was symmetric around the slice that contained the probe. However, the clearance rate constant for ethanol obtained in the striatal study (2.0 ± 0.3 min^{-1}) was higher than expected if ethanol is only cleared from the brain tissue by loss to blood flow. It is probable that our clearance estimate in this case was influenced by loss of ethanol from the tissue during the extensive cryostat slicing method. In a subsequent study in rat hippocampus we modified the procedure to dissect the tissue surrounding the probe before freezing the tissue fragment in liquid nitrogen. In this case, the clearance rate constant we obtained was 0.43 ± 0.07 min^{-1}, a value considerably lower than the one we obtained in our striatal study. Additional studies are required to obtain a better value for the clearance of ethanol *in vivo* during microdialysis.

When the required experimentally derived parameters for microdialysis of ethanol in the striatum are obtained, the distribution of ethanol in the extracellular space can be predicted. This predicted radial profile for ethanol at steady-state is shown in Figure 12.7. The profile shown is for the conditions specified in the striatal experiment and is not applicable to other conditions of flow rate or probe composition or geometry. It should also be emphasized that the profile is averaged across the axial dimension of the probe. In the tissue there will exist a series of profiles that will fall either above or below the average shown in Figure 12.7. In this way, a range of extracellular concentrations of ethanol can be predicted for a particular set of

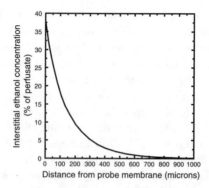

FIGURE 12.7 Predicted steady-state tissue concentration profile for ethanol perfused through microdialysis probe in rat striatum. Application of steady-state microdialysis theory of Bungay et al. (1990) provides a way to estimate extracellular concentrations of ethanol at various distances from probe–tissue interface. Profile shown represents an averageethanol concentration over the axial dimension. Prediction is valid only for probe geometry and composition that were used to derive theoretical parameters (Gonzales et al., 1998). In this case, probe parameters were 3.0 mm long, 270 μm diameter, 2.7 μl/min flow rate, regenerated cellulose membrane. The estimated profile suggests that, under these conditions, concentration of ethanol drops to approximately 35% of perfusate concentration at probe–tissue interface, and concentrations greater than 1% of perfusate extend up to 500 μm from probe membrane.

experimental conditions. Another profile was recently generated from experiments in hippocampus, and this is shown in Chapter 11.

Although the tissue concentration profiles generated in this way will be specific to the experimental conditions used, there are some general implications of the studies we have carried out so far. First, we now have firm experimental evidence that ethanol is highly diffusible within brain tissue, and the values we obtained for D_t are consistent with the existence of a transcellular diffusion pathway for ethanol as well as through the extracellular space. Second, the clearance of ethanol from brain tissue *in vivo* is consistent with removal to the blood and is relatively low compared with neurotransmitters that are actively metabolized or transported into cells. The combination of high tissue diffusivity for ethanol and low clearance lead to the prediction that the penetration of ethanol into the extracellular space under steady-state conditions is relatively deep (at least 0.5 mm from the probe–tissue interface). Therefore, perfusion of ethanol through a probe will likely affect cellular activity much farther from the probe than just the local area that is being sampled by the probe. For example, an interneuron may be affected by ethanol delivered through a probe that then can alter cellular activity of the neuron releasing a neurotransmitter that can be detected by the probe. In this case, any effect of ethanol on dialysate levels of a neurotransmitter may be an indirect effect due to a local circuit rather than a direct effect on the nerve terminals. Finally, it should be emphasized that perfusion of high concentrations of ethanol through a microdialysis probe likewise will generate high concentrations of ethanol within the tissue particularly within a few microns of the probe-tissue interface. Although the values will depend on the experimental conditions, the predictions we have made thus far suggest that under conditions commonly used in microdialysis experiments with ethanol, the tissue will be exposed to ethanol concentrations approximately one third of the perfused concentrations. Therefore, perfusion of concentrations of ethanol through a microdialysis probe greater than 200 mM should be interpreted with caution, and may represent pharmacological conditions that will not be applicable to a conscious behaving rat.

12.5 FUTURE DIRECTIONS AND CONCLUSIONS

Microdialysis has been and promises to be a widely used technique in alcohol research and in neuroscience research in general. However, the promise of the technique is as yet unfulfilled due to the lack of a fully validated theoretical model that will put the technique in the category of a fully characterized quantitative method. This chapter has presented a brief overview of the major concepts, difficulties, and applications of quantitative microdialysis to problems in alcohol research. It should be clear from this chapter that microdialysis will continue to be a useful experimental tool to study pharmacokinetics and pharmacodynamics of ethanol. No other technique is currently available to monitor ethanol levels in the brain of an experimental subject that will exhibit ethanol-related behaviors, such as operant responding or self-administration (see Chapters 9 and 10). It is possible that an enzyme-linked probe that is selective for ethanol analysis will be developed for use *in vivo*, but such a device is not available at present. Additionally, microdialysis can be used in combination with other techniques (such as single-cell electrophysiological recording) that can result in new powerful approaches for analysis of brain function (see Chapter 11). Furthermore, it deserves mention that microdialysis allows sampling of a variety of small molecules.

Other *in vivo* techniques exist that can be used to obtain specific neurochemical information, notably *in vivo* electrochemistry (Garris and Wightman, 1995; Gerhardt, 1995; Gonon, 1995). The major advantages of *in vivo* electrochemistry over microdialysis are that it can be used to monitor neurochemical events at much shorter time scales (millisecond range) and the probes used are much smaller (tens of microns), so they cause less tissue damage. However, *in vivo* electrochemistry also has significant disadvantages that microdialysis does not. For example, microdialysis sampling can take place over longer periods of time (days rather than hours). There are still significant issues with respect to chemical specificity of the signals obtained with electrochemical techniques, and electrochemistry can be used only with compounds that are electrochemically active. Finally, the current concentration detection limits of *in vivo* electrochemistry for an analyte like dopamine are generally not as sensitive as can be obtained with microdialysis combined with a high sensitivity technique like HPLC with electrochemical detection. For example, the current ability to detect changes in extracellular dopamine concentration with fast scan cyclic voltammetry is in the range of 50 nM (Kilpatrick et al., 2000). The ethanol-induced changes in extracellular concentration of dopamine that have been published are in the range of a few nM or less. In fact, Samson et al. (1997) were not able to detect ethanol-induced increases in extracellular dopamine using *in vivo* voltammetry. Therefore, the use of voltammetry may not be applicable to the study of ethanol-induced changes in extracellular dopamine.

The major drawback to the technique of microdialysis, whether quantitative or qualitative, is probably related to the probe-induced tissue trauma associated with the use of probes that are several hundred microns in diameter. The probe-induced damage could partly explain some of the discrepancies between microdialysis theory and experimental results. The tissue trauma inherent in the technique is likely to be difficult to control until smaller probes can be fabricated routinely. As discussed in section 12.2.3, the issue of tissue trauma has only recently been incorporated into theoretical models. Clearly, additional theoretical and experimental work is necessary to fully characterize the influence of tissue factors on the quantitative results obtained with microdialysis. However, it should be emphasized that reproducible and interpretable results with quantitative microdialysis have been obtained with the present technology and probe sizes. Therefore, the experimenter who uses this technique needs to maintain caution regarding the generality of results to healthy tissue *in vivo* and the accuracy of any quantitative conclusions.

There is a general trend in analytical chemistry toward miniaturization of instruments, and this trend has already proven to be advantageous to microdialysis experiments. For example, capillary electrophoresis with laser-induced fluorescence detection allows the quantitation of analytes in samples as small as a few nanoliters (Lada and Kennedy, 1995; Bert et al., 1996a; Bert et al., 1996b; Lada and Kennedy, 1996; Shear et al., 1996; Lada et al., 1997; Gostkowski et al., 1998). These small sample sizes provide an opportunity for more rapid sampling during dialysis experiments, and this has already been demonstrated with sampling intervals of less than 10 sec (Lada et al., 1998). Thus, the application of microdialysis is poised to enter a regime that is at least an order of magnitude closer to what is needed to relate *in vivo* dialysis measurements to synaptic transmission. With these technical improvements already present, and additional technical breakthroughs sure to come in the near future, we can predict that quantitative microdialysis will remain a very useful technique for understanding ethanol pharmacodynamics *in vivo* for years to come.

ACKNOWLEDGMENTS

This work was supported by grants from the National Institute on Alcohol Abuse and Alcoholism (AA11852, AA08484, AA07471), the Alcoholic Beverage Medical Research Foundation, and the Texas Commission on Alcohol and Drug Abuse. D.L.R. was supported by a training grant from National Institute on Alcohol Abuse and Alcoholism (AA07573). D.L.R. and A.T. were supported by a fellowship provided by the Waggoner Center for Alcohol and Addiction Research.

REFERENCES

Amberg G and Lindefors N (1989) Intracerebral microdialysis: II. Mathematical studies of diffusion kinetics. *J Pharmacol Methods* 22: 157-183.

Benjamin D, Grant ER and Pohorecky LA (1993) Naltrexone reverses ethanol-induced dopamine release in the nucleus accumbens in awake, freely moving rats. *Brain Res* 621: 137-140.

Bert L et al. (1996a) High-speed separation of subnanomolar concentrations of noradrenaline and dopamine using capillary zone electrophoresis with laser-induced fluorescence detection. *Electrophoresis* 17: 523-525.

Bert L et al. (1996b) Enhanced temporal resolution for the microdialysis monitoring of catecholamines and excitatory amino acids using capillary electrophoresis with laser-induced fluorescence detection analytical developments and *in vitro* validations. *J Chromatogr A* 755: 99-111.

Blanchard BA et al. (1993) Sex differences in ethanol-induced dopamine release in nucleus accumbens and in ethanol consumption in rats. *Alcohol Clin Exp Res* 17: 968-973.

Bungay PM et al. (2001) Probe calibration in transient microdialysis *in vivo*. *Pharm Res* 18: 361-366.

Bungay PM and Gonzales RA (1996) Pressure-enhanced delivery of solutes via microdialysis. *Soc Neurosci Abstracts* 22:2076.

Bungay PM, Morrison PF and Dedrick RL (1990) Steady-state theory for quantitative microdialysis of solutes and water *in vivo* and *in vitro*. *Life Sci* 46: 105-119.

Bungay PM et al. (1999) Model for microdialysis probe implantation trauma employing a layer of tissue with altered properties adjacent to the probe. In: *Monitoring Molecules in Neuroscience, Proceedings of the 8th International Conference on In Vivo Methods* (Rollema H et al., Eds.), pp. 57-58. Newark: The State University of New Jersey.

Crippens D et al. (1999) Gender differences in blood levels, but not brain levels, of ethanol in rats. *Alcohol Clin Exp Res* 23: 414-420.

Di Chiara G and Imperato A (1988) Drugs abused by humans preferentially increase synaptic dopamine concentrations in the mesolimbic system of freely moving rats. *Proc Natl Acad Sci USA* 85: 5274-5278.

Dykstra, KH et al. (1993) Microdialysis study of zidovudine (AZT) transport in rat brain. *J Pharmacol Exp Ther* 267:1227-1236.

Dykstra, KH et al. (1992) Quantitative examination of tissue concentration profiles associated with microdialysis. *J Neurochem* 58: 931-940.

Ferraro TN et al. (1990) Continuous monitoring of brain ethanol levels by intracerebral microdialysis. *Alcohol* 7: 129-132.

Garris PA and Wightman RM (1995) Regional differences in dopamine release, uptake, and diffusion measured by fast-scan cyclic voltammetry. In: *Neuromethods, Vol. 27: Voltammetric Methods in Brain Systems* (Boulton AA, Baker GB and Adams RN, Eds.), pp 179-220. Totowa: Humana Press.

Gerhardt GA (1995) Rapid chronocoulometric measurements of norepinephrine overflow and clearance in CNS tissues. In: *Neuromethods, Vol. 27: Voltammetric Methods in Brain Systems* (Boulton AA, Baker GB and Adams RN, Eds.), pp 117-151. Totowa: Humana Press.

Gonon F (1995) Monitoring dopamine and noradrenaline (norepinephrine) release in central and peripheral nervous systems with treated and untreated carbon-fiber electrodes. In: *Neuromethods, Vol. 27: Voltammetric Methods in Brain Systems* (Boulton AA, Baker GB and Adams RN, Eds.), pp 153-177. Totowa: Humana Press.

Gonzales RA et al. (1998) Quantitative microdialysis of ethanol in rat striatum. *Alcohol Clin Exp Res* 22: 858-867.

Gostkowski ML, Wei J and Shear JB (1998) Measurements of serotonin and related indoles using capillary electrophoresis with multiphoton-induced hyperluminescence. *Anal Biochem* 206: 244-250.

Imperato A and Di Chiara G (1986) Preferential stimulation of dopamine release in the nucleus accumbens of freely moving rats by ethanol. *J Pharmacol Exp Ther* 239: 219-228.

Jacobson I, Sandberg M and Hamberger A (1985) Mass transfer in brain dialysis devices: a new method for estimation of extracellular amino acid concentration. *J Neurosci Methods* 15: 263-268.

Justice JB Jr. (1993) Quantitative microdialysis of neurotransmitters. *J Neurosci Methods* 48: 263-276.

Kenakin T (1997) Statistical assessment of biological significance from data. In: *Pharmacological Analysis of Drug-Receptor Interaction, 3rd ed.*, pp 198-241. Philadelphia, PA: Lippincott-Raven.

Kehr J (1993) A survey on quantitative microdialysis: theoretical models and practical implication. *J Neurosci Methods* 48: 251-261.

Kiianmaa K et al. (1995) Effect of ethanol on extracellular dopamine in the nucleus accumbens of alcohol-preferring AA and alcohol-avoiding ANA rats. *Pharmacol Biochem Behav* 52: 29-34.

Kilpatrick MR et al. (2000) Extracellular dopamine dynamics in rat caudate-putamen during experimenter-delivered and intracranial self-stimulation. *Neurosci* 96: 697-706.

Lada MW and Kennedy RT (1995) Quantitative *in vivo* measurements using microdialysis on-line with capillary zone electrophoresis. *J Neurosci Methods*, 63: 147-152.

Lada MW and Kennedy RT (1996) Quantitative *in vivo* monitoring of primary amines in rat caudate nucleus using microdilaysis coupled by a flow-gated interface to capillary electrophoresis with laser-induced fluorescence detection. *Anal Chem* 68: 2790-2797.

Lada MW, Vickroy TW and Kennedy RT (1997) High temporal resolution monitoring of glutamate and aspartate *in vivo* using microdialysis on-line with capillary electrophoresis with laser-induced fluorescence detection. *Anal Chem* 69: 4560-4565.

Lada MW, Vickroy TW and Kennedy RT (1998) Evidence for neuronal origin and metabotropic receptor-mediated regulation of extracellular glutamate and aspartate in rat striatum *in vivo* following electrical stimulation of the prefrontal cortex. *J Neurochem* 70: 617-625.

Lindefors N, Amberg G and Ungerstedt U (1989) Intracerebral microdialysis: I. Experimental studies of diffusion kinetics. *J Pharmacol Methods* 22: 141-156.

Lönnroth P, Jansson PA and Smith U (1987) A microdialysis method allowing characterization of intercellular water space in humans. *Am J Physiol* 253: E228-E231.

Lu Y, Peters JL and Micheal AC (1998) Direct comparison of the response of voltammetry and microdialysis to electrically evoked release of striatal dopamine. *J Neurochem* 70: 584-593.

Ludvig, N et al. (2001) Evidence for the ability of hippocampal neurons to develop acute tolerance to ethanol in behaving rats. *Brain Res* 900: 252-260.

Morrison PF et al. (1991) Quantitative microdialysis: analysis of transients and application to pharmacokinetics in brain. *J Neurochem* 57: 103-119.

Nurmi M, Kiianmaa K and Sinclair JD (1994) Brain ethanol in AA, AN, and Wistar rats monitored with one-minute microdialysis. *Alcohol* 11: 315-321.

Nurmi M, Kiianmaa K and Sinclair JD (1999) Brain ethanol levels after voluntary ethanol drinking in AA and Wistar rats. *Alcohol* 19: 113-118.

Olson RJ and Justice JB Jr. (1993) Quantitative microdialysis under transient conditions. *Anal Chem* 65: 1017-1022.

Parsons LH and Justice JB Jr. (1992) Extracellular concentration and *in vivo* recovery of dopamine in the nucleus accumbens using microdialysis. *J Neurochem* 58: 212-218.

Parsons LH and Justice JB Jr. (1994) Quantitative approaches to *in vivo* brain microdialysis. *Crit Rev Neurobiol* 8: 189-220.

Peters JL and Michael AC (1998) Modeling voltammetry and microdialysis of striatal extra-cellular dopamine: the impact of dopamine uptake on extraction and recovery ratios. *J Neurochem* 70: 594-603.

Pettit HO and Justice JB Jr. (1991) Procedures for microdialysis with smallbore HPLC. In: *Microdialysis in the Neurosciences* (Robinson TE and Justice JB, Jr., Eds.), pp 117-153. New York: Elsevier.

Robinson DL et al. (2000) Quantification of ethanol concentrations in the extracellular fluid of the rat brain: *in vivo* calibration of microdialysis probes. *J Neurochem* 75: 1685-1693.

Samson HH et al. (1997) The effects of local application of ethanol in the n. accumbens on dopamine overflow and clearance. *Alcohol* 14: 485-492.

Shear JB, Brown EB and Webb WW (1996) Multiphoton-excited fluorescence of fluorogen-labeled neurotransmitters. *Anal Chem* 68: 1778-1783.

Smith AD and Justice JB Jr. (1994) The effect of inhibition of synthesis, release, metabolism and uptake on the microdialysis extraction fraction of dopamine. *J Neurosci Methods* 54: 75-82.

Vinson PN and Justice JB (1997) Effect of neostigmine on concentration and extraction fraction of acetylcholine using quantitative microdialysis. *J Neurosci Methods* 73: 61-67.

Wozniak KM et al. (1991) Focal application of alcohols elevates extracellular dopamine in rat brain: a microdialysis study. *Brain Res* 540: 31-40.

Yan QS et al. (1996) Focal ethanol elevates extracellular dopamine and serotonin concentrations in the rat ventral tegmental area. *Eur J Pharmacol* 310: 49-57.

Yim HJ and Gonzales RA (2000) Ethanol-induced increases in dopamine extracellular concentration in rat nucleus accumbens are accounted for by increased release and not uptake inhibition. *Alcohol* 22: 107-115.

Yim HJ et al. (2000) Dissociation between the time course of ethanol and extracellular dopamine concentrations in the nucleus accumbens after a single intraperitoneal injection. *Alcohol Clin Exp Res* 24: 781-788.

Yim HJ et al. (1998) Comparison of local and systemic ethanol effects on extracellular dopamine concentration in rat nucleus accumbens by microdialysis. *Alcohol Clin Exp Res* 22: 367-374.

Yoshimoto K et al. (1991) Alcohol stimulates the release of dopamine and serotonin in the nucleus accumbens. *Alcohol* 9: 17-22.

Yoshimoto K and Komura S (1993) Monitoring of ethanol levels in the rat nucleus accumbens by brain microdialysis. *Alcohol* 28: 171-174.

Zhao Y, Liang X and Lunte CE (1995) Comparison of recovery and delivery *in vitro* for calibration of microdialysis probes. *Anal Chim Acta* 316: 403-410.

13 Structural Magnetic Resonance Imaging (MRI) of the Human Brain

Terry L. Jernigan

CONTENTS

13.1 INTRODUCTION: MODERN *IN VIVO* IMAGING TECHNIQUES OPEN A WINDOW TO THE BRAIN

The availability of *in vivo* brain imaging techniques has led to dramatic advances in neuroanatomical investigation of psychiatric disorders. Until less than three decades ago, evidence for brain abnormalities in psychiatric disorders came almost entirely from autopsy studies. While neuropathological examination was, and remains, a powerful investigative tool, the limitations of the approach are particularly severe for the study of neuropsychiatric disorders because of their early onset and decades-long duration. Magnetic resonance imaging (MRI) provides the opportunity to study living patients at various points in the course of the illness as well as before and after treatment. The need for *in vivo* methods is particularly acute in the study of alcohol effects, both those associated with prenatal exposure and those resulting from chronic abuse in adults. In the former case, the full implications of fetal alcohol exposure must be defined in terms of interactions between effects of early alcohol exposure and protracted processes of brain maturation. These are best described with information about the nature of the brain anomalies present at multiple points during development, information that would be very difficult to obtain with autopsy material. Similarly, given the long histories of alcohol dependence with which many patients seeking treatment present, there is a need to separate the effects of duration of alcoholic drinking from those of normal aging, and to describe any interactions between these effects. Again, the availability of an informative, noninvasive method for examining the human brain has made the study of such interactions feasible.

13.2 STRUCTURAL MAGNETIC RESONANCE IMAGING (SMRI) METHODS

13.2.1 Physical Basis for sMRI

MRI of the type described here is possible because the abundant protons in our body tissues, when placed in a magnetic field, behave like small magnets. MRI of the brain involves creating a strong static magnetic field into which the subject is placed for imaging. Field strengths of magnets used for MRI vary, but it is particularly common for clinical imagers today to create a 1.5 Tesla magnetic field. Protons tend to align parallel to this applied magnetic field, much as small bar magnets would. In reality, the protons resemble spinning tops precessing around the axis of the static field (Cohen and Bookheimer, 1994, Figure 13.1). They may be oriented parallel to, or in the

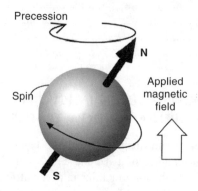

FIGURE 13.1 The behavior of protons in a static magnetic field (Cohen and Bookheimer, 1994).

opposite orientation to, the magnetic field, but, because they are more likely to align parallel to the field, there is net magnetization in the tissues along the long (z) axis of the static magnetic field. The rate of precession of the protons is determined by the field strength, thus all of the protons precessing in a homogenous magnetic field will precess at the same rate. This precession results in a rotating magnetic field at right angles (orthogonal) to the long axis of the applied static field. In other words, there is, for each proton, a rotating magnetic moment in the x-y plane. However, the spins of the protons are at random phases; so, at equilibrium, there is no net magnetization in the x-y plane of the field, because the rotating components cancel each other out. The MR signal is created by applying a radiofrequency (RF) pulse orthogonal to the long (z) axis of the static field and at the precession frequency of the protons. This pulse creates a rotating magnetic field in the x-y plane by inducing greater phase coherence among the individual rotating fields of the protons. In other words, the RF pulse creates net magnetization in the x-y plane where there was no net magnetization before the application of the RF pulse. The resulting magnetization produces an electromagnetic signal that can be detected as the MR signal.

Much of the useful information in the MR signal can be extracted by modeling the decay (or relaxation) of the signal. Two important parameters in the model are the exponential time constants that describe what is referred to as longitudinal relaxation rate, or T1, and transverse relaxation rate, or T2. The former relates to the rate at which equilibrium magnetization occurs when the protons are placed in the static magnetic field. The latter refers to the loss of signal (i.e., magnetization in the x-y plane), associated with loss of phase coherence among the rotating fields of the protons, after the RF pulse is discontinued. This transverse relaxation rate, or T2, is influenced by several different factors, including interactions among the protons, various other sources of inhomogeneity in the magnetic field, and diffusion (or motion). Different pulse sequences result in differing contributions of these different factors to the MR signal.

Thus far, the process by which an MR signal is produced within an ensemble of protons has been described. However, MR images would be of little interest if they provided only information about the nature of the composite MR signal produced by the body's protons. To reveal the internal structure of the body, the MR signal from different locations in the body must be isolated from each other. This is possible because, as mentioned previously, the precession frequency of protons is determined by field strength. To isolate MR signals from different spatial locations in the body, magnetic field gradients are introduced across the field of view. This results in variability in the precession frequency across the field of view, due to the variation in field strength, and the signal from different points along the gradient can therefore be isolated by frequency analysis.

13.2.2 sMRI Provides Multiple Views
of the Underlying Anatomy

Underlying any MR image is a two-dimensional matrix of numbers. The information in the matrix is visualized by assigning gray levels or colors to the values in the matrix and displaying the information as an image. Whether a particular kind of

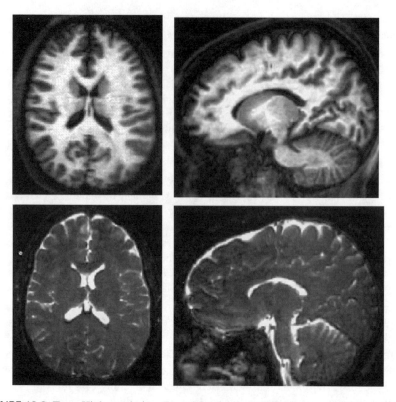

FIGURE 13.2 Top: High resolution T1-weighted spiral FSE images exhibiting excellent tissue contrast. Bottom: High resolution T2-weighted spiral FSE images. (Wong et al., 2000).

anatomical information is detectable visually in the image depends upon both the degree to which that information is present within the numbers themselves and the degree to which it is preserved by the gray scaling or color coding used to display the image data. As described above, the MR signal is complex, and different pulse sequences produce MR signals more or less strongly influenced by such factors as T1, T2, etc. Because different kinds of biological information are represented in different components of the MR signal, and because the different components can be differentially weighted by varying the pulse sequence, many different views of the underlying anatomy are possible with MRI. Generally speaking, T1-weighting of images (such as those in the top panels of Figure 13.2) results in good contrast between the gray and white matter of the brain. Strong T2-weighting (bottom panels of Figure 13.2) yields strong contrast between brain and CSF structures, and better sensitivity to the subtle increases in tissue water content that can occur in association with ischemia, gliosis, edema, demyelination, and other tissue abnormalities.

The level of anatomical detail possible with MRI has increased rapidly over the last decade. It is now routine to acquire full-brain image volumes with 1 mm resolution in all three axes, and specialized procedures permit even higher spatial resolution in more focused fields of view. Many sMRI procedures also afford excellent tissue contrast (see

Figure 13.2). However, in spite of the remarkable visual quality of these images, accurate quantitation and interpretation of the information they provide remain difficult problems.

13.2.3 THE INTERPRETATION AND QUANTITATION OF IMAGING SIGNALS

In the early years after the introduction of MRI, much clinical research was needed to describe the features and anomalies that could be detected visually with this new technique. Modifications to the acquisition protocols were devised to improve tissue and lesion contrast and thereby enhance the detectability of such features. This important work continues apace, and new methods for extracting novel, biologically informative signals from the brain are constantly emerging. However, investigators soon became dissatisfied with the limitations of purely visual, and necessarily subjective, interpretations of the imaging data. This was nowhere a more significant problem than in the investigation of neuropsychiatric disorders in which the abnormalities, when they are present, are rarely focal and are often very subtle. It was soon surmised that the inherently numerical nature of the imaging data could be further exploited by developing and applying more sensitive quantitative methods.

Although the merits of objective quantitative methods for characterizing brain structure with MRI are beyond dispute, a number of significant challenges are nevertheless associated with their application. MRI produces large matrices of signal values in which each value represents the estimated signal from a volume unit (voxel) within the imaged object (or field of view). The variability in voxel signal values reflects the variable characteristics of the underlying tissues. As described above, visualization of the structure is possible by coding the signal values with a gray scale or a color scale. However, while an expert observer can readily identify the underlying structure, the images contain, in addition to the information of interest, considerable signal fluctuation that is artifactual, or irrelevant. For example, signals from structures of the brain itself are accompanied in the images by signals from cranial and extracranial structures (such as skull, eyes, facial tissues, etc.) that are themselves of little interest to neuroscientists. However, these signals can, in some cases, be numerically similar to brain signals and can therefore contaminate measurements of brain structures. Furthermore, due to imperfect magnetic field gradients, motion, or magnetic properties of the tissues themselves, the signals can be distorted, leading to subtle artifacts that can nonetheless significantly degrade measurement validity. For example, most MRI datasets (3-D matrices of signal values) exhibit low frequency signal drift across the field of view that results in slightly different values for the same tissue in different parts of the brain. When attempting to detect subtle tissue changes or abnormalities, this fluctuation can introduce an unacceptable level of noise. Methods for addressing these and other problems have been developed and will be outlined below.

The focus of this chapter is on the use of MRI for the study of brain structure. Most of the studies to be reviewed here involve MR morphometric techniques; i.e., they have examined the effects of alcohol exposure on brain morphology with quantitative measures of specific brain structures. The impetus for these studies came from earlier observations with x-ray computed tomography, particularly of patients with histories of chronic alcoholism. Computed tomography (CT) of the brain yields only limited contrast among the soft tissues, but low-density cerebrospinal fluid (CSF) within the cerebral ventricles, the subarachnoid spaces, and within areas of focal damage, is well visualized. Although chronic alcoholic patients rarely have focal findings on CT, even subjective (visual) evaluation of CT in these patients often yields reports of enlarged CSF spaces (see Figure13.3). Early quantitative studies with CT quickly confirmed the association of alcoholic drinking with increased CSF spaces and these findings were interpreted as alcohol-related atrophy of the brain (Bergman et al., 1980; Ron et al., 1980; Cala and Mastaglia, 1981; Jernigan et al., 1982; Wilkinson, 1982); in other words, it was assumed that loss or shrinkage of brain tissue had resulted in passive expansion of the cerebral ventricles and subarachnoid spaces. Unfortunately, the limited tissue contrast afforded by CT precluded further studies with this method to determine which tissues and structures exhibited the most volume loss in alcoholic patients. Later MRI studies exploited the improved tissue contrast and spatial resolution to extend these early CT studies by estimating the volumes of individual brain tissues and structures. In the following section, the basic methodological requirements for such studies are outlined.

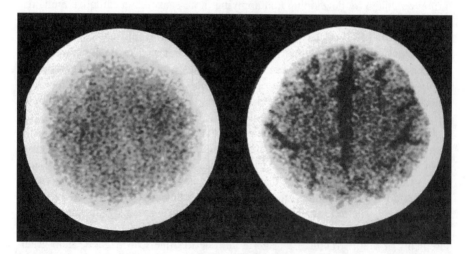

FIGURE 13.3 Sections from x-ray computed tomography examinations of two individuals. The sections are both from near the vertex (through the cerebral cortex). Section on left is from a normal control and shows little CSF in cortical sulci. Section on right is from a chronic alcoholic of similar age, and shows the increase in sulcal CSF often reported in these patients.

Specific methods for volumetric studies with MRI vary across laboratories, but most involve a set of important steps listed below:

- *Isolation of brain from nonbrain areas within the images (sometimes referred to as stripping)* — A surprisingly challenging problem is the separation of the brain from adjacent structures. While trained anatomists can generally distinguish the boundaries of the brain fairly reliably, the process of "tracing" these boundaries within each imaged brain section can be tedious and time consuming. Also, within sections at the edges of the brain, the problem is complicated by the presence of overlapping structures with similar signal to brain. Most laboratories have adopted substantially automated methods that exploit the change in signal values at the interfaces between the cerebral cortex, the adjacent subarachnoid space, and the cranium; however, in most cases some human supervision (or editing) of the process is necessary to achieve results acceptable for volumetric studies.

- *Filtering or other methods for correcting signal drift in image values* — As described above, the signal values for a given tissue sometimes differ in different parts of the image due primarily to magnetic field inhomogeneity. A variety of bias-correction methods have been employed (Arnold et al., 2001) to reduce these effects, most of which produce similar results. Though recent improvements in instrumentation have mitigated this problem to some extent, most investigators still find that better tissue segmentation results are obtained after processing the image volumes to further reduce drift artifacts.

- *Identification of voxels within different tissues (usually white matter, gray matter and CSF)* — Since the ultimate goal of these methods is to estimate the size of specific structures, and since a structure is usually defined in part by its tissue type, it is important to operationalize the definition of the different tissues. For example, when attempting to estimate the volume of the cerebral cortex, one generally intends to include the gray matter of the cortical ribbon while excluding the underlying white matter. Automated, or semi-automated, tissue classification is therefore an important step in the process of extracting a usable 3D model of the brain's structure from a raw MR image volume. The utility of efficient methods for tissue segmentation of MR images has been widely acknowledged (Jernigan and Ostergaard, 1993; Caviness et al., 1995; Collins et al., 1995; Andreasen et al., 1996), and several approaches have emerged see (Bezdek et al., 1993; Bonar et al., 1993; Cagnoni et al., 1993; Clarke et al., 1993; Fletcher et al., 1993; Clarke et al., 1994; Jackson et al., 1994; Kao et al., 1994; Kennedy et al., 1994; Zijdenbos and Dawant, 1994; Collins et al., 1995; Simmons et al., 1996; Vaidyanathan et al., 1997; Vinitski et al., 1997) for reviews. However, the basic rationale underlying tissue segmentation merits some scrutiny. Most schemes attempt either a) to classify each voxel as gray matter, white matter, or CSF (sometimes additional categories, such as lesion or tumor are included), or b) to assign some portion of each voxel to each of these categories. The latter are sometimes referred to as "partial-volume" methods and are inherently appealing because it is known that many voxels span the boundaries of gray, white, and CSF structures, and thus contain a mix of tissues. Although a more accurate volume estimate of a given tissue is obtained from a partial-volume method than a classification method, in some cases the simpler structure of the segmented image volume resulting from a classification method is of greater value to the investigator than the increase in sensitivity gained by a partial-volume method. If a classification method assigns each voxel to the tissue category that represents the largest percentage of its volume, an equally accurate partial-volume method affords little increase in sensitivity when the voxel size is small relative to the sizes of the target structures. However, when the target structure has an expected volume that is a small multiple of the voxel size, the advantages of partial-volume methods can be considerable. Since most morphometric studies reviewed below examined brain structures hundreds to thousands of times larger than the voxels, it is not surprising that most of these studies used voxel classification methods.

From a practical point of view, a much more significant challenge to tissue segmentation occurs in the form of signal inhomogeneity within tissue classes. As mentioned above, non-biological sources of signal non-uniformity, such as RF inhomogeneity, contribute substantially to this variability. However, biological sources also exist. For example, even to the naked eye, the MR signal is not uniform across regions deemed by anatomical nomenclature to be "white matter" regions. This is because the white matter is, to some extent, biochemically heterogeneous. Similar heterogeneity is also present within the cerebral gray matter. With most MR acquisition sequences, this results in some fully-volumed voxels within the "white matter" regions having signal characteristics identical to those of some fully-volumed voxels within the "gray matter" regions. When this occurs because of variability within a tissue class in the biochemical composition of the voxels, it is not always clear what is meant by "white matter" and "gray matter" (i.e., is it defined by location or by MR signal value). There also exists variation across individual brains in the signal characteristics of the tissues, and in the tissue contrast, and this variation is by no means limited to "pathological" cases. These factors considerably complicate the already formidable task of validating tissue classification methods, and readers of the MR morphometry literature should consider them when interpreting the results.

Most methods identify tissues by first defining characteristic signal values for each of the tissue types to be distinguished. Supervised methods generally require that an operator select regions within the tissues of interest, and then these values are used to "seed" the segmentation algorithm. In other words, an operator designates certain voxels to be sampled for an estimate of the characteristic value of each tissue. A few methods "seed" the algorithm with voxel values selected using an automated method. Other methods, not requiring supervision, estimate the characteristic tissue values by assuming that the histogram of signal values from the matrix consists of overlapping distributions from a given number of tissues with distinct values. By making assumptions about the underlying distributions, the best fit to the model can be determined statistically and the characteristic value of each tissue can be estimated from the histogram. With either approach, after the values for the different tissues have been defined, each voxel can either be classified based on the similarity of its signal value to these tissue values, or each voxel's tissue composition can be estimated using a linear model.

Most of these tissue segmentation methods produce reasonable, and similar, results when applied to human MRI data. They also produce similar results when tested on phantom data (images of models constructed with simulated "tissues"), and they produce highly reproducible results (particularly the more highly automated ones, of course) when applied repeatedly to the same dataset. Unfortunately, few have been validated on serial MRI examinations in the same individuals. As mentioned above, the assumption on which these tissue segmentation methods are based, namely the assumption that the brain consists of a discrete number of tissue types with uniform values within each type, is not strictly true. Therefore, the results of all of the methods contain errors. Nevertheless, the results of different methods applied to MRI from young healthy volunteers are remarkably similar, and have good face validity (i.e., the segmentations look right). Unfortunately, there is ample evidence that the signal values of brain tissues are affected by factors associated with aging, and by toxicity and disease. This raises concern about the effects of such factors on results of tissue segmentation in older individuals and patients, since violation of the segmentation models may be more severe in these cases. An association of segmentation bias with diagnosis or other clinical variables can lead to significant confounding of the experimental effects.

- *Delineation of the boundaries of different brain structures of interest* — Because most MR morphometry studies of alcohol effects have focused on structure volumes, the accurate delineation of the boundaries of target structures has been of utmost importance. If, for example, one is attempting to measure the volume of the amygdala in an attempt to detect any alcohol-related loss of volume in that structure, it is necessary to define the contour that separates the amygdala from adjacent entorhinal cortex, basal forebrain, hippocampus, and white matter. Any inconsistency in setting these boundaries can result in considerable unreliability in the estimated amygdalar volumes. Because of the difficulty involved, most investigators have relied upon manual delineation of boundaries by trained anatomists. Nevertheless, for many brain structures of interest, there are few gross morphological features visible on MRI that can be used to define their boundaries. For this reason, precise criteria must be developed, sometimes involving the positions of nearby anatomical landmarks, to guide operators in setting reliable boundaries. Often, it is not possible to define boundaries for a brain structure following cytoarchitectural conventions. In these cases, there are two goals: first, to define in a highly consistent way the boundaries of a region that includes the structure and, second, to ensure that the structure of interest contributes a sufficiently large percentage of the total volume of that region that any variability in the volume of the structure is likely to make up the majority of the variability in the volume of the total region.

When reporting the results of volumetric studies, investigators usually report both the specific criteria used to define boundaries between structures and provide estimates of the reliability with which human operators can reproduce the boundaries using those criteria.

- *Computation of volume estimates (and correction for overall cranial volume)* — Once the boundaries have been defined, structural volumes can be estimated by summing the voxels, or voxel tissue proportions, within the boundaries, usually across the imaged brain sections containing the structures. Most investigators are well aware that these estimates of raw volume show very wide variability across individuals. For example, raw volumes of the caudate nuclei of two adult individuals can easily vary by a factor of nearly two. Much of this variation appears simply to be related to body size; larger people tend to have larger heads (and brains) than do smaller people. Consistent with this, women have smaller heads than do men. Furthermore, there appears to be additional individual difference variability in head size, just as there is in nose size. This large degree of individual difference variability in head size, which is also present in volumes of individual brain structures, is a problem when one is attempting to infer volume *loss* from volume. When there is good reason to believe that the individuals studied had normal brain development (i.e., that any loss occurred after the brain reached its full size), it is often helpful to correct the raw volumes for overall cranial volume. The logic here is that the cranial volume indexes the premorbid brain size, and that any reduction in the volume of a brain structure relative to the size of the cranial volume is likely to reflect volume loss in that structure. Frequently, the volumes are expressed as proportions of cranial volume. Most results of volumetric studies in adult alcoholics and their controls are reported in this way. Obviously, however, the same logic cannot be applied to estimate the degree of brain hypoplasia. In this case, the cranial volume is itself reduced as a result of the pathological process.

13.2.4 WITHIN-SUBJECT (TEMPORAL) VARIABILITY: CHRONICITY AND DEVELOPMENTAL CHANGE

Most MRI studies of alcohol effects have been cross-sectional, controlled group studies. In these studies, individuals of widely varying age and with widely varying levels of alcohol exposure have been examined. Furthermore, in studies of chronic alcoholics, though most subjects have been detoxified when imaged, they may have varied considerably in terms of the time elapsed between the cessation of drinking and the imaging examination. The interpretation of the results of these studies is complicated by the fact that any alcohol effects present are occurring in the context, not only of substantial individual variability, but also on a background of dynamic change in brain structure associated with processes of brain maturation, involution, and structural restitution (in the case of recovering alcoholics). Of particular concern is the possibility that such processes may interact with the alcohol-induced effects. As an example, consider the relationship between aging and cumulative alcohol exposure in chronic alcoholic patients. It is known that there are brain structural changes that accompany aging in the absence of alcohol abuse (see Jernigan et al., 2001 for a recent summary from the author's laboratory). However, on average, patients with more extensive alcohol histories are older. To fully understand the effects of alcohol exposure one should take care to model not only the effects of diagnosis, but also those of aging, cumulative alcohol intake, and the interactions between these factors.

A similar dilemma exists when studying the effects of fetal alcohol exposure on brain development. Most studies include children or adolescents of different ages; however, the brain's structure changes substantially during normal development and the changes continue well into adulthood (Jernigan et al., 1991c; Sowell and Jernigan, 1998; Sowell et al., 1999b; Sowell et al., 1999a). Again, a full description of the effects of fetal alcohol exposure on brain maturation should include estimates of the effects of varying levels of alcohol exposure on the entire process of brain maturation, as well as the interaction of these factors.

As already mentioned, a significant advantage of MRI over other methods for examining brain structure is the possibility of studying individuals repeatedly over time. This is a particularly important advantage in the study of alcohol effects, for reasons given above. Longitudinal studies employing within-subject methods could be particularly helpful in isolating the effects of different factors. Unfortunately these studies are rare.

13.2.5 VOXEL-BASED ANALYSES OF BRAIN DYSMORPHOLOGY

In recent years, methods have emerged that attempt to characterize brain morphological effects on MRI on a voxel-by-voxel basis. These methods have the potential to provide a more detailed description of the shape anomalies produced by hypoplasia or degeneration in the brain than the conventional volumetric techniques described above. Although voxel-based approaches sometimes require the identification of anchoring landmarks within the brain, they obviate the need to delineate the complete outside boundaries of structures using conventional anatomical criteria. The desired results of these methods are statistical maps of the brain revealing the pattern of structural anomaly present in a group of patients (relative to a control group) with high spatial resolution. Because the precise anatomical plane of section varies in MRI volumes, and because there is considerable normal subject-to-subject variability in brain morphology, these methods require that the datasets across individuals be registered to each other accurately, and that the variability in each voxel location associated with group membership be compared statistically with that occurring within the experimental groups.

As in the case of the more conventional morphometric techniques described above, a significant challenge to these methods is the large degree of variability in head size. One approach to this problem is to "size normalize" the data in the MR images before examining the distribution of shape anomalies. This results in considerably more precision in the maps. Normalizing the maps using a strictly linear reduction or expansion of the voxels leads to a reasonably straightforward distribution for subsequent shape analysis, but there is still considerable variability in the

MR signals at a given voxel location due to individual differences in brain shape. Voxel-based approaches to describing the abnormalities resulting from degenerative processes in patients with adult-onset disorders have generally used such linear methods for size normalization. Here, an assumption that the overall size of the brain (or other structure to be examined) is unrelated to clinical status seems reasonable, and, in fact, this assumption can be examined empirically by comparing the groups first on these measures before performing voxel-based shape analysis. Under these circumstances, a voxel-wise map of the variability associated with clinical group (taking into account the within-group variability) should reveal where in the brain or structure the greatest morphological anomalies are present.

It should be noted that the effect of size-normalization must be considered carefully when using voxel-based approaches to compare morphology in groups that differ in overall brain, or structure, size. Imagine, for example, that a brain structure is smaller in a clinical group than in controls, and further, that this size reduction is entirely due to a reduction of the more anterior parts of the structure. Assuming that the structures are properly aligned in a standard anatomical space, voxel-wise statistical maps (comparing the groups) would be expected to reveal anomalies in the anterior, but not the posterior, parts. However, if within-group size variability is great, the resulting noise at each voxel that is due to this factor may reduce power. On the other hand, if a size normalization procedure were employed to render each structure equal in size, the statistical maps would be expected to show anomalies throughout the extent of the structure, with anterior portions exhibiting anomalies in one direction and posterior portions exhibiting anomalies in the opposite direction. In other words, the maps would show that when size is held constant, the smaller anterior portions are *relatively* smaller, but the posterior portions are also *relatively* larger in the clinical group. This is strictly true, but in this case it should probably not be interpreted to mean that the posterior as well as the anterior portions are *abnormal*. This is just one example of the interpretive difficulties that arise when examining morphology using voxel-based maps in subjects in whom systematic differences in size of the structures are present. Note that similar problems could occur in mapping age-effects on tissue composition in samples within which brain size varies systematically with age.

Given these caveats, it is clear that mapping, using voxel-based methods, of morphological anomalies in groups like FAS subjects with known overall brain hypoplasia must be approached very cautiously. However, such methods have considerable potential to reveal differences not apparent in more conventional analyses; and their application is substantially more efficient.

13.3   APPLICATIONS OF sMRI IN ALCOHOL RESEARCH

13.3.1   STUDIES OF BRAIN STRUCTURE IN CHRONIC ALCOHOLICS

As described above, early CT studies suggested that many chronic alcoholic patients had suffered brain volume loss as indexed by increased CSF. MRI was the first method with which the volumes of individual brain structures could be estimated in living individuals. This made it possible to survey the regional pattern of brain volume loss in chronic alcoholic patients to determine whether some structures appeared to be more vulnerable than others to the effects of alcohol abuse. In the author's laboratory, volumetric methods of the type described in the previous section were developed for application to MRI. The first study examined the effects of normal aging, uncomplicated by substance abuse or other significant neurological or neuropsychiatric conditions (Jernigan et al., 1991d). This study revealed that, over the range from 30 to 79 years, age-related volume loss occurred in cerebral gray matter, in subcortical as well as cortical structures, and that the effects were best measured by correcting the raw volumes for the volume of the cranial vault. This early MR morphometric study of normal individuals produced data that were used to derive measures of brain structural volumes that were corrected both for cranial volume and for the effects of "normal aging." These measures were used to examine the volumes of subcortical brain structures and cortical regions in a group of 28 chronic alcoholic patients and matched controls (Jernigan et al., 1991e). Regarding the CSF-filled structures, the results confirmed those of several earlier CT studies (e.g., Jernigan et al., 1982) showing modest increases in ventricular size, but striking increases in subarachnoid spaces. As the subarachnoid spaces are adjacent to the cerebral cortex, while the ventricles lie deep in the center of the brain, this pattern had been interpreted by many to mean that the cerebral cortex might be particularly vulnerable to alcohol-related damage, whereas the effects on subcortical structures might be less severe. The study by Jernigan et al. (1991e) revealed that indeed the cerebral cortex was reduced in volume in chronic alcoholic patients, however, there were losses in subcortical gray matter structures (e.g., basal ganglia and diencephalon) that were at least as striking as those in cortex. Furthermore, while the losses in gray matter structures throughout the brain were substantial, they did not seem entirely to explain the dramatic increases in subarachnoid CSF. The authors pointed out that some of the volume loss was probably in white matter, and, in fact, their study revealed significant abnormality of the cerebral white matter in the form of increased volume of white matter with high signal on MRI (Jernigan et al., 1991a; Jernigan et al., 1991e). Whatever the sources of brain volume loss in the patients of this study, there did seem to be a moderate correlation between the degree of loss (as reflected in increased CSF) and the severity of their performance decrements on tests of memory, attention, and speed of information processing.

Reports that appeared soon after this one substantially extended the findings, revealing in a larger sample of 49 recently detoxified patients significantly decreased volumes of white matter as well as gray matter in cortical regions (Pfefferbaum et al., 1992, 1993). Also, an interaction between the effects of age and those of alcoholism clearly seemed to be observable in this study. The older patients showed considerably more volume loss relative to their controls than did the younger patients. Surprisingly, within this sample, there was little association between severity of alcohol history and age; therefore, the authors interpreted their results to suggest that the older brain was more vulnerable to the effects of alcohol. Although, in this study by Pfefferbaum and associates, as well as in the earlier study by Jernigan et al., there was some minor variability in the amount of loss observed in different cortical regions, the evidence for greater vulnerability of specific cortical regions was not strong. However, it should be noted that, in both studies, the anatomical boundaries between cortical regions were set using stereotactic criteria, rather than conventional sulcal landmarks, and did not specifically delineate the cortical lobules or gyri.

Several subsequent morphometric studies have focused on structures of the temporal lobe (and in particular the hippocampus) in chronic alcoholic patients. Sullivan and associates (Sullivan et al., 1995, 1996a) reported on specific measures of the volume of the anterior and posterior segments of the hippocampus. Measures of gray and white matter volumes of the temporal lobe (excluding hippocampus) were also available. They observed that the anterior hippocampus, like temporal lobe gray matter, was reduced in the alcoholic patients, and that the losses in both areas were greater in the alcoholics than would be expected from the subjects' ages. The results of analyses to examine the relationship of these losses to alcohol history suggested little or no association, providing further evidence for an increase in brain vulnerability to alcohol with aging. Of particular interest were the comparisons by this research group of patients with and without histories of alcohol withdrawal seizures. Although those patients with no history of seizures showed hippocampal and temporal lobe gray matter losses comparable to those with seizure histories, the latter had significantly less temporal lobe white matter. This suggests that subsequent investigation of the mechanisms by which alcohol withdrawal induces seizures should consider a possible role for the white matter of the temporal lobe.

A recent MR morphometric study also focused on the hippocampus. Beresford et al. (1999) examined the volumes of the hippocampus and the pituitary in alcohol-dependent subjects seeking evidence that the hippocampal damage is mediated via the hypothalamic-pituitary-adrenal axis. They reasoned that if the hippocampal damage were so mediated, the pituitary might be enlarged in association with hypercortisolemia. Indeed, consistent with this hypothesis, their study showed that the ratio of hippocampal volume to pituitary volume was significantly reduced in a small group of alcohol-dependent individuals relative to controls.

Two recent studies by Pfefferbaum, Sullivan, and their associates (Pfefferbaum et al., 1997; Sullivan et al., 1998) have examined in greater detail the regional pattern of gray and white matter losses in chronic alcoholics. In a comparison of younger and older alcoholics (Pfefferbaum et al., 1997), they found evidence that the older patients exhibited disproportionate damage to gray and white matter of the frontal lobes, while in younger patients the cortical losses were more uniform. In a separate study (Sullivan et al., 1998), the regional pattern of volume loss in chronic alcoholics was compared to that in schizophrenic patients. The latter had somewhat greater loss of cortical gray matter overall, but disproportionate loss in more anterior areas. In contrast, the cortical gray matter losses in the alcoholics were more uniform across the cortex. White matter losses, which were more severe overall in alcoholics than in schizophrenics, were greatest in frontal and posterior temporal areas.

A final MR morphometric study of alcoholics examined the cerebellar vermis and hemispheres (Sullivan et al., 2000). Alcoholic patients with and without Korsakoff's syndrome were studied. Gray matter abnormalities of the cerebellar hemispheres were observed in both groups of alcoholics, and abnormalities in both gray and white matter tissue compartments were present in the Korsakoff patients. There was also evidence that the degree of volume loss in anterior vermal white matter was related to degree of ataxia within the alcoholic patients.

An early study applying voxel-based, rather than morphometric, methods was conducted by Pfefferbaum and associates (1996), who compared corpus callosum shapes in older alcoholics and older controls. Their analysis was performed on the averaged midsaggital sections through the corpus callosum and revealed significant reductions in the midbody area of the callosum in the alcoholics, with relative sparing of genu and splenium.

In summary, results of MR morphometric studies of chronic alcoholic patients suggest that there are volume losses in the cerebral cortex, the underlying cerebral white matter, the cerebellum, and the midline basal ganglia, limbic, and diencephalic structures. In addition, there appear to be some MR signal alterations in the remaining cerebral white matter. Older patients appear to have disproportionate, more frontally distributed losses, suggesting that the older brain, and particularly the older frontal lobes, may be more vulnerable to alcohol effects. These studies inform hypotheses about the pathogenesis of alcohol-related neurodegeneration, and they demonstrate that it is feasible to monitor these changes in brain structure during life. Nevertheless, there is much work left to be done to reveal the relevant neurodegenerative mechanisms and clinical implications associated with these phenomena.

13.3.2 STUDIES OF EFFECTS OF ABSTINENCE ON BRAIN STRUCTURAL ABNORMALITIES

Given the early CT findings of increased CSF in alcoholics, the presumption that such increases reflected loss of brain parenchyma, and the prevalent view at that time that little brain regeneration occurred; it was surprising to many that subsequent CT studies of recovering patients suggested that the CSF spaces were smaller after successful treatment (Carlen et al., 1978; Artmann et al., 1981). The reversibility of CSF changes with abstinence was subsequently confirmed in MRI studies (Schroth et al., 1988; Zipursky et al., 1989). An MR morphometry study comparing abstaining to relapsing alcoholics 3 months after baseline revealed significant increases in cerebral white matter volumes in the abstaining group (Shear et al., 1994). Pfefferbaum et al. (1995) examined changes on MRI over a period of about 3 weeks in 58 patients, and over a longer period in a smaller group. This study was particularly informative because normal controls were also followed with MRI. The findings suggested that the short-term reductions in CSF spaces observed in abstaining patients were associated with increases in gray matter, and in the longer-term follow-up, white matter volume was observed to decrease in relapsers. Taken together, these studies would seem to confirm alcohol-related shrinkage of brain tissue in many chronic alcoholics that is to some substantial degree reversible with abstinence. Some have suggested that the changes might reflect rehydration of the brain associated with abstinence. Mann and his associates (1993) tested this hypothesis directly by examining CT values of tissue in recovering alcoholics. Rehydration of the brain in association with abstinence would be expected to decrease CT values; however, these authors reported slightly increased values in the abstaining patients in association with the expected reductions in CSF volumes. Given these findings, it seems likely that, rather than rehydration, neuroplastic changes, perhaps trophic increases in cell size, remyelination, or growth of neuronal processes, may lead to the observed tissue volume increases.

13.3.3 Studies of Brain Structure in Alcoholic Amnesics

A rare syndrome of anterograde and retrograde amnesia sometimes occurs in chronic alcoholics in the context of the Wernicke-Korsakoff syndrome. Autopsy studies have generally implicated diencephalic structures (particularly mammillary bodies and medial thalamus) in the pathogenesis of the memory deficits of these patients (Victor et al., 1971; Mair et al., 1979; Mayes et al., 1988); however, prior to the advent of MRI, it was not possible to examine these structures directly in living patients. Because the mammillary bodies are particularly often involved in Wernicke-Korsakoff cases, and because these structures are well visualized in midsaggital MR images, they were the objects of scrutiny in several early MRI studies. Most of these studies employed subjective evaluations of the degree of mammillary body damage. Numerous reports confirmed the high incidence of mammillary body damage in alcoholic amnesics (Charness and DeLaPaz, 1987; Squire et al., 1990; Blansjaar et al., 1992; Davila et al., 1994; Shear et al., 1996); but they also documented an increase in the incidence of similar damage in nonamnesic alcoholics (Blansjaar et al., 1992; Davila et al., 1994; Shear et al., 1996). Furthermore, there were some reports of an absence of any visually apparent mammillary body damage in some alcoholic amnesics (Shear et al., 1996). In a recent study, a high-resolution MRI protocol and morphometric techniques were employed for estimating the volume of these small, midline structures (Sullivan et al., 1999). Mammillary body volumes were compared in demented alcoholics, amnesic alcoholics, nonamnesic alcoholics, and matched controls. The demented and amnesic patient groups were very small, but the results suggest that mammillary body volume loss, while present in nonamnesic alcoholics, is nevertheless more severe in amnesic, and even more severe in demented alcoholic patients. Overall, the results of this study suggested that both degree of mammillary body damage and cognitive impairment within nonamnesic alcoholics is on a continuum with that seen in Wernicke-Korsakoff syndrome. These authors also examined the relationship between a history of Wernicke's encephalopathy and the degree of mammillary body damage. They found an association of both with declarative memory impairment across the groups of alcoholics, and interpreted this as evidence for a distinct form of neuropathology related to Wernicke's encephalopathy.

Jernigan et al. (1991b) did not examine mammillary bodies, but did estimate the volumes of a number of other brain structures in their study of alcoholic amnesia. Eight alcoholic amnesics were compared with matched groups of nonamnesic alcoholics and normal controls. The volumes of CSF structures, subcortical gray matter structures, and regions of the cerebral cortex were examined. Moderate increases in subarachnoid CSF volumes, roughly comparable in degree, were present in both alcoholic groups. In contrast, the increases in ventricular CSF were only modest in the nonamnesic patients, but were significantly more severe in the amnesics. Circumscribed gray matter losses were present in the nonamnesic patients in diencephalon and ventral cortex. In contrast, the amnesic patients had considerably more extensive gray matter loss. In particular, their losses in the anterior diencephalon (which did not include mammillary bodies), the limbic structures on the mesial surface of the temporal lobes, and orbito-frontal cortex, were significantly greater than those of the nonamnesic patients. This suggests that damage in structures other than the mammillary bodies and midline thalamus may play a role in the amnesia of Wernicke-Korsakoff syndrome. Some other investigators (Sullivan et al., 1996b), but not all (Squire et al., 1990), also have observed mesial temporal lobe damage on MRI in alcoholic amnesics.

13.3.4 Studies of the Effects of Fetal Alcohol Exposure on Brain Structure

Jones et al. (1973) were the first to describe a syndrome of facial dysmorphology, growth retardation, mental retardation, and other neurological symptoms in children who had suffered from severe prenatal exposure to alcohol. A later autopsy study (Jones and Smith, 1975) revealed a small, abnormal brain lacking the corpus callosum (the largest interhemispheric fiber tract in the brain) in a case with this syndrome. Most children with severe fetal alcohol exposure have a relatively normal life expectancy and, fortunately, do not come to autopsy during their development. However, such children can and have been examined with MRI in attempts to define the prevalence and pattern of brain dysmorphology associated with fetal alcohol exposure.

In an early case study, two young patients with severe fetal alcohol syndrome (FAS) were examined with MR morphometric techniques (Mattson et al., 1992). The brains of these two patients were compared with those of a group of age-matched controls, and a small group of children with Down Syndrome. On visual inspection, it was apparent that the corpus callosum was absent in one of the FAS patients, and moderately hypoplastic in the other. Volumetric analyses revealed severe overall brain size reduction, affecting both cerebrum and cerebellum, and suggested disproportionate reduction in the volume of subcortical gray matter structures, such as basal ganglia and diencephalon. Data for two additional individuals were added in a second preliminary report (Mattson et al., 1994). These two young subjects had histories of severe prenatal alcohol exposure, mental retardation, and behavioral problems, but they lacked the facial features required for the diagnosis of FAS. These youngsters also had significantly reduced cerebral and cerebellar volumes, and again, there was evidence for disproportionate hypoplasia of the basal ganglia. In another study by this group, measurements of the cerebellar vermis in children with severe prenatal alcohol exposure (FAS and nonFAS) were compared with those in normal controls (Sowell et al., 1996). The children with fetal alcohol exposure had reduced anterior vermal size, but no significant reductions in posterior vermal areas.

A subsequent report (Mattson et al., 1996) described morphometry results in an additional six children with FAS and seven control children, none of whom had agenesis of the corpus callosum, though the callosum may have been thin. The study focused on overall brain hypoplasia and on the volumes of cortical and subcortical structures. Again, there was evidence for cerebral and cerebellar hypoplasia. Volumes of subcortical structures were expressed as proportions of the total cranial volume to determine whether the reductions in these structures were commensurate with or disproportionate to overall cerebral hypoplasia. The basal ganglia proportion was significantly reduced in the FAS subjects. In a secondary analysis, volumes of the caudate nuclei and the lenticular nuclei (components of the basal ganglia) were examined separately. The basal ganglia effects appeared to be particularly severe in caudate nuclei, and there was some indication that the caudate nuclei might be more severely hypoplastic than the lenticular nuclei (though the region by group interaction failed to reach significance).

The data published in the latter report were generally consistent with those from another MRI report that appeared about the same time (Johnson et al., 1996), in which visually identifiable anomalies on MRI were noted in a group of nine cases of FAS. These authors stressed the prevalence in the disorder of midline abnormalities, i.e., in corpus callosum, cavum septi pellucidum, and brainstem. Hippocampal and amygdalar volumetry was performed using MRI in a separate study (Riikonen et al., 1999) of 11 FAS individuals. There was some evidence for disturbed asymmetry of the hippocampus (with relatively smaller left hippocampi) in FAS subjects relative to controls. Surprisingly, in the latter report, the authors described findings they interpreted as indicative of atrophy; i.e., loss of brain tissue.

Recently, a more comprehensive MR morphometric study compared 14 FAS individuals, eight subjects with severe prenatal alcohol exposure (but without facial dysmorphology), and 41 controls (Archibald et al., 2001). No subjects with agenesis of the corpus callosum were included. Detailed anatomical analyses of different tissues and structures within cerebrum, cerebellum, and brainstem were performed (see Figure 13.4). The results in the FAS subjects were generally consistent with, and substantially extended, previous observations; however, they provided no evidence for atrophy (as indexed by CSF increases). Instead there was evidence for a specific pattern of hypoplasia, affecting the cerebrum and brain stem, but perhaps affecting the cerebellum even more dramatically. The results also indicated that within the cerebrum, white matter was more severely hypoplastic than was gray matter, and that the parietal lobe was disproportionately affected overall. There was again definite evidence for disproportionately reduced size of the caudate nuclei in FAS subjects; but, in contrast, the hippocampus appeared to be significantly spared. In general, brain morphology in the alcohol-exposed subjects without facial dysmorphology was more similar to that in controls than to FAS. However, *post hoc* analyses focusing on those regions most abnormal in the FAS subjects, namely caudate nuclei and parietal lobes, produced some evidence for modest volume reductions in this less affected group.

An example of a very interesting study of fetal alcohol effects using a voxel-based approach is that of Sowell et al. (2001). The subjects in this study included many of the fetal alcohol-exposed subjects and controls examined in the study by Archibald et al. (2001). A particularly interesting analysis attempted to map the pattern of hypoplasia as represented at the cortical surface. For this analysis, the authors defined the surface of the cerebral cortex for each subject and aligned the brain volumes so that they were registered based on the positions of principal cortical sulci. They then determined the distance from each of 65,536 points on the cortical surface to a point deep in the brain on the midline (at the decussation of the anterior commissure). They refer to these values as differences from center

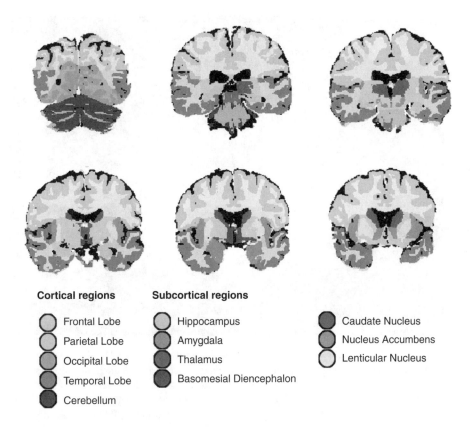

Cortical regions

- Frontal Lobe
- Parietal Lobe
- Occipital Lobe
- Temporal Lobe
- Cerebellum

Subcortical regions

- Hippocampus
- Amygdala
- Thalamus
- Basomesial Diencephalon

- Caudate Nucleus
- Nucleus Accumbens
- Lenticular Nucleus

FIGURE 13.4 A set of "processed" MR images from a single individual studied in an MR morphometry study performed in the author's laboratory. The different regions and structures examined are color-coded to show the structural boundaries. (Figure also shown in color insert following page 202.)

(DFC). Figure 13.5 shows the statistical map comparing the alcohol-exposed to the control subjects in terms of these DFC values. Note that no size-normalization was applied, as the purpose was to map the pattern of brain shape anomaly that results from the hypoplasia. There appear to be large reductions in DFC in parietal areas bilaterally, which could be expected, given the disproportionate hypoplasia in parietal lobe observed by Archibald et al. (2001). However, the pattern shown within this area may provide clues about the specific gyri most affected. Furthermore, there is an area of reduced DFC visible in the ventral part of the frontal lobe, in the left hemisphere, which was unexpected. The only measures of frontal lobe available in the earlier MR morphometric study included the entire lobe. A focal decrease in volume in the area implicated by the DFC statistical map may easily have been masked in this study using large regions of interest. Thus the voxel-based approach has in this case raised an important question about the effects of prenatal alcohol exposure on ventral structures within frontal lobe.

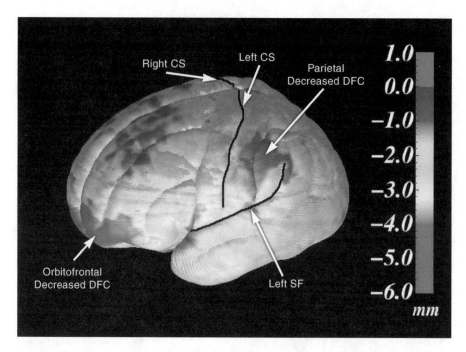

FIGURE 13.5 DFC group-difference maps showing differences in DFC (in mm) between the alcohol-exposed and control subjects according to the color bar on the right. Note negative effects in nearly all regions, most prominent in parietal and orbital frontal regions, i.e., DFC is greater in controls than in alcohol-exposed patients, with regional patterns of DFC reduction up to 8 mm. While only left hemisphere is fully visualized here, DFC reduction in both parietal and orbital frontal regions is bilateral. Left and right central sulcus (CS) and left hemisphere Sylvian fissure (SF) are highlighted in black. (Figure also shown in color insert following page 202.)

13.4 EVALUATION OF sMRI METHODS AND IMPLICATIONS FOR FUNCTIONAL IMAGING STUDIES

It is clear that MRI and MR morphometry have provided much important information about the effects of prenatal alcohol exposure, chronic alcoholism, successful alcohol treatment, and alcohol-related illnesses on the structure of the human brain. There remains little doubt that substantial dysmorphology exists in a significant proportion of individuals affected by alcohol abuse.

Conventional morphometric methods have focused on sizes of brain structures, and the consistency of the results suggest that most investigators are measuring brain volumes with reasonable accuracy. However, as described above, most volumetric techniques make the assumption that the different tissues are reasonably homogenous in signal value, and that the tissue values themselves are relatively independent of diagnosis. This may not be true. For example, it is possible that widespread effects of alcohol abuse on white matter biochemistry could lead to alteration of the MR signal of white matter in some alcoholics. Such signal changes could result in segmentation biases, particularly with unsupervised segmentation methods, resulting in apparent tissue volume changes that were in fact the result of white matter signal alteration. Further study of tissue characteristics in alcohol-exposed human populations and animal models could help to resolve such uncertainty.

Additional concerns arise concerning the use of voxel-based methods, particularly regarding the effects of spatial normalization. It should be noted that most of the voxel-based image processing methods were developed for use with functional imaging. The goal was originally to devise methods for intersubject averaging that would best bring functionally homologous brain structures into spatial alignment, so that modal brain maps for particular functions could be defined. The early so-called "affine" methods were superceded by new methods that involve nonlinear warping of the brain image data to minimize the structural discrepancies in size and shape among individuals. This is a reasonable approach if one can make the assumption that individual differences in brain size and shape are relatively independent of functional characteristics of the structures. The use of voxel-based analysis for mapping structural differences has, in some cases, been bootstrapped from these methods. The proper interpretation of maps of sMRI signal anomaly that compare groups that may differ substantially in brain size and disease related dysmorphology, particularly when intensive spatial normalization has occurred, is a complex subject. There is a danger, as described briefly above, that interactions between the effects of diagnosis and the effects of spatial normalization may produce apparent abnormalities in the maps that are remote in location from the sites of the actual structural abnormalities present in the clinical group.

Bookstein (1996) was among the first to propose mathematical approaches to characterizing dysmorphology (Bookstein, 1996). He and his associates have attempted to define the pattern of callosal shape abnormality on mid-saggital MR images using a thin-plate spline method. He has also recently cautioned against the use of voxel-based methods for group comparisons of brain morphology, citing problems resulting from the effects of the spatial normalization (Bookstein, 2001).

New functional imaging modalities, described in the next chapter, have the potential to add critical information about the functional implications of the anatomical changes described in this chapter, as well as to reveal brain functional aspects of alcohol abuse that are independent of measurable anatomical changes. However, it should be noted that in virtually all functional imaging, structural imaging is used in the image analysis path to localize the effects. Functional imaging modalities, such as fMRI, usually involve averaging across individuals, and they produce maps of the effects observed, plotted in anatomically standardized coordinates. Producing these maps requires several complex postprocessing steps. Structural MR images are sometimes processed with automated algorithms to define the different tissues. Size normalization is almost invariably performed. Furthermore, to further reduce noise associated with individual morphological variability, nonlinear warping methods are applied to attempt to bring homologous structures in different individuals into better spatial registration. These processes are usually applied before group differences are tested. Most of the methods for analyzing functional imaging data in common use today were designed making the assumption that morphological variability is independent of any of the experimental variables. The implications of systematic morphological effects on the outcome of these analyses are poorly under-stood, but there is no doubt that the results are affected by them to some extent. The results of an fMRI study, the data from which were analyzed with any of a number of common analysis packages, comparing groups of FAS subjects and normal con-trols, would, by themselves, be ambiguous. It would be unclear in many cases whether any group differences in brain activation observed were due entirely to morphological differences, caused by the same changes reflected in anatomical changes, or independent of any morphological changes. Furthermore, it would be difficult to conclude that the effects were in the structures to which they would be assigned on the basis of the standardized coordinates, since it would not be certain that the warping algorithms would affect subjects in the two groups similarly. These are important problems that must be addressed as functional imaging modalities are applied more widely to investigate the effects of alcoholism.

13.5 FUTURE DIRECTIONS

It is virtually a certainty that the spatial resolution and anatomical precision of MRI will continue to improve. This will provide even more detailed information about the neuroanatomical effects of alcohol abuse. One of the major shortcomings of the anatomical work reviewed above is the paucity of information provided about spe-cific effects on white matter. The importance of this limitation is underscored by the evidence that, in both FAS and chronic alcoholism, effects on white matter are among the most pronounced. Diffusion tensor imaging (DTI), to be discussed in the next chapter, promises to add vital information about the location of individual fiber tracts and about their structural integrity in these populations. Other new imaging modalities will provide additional information about the structure and biochemistry of brain tissues.

One of the greatest challenges facing investigators in the future is that of inte-grating the vast amounts of information obtained with brain imaging. New methods are needed for analyzing these huge datasets to allow us to extract the information that is most relevant to the pathogenesis and resulting dysfunction of alcohol-related disorders. These methods must yield an accurate characterization of brain structural anomaly in individual subjects; they must be efficient; and they must provide the means for defining the relationship between anatomical factors and the other effects under investigation. This work represents a branch of a new field referred to as neuroinformatics. To succeed in developing these new modeling techniques, investi-gators must form teams that include, in addition to neuroscientists and behavioral scientists, mathematicians and information scientists who bring knowledge of relevant recent developments in those fields. The urgency of this problem can hardly be overemphasized. While numerous sophisticated informatics tools exist for accom-plishing some or all of the processes involved in basic brain mapping; these are frequently inadequate for use in investigations of clinical problems. There is a serious need for appropriate data reduction and data analysis schemes with sufficient com-plexity to model even the major known factors contributing to the results of functional imaging in patients with brain dysfunction. This is probably the most significant problem reducing the yield of neuroimaging investigations of alcohol abuse today.

ACKNOWLEDGMENTS

The author would like to acknowledge the support of the Medical Research Service of the Department of Veterans Affairs and the National Institute on Alcohol and Alcohol Abuse. L. Jay Starkey made a number of significant contributions to the preparation of this chapter.

REFERENCES

Andreasen NC et al. (1996) Automatic atlas-based volume estimation of human brain regions from MR images. *J Comput Assist Tomogr* 20:98-106.

Archibald SL et al. (2001) Brain dysmorphology in individuals with severe prenatal alcohol exposure. *Dev Med Child Neurol* 43:148-154.

Arnold J et al. (2001) Qualitative and quantitative evaluation of six bias-correction algorithms. *Neuroimage.* (In press).

Artmann H et al. (1981) Reversible enlargement of cerebral spinal fluid spaces in chronic alcoholics. *Am J Neuroradiol* 2:23-27.

Beresford T et al. (1999) Hippocampal to pituitary volume ratio: a specific measure of reciprocal neuroendocrine alterations in alcohol dependence. *J Stud Alcohol* 60:586-588.

Bergman H et al. (1980) Computed tomography of the brain, clinical examination and neuropsychological assessment of a random sample of men from the general population. *Acta Psychiatr Scand Suppl* 286:47-56.

Bezdek JC, Hall LO and Clarke LP (1993) Review of MR image segmentation techniques using pattern recognition. *Med Phys* 20(4):1033-1048.

Blansjaar BA et al. (1992) Similar brain lesions in alcoholics and Korsakoff patients: MRI, psychometric and clinical findings. *Clin Neurol Neurosurg* 94:197-203.

Bonar DC et al. (1993) Graphical analysis of MR feature space for measurement of CSF, gray-matter and white-matter volumes. *J Comp Assisted Tomog* 17:461-470.

Bookstein FL (1996) Endophrenology: new statistical techniques for studies of brain form. Life on the hyphen in neuro-informatics. *Neuroimage* 4:S36-38.

Bookstein FL (2001) Voxel-based morphometry should not be used with imperfectly registered images. *Neuroimage.*

Cagnoni S et al. (1993) Neural network segmentation of magnetic resonance spin echo images of the brain. *J Biomed Eng* 15:355-362.

Cala LA and Mastaglia FL (1981) Computerized tomography in chronic alcoholics. *Alcohol Clin Exp Res* 5:283-294.

Carlen PL et al. (1978) Reversible cerebral atrophy in recently abstinent chronic alcoholics measured by computed tomography scans. *Science* 200:1076-1078.

Caviness VS, Jr. et al. (1995) Advanced application of magnetic resonance imaging in human brain science. *Brain Dev* 17:399-408.

Charness ME and DeLaPaz RL (1987) Mamillary body atrophy in Wernicke's encephalopathy: Antemortem identification using magnetic resonance imaging. *Ann Neurol* 22:595-600.

Clarke LP et al. (1993) MRI: stability of three supervised segmentation techniques. *Mag Res Im* 11:95-106.

Clarke LP et al. (1994) MRI segmentation: methods and applications. *Mag Res Im* 13 (3):343-368.

Cohen MS and Bookheimer SY (1994) Localization of brain function using magnetic resonance imaging. *Trends in Neurosciences* 17:268-277.

Collins DL et al. (1995) Automatic 3-D model-based neuroanatomical segmentation. *Hum Brain Map* 3:190-208.

Davila MD et al. (1994) Mammillary body and cerebellar shrinkage in chronic alcoholics: An MRI and neuropsychological study. *Neuropsychology* 8:433-444.

Fletcher LM, Barsotti JB and Hornak JP (1993) A multispectral analysis of brain tissues. *Mag Res Med* 29:623-630.

Jackson EF, Narayana PA and Falconer JC (1994) Reproducibility of nonparametric feature map segmentation for determination of normal human intracranial volumes with MR imaging data. *J Mag Res Im* 4:692-700.

Jernigan TL and Ostergaard AL (1993) Word priming and recognition memory are both affected by mesial temporal lobe damage. *Neuropsychology* 7:14-26.

Jernigan TL, Butters N and Cermak LS (1991a) Studies of brain structure in chronic alcoholism using magnetic resonance imaging. *National Institute on Alcohol Abuse and Alcoholism Research Monograph 21: Imaging in Alcohol Research*. Rockville, MD: U.S. Department of Health and Human Services.

Jernigan TL et al. (1991b) Magnetic resonance imaging of alcoholic Korsakoff patients. *Neuropsychopharmacology* 4:175-186.

Jernigan TL et al. (1991c) Maturation of human cerebrum observed *in vivo* during adolescence. *Brain* 114:2037-2049.

Jernigan TL et al. (1982) CT measures of cerebrospinal fluid volume in alcoholics and normal volunteers. *Psych Res* 7:9-17.

Jernigan TL et al. (1991d) Cerebral structure on MRI, Part I: Localization of age-related changes. *Biol Psych* 29:55-67.

Jernigan TL et al. (2001) Effects of normal aging on tissues and regions of the cerebrum and cerebellum. *Neurobiology of Aging*. (In press)

Jernigan TL et al. (1991e) Reduced cerebral grey matter observed in alcoholics using magnetic resonance imaging. *Alcohol Clin Exp Res* 15:418-427.

Johnson VP et al. (1996) Fetal alcohol syndrome: Craniofacial and central nervous system manifestations. *Am J Med Genet* 61:329-339.

Jones KL and Smith DW (1975) The fetal alcohol syndrome. *Teratology* 12:1-10.

Jones KL et al. (1973) Pattern of malformation in offspring of chronic alcoholic mothers. *Lancet* 1:1267-1271.

Kao YH et al. (1994) Dual-echo MRI segmentation using vector decomposition and probability techniques: a two-tissue model. *Mag Res Med* 32:342-357.

Kennedy DN et al. (1994) MRI-based topographic segmentation. In: *MRI-Based Topographic Segmentation*, pp 201-208: Academic Press, Inc.

Mair WG, Warrington EK and Weiskrantz L (1979) Memory disorder in Korsakoff's psychosis: a neuropathological and neuropsychological investigation of two cases. *Brain* 102:749-783.

Mann K et al. (1993) The reversibility of alcoholic brain damage is not due to rehydration: a CT study. *Addiction* 88:649-653.

Mattson SN et al. (1996) A decrease in the size of the basal ganglia in children with fetal alcohol syndrome. *Alcohol Clin Exp Res* 20:1088-1093.

Mattson SN et al. (1994) A decrease in the size of the basal ganglia following prenatal alcohol exposure: a preliminary report. *Neurotoxic & Teratol* 16:283-289.

Mattson SN et al. (1992) Fetal alcohol syndrome: A case report of neuropsychological, MRI, and EEG assessment of two children. *Alcohol Clin Exp Res* 16(5):1001-1003.

Mayes A et al. (1988) Locations of lesions in Korsakoff's syndrome: Neuropsychological and neuropathological data on two patients. *Cortex* 24(3):367-88.

Pfefferbaum A et al. (1996) Thinning of the corpus callosum in older alcoholic men: a magnetic resonance imaging study. *Alcohol Clin Exp Res* 20:752-757.

Pfefferbaum A et al. (1997) Frontal lobe volume loss observed with magnetic resonance imaging in older chronic alcoholics. *Alcohol Clin Exp Res* 21:521-529.

Pfefferbaum A et al. (1993) Increase in brain cerebrospinal fluid volume is greater in older than in younger alcoholic patients: a replication study and CT/MRI comparison. *Psych Res* 50:257-274.

Pfefferbaum A et al. (1995) Longitudinal changes in magnetic resonance imaging brain volumes in abstinent and relapsed alcoholics. *Alcohol Clin Exp Res* 19:1177-1191.

Pfefferbaum A et al. (1992) Brain gray and white matter volume loss accelerates with aging in chronic alcoholics: a quantitative MRI study. *Alcohol Clin Exp* Res 16:1078-1089.

Riikonen R et al. (1999) Brain perfusion SPECT and MRI in foetal alcohol syndrome. *Dev Med Child Neurol* 41:652-659.

Ron MA, Acker W and Lishman WA (1980) Morphological abnormalities in the brains of chronic alcoholics — a clinical psychological and computerized axial tomographic study. *Acta Psychiatr Scand Suppl* 286:41-46.

Schroth G et al. (1988) Reversible brain shrinkage in abstinent alcoholics, measured by MRI. *Neuroradiology* 30:385-389.

Shear PK, Jernigan TL and Butters N (1994) Volumetric magnetic resonance imaging quantification of longitudinal brain changes in abstinent alcoholics [published erratum appears in *Alcohol Clin Exp Res* 1994 Jun;18(3):766]. *Alcohol Clin Exp Res* 18:172-176.

Shear PK et al. (1996) Mammillary body and cerebellar shrinkage in chronic alcoholics with and without amnesia. *Alcohol Clin Exp Res* 20:1489-1495.

Simmons A et al. (1996) Simulation of MRI cluster plots and application to neurological segmentation. *Mag Res Im* 14(1):73-92.

Sowell ER and Jernigan TL (1998) Further MRI evidence of late brain maturation: Limbic volume increases and changing asymmetries during childhood and adolescence. *Devel Neuropsych* 14:599-617.

Sowell ER et al. (1999a) *In vivo* evidence for post-adolescent brain maturation in frontal and striatal regions [letter]. *Nat Neurosci* 2:859-861.

Sowell ER et al. (1996) Abnormal development of the cerebellar vermis in children prenatally exposed to alcohol: size reduction in lobules I-V. *Alcohol Clin & Exper Res* 20:31-34.

Sowell ER et al. (1999b) Localizing age-related changes in brain structure between childhood and adolescence using statistical parametric mapping. *Neuroimage* 9:587-597.

Sowell ER et al. (2001) Regional brain shape abnormalities persist into adolescence after heavy prenatal alcohol exposure. (Submitted).

Squire LR, Amaral DG and Press GA (1990) Magnetic resonance imaging of the hippocampal formation and mammillary nuclei distinguish medial temporal lobe and diencephalic amnesia. *Neuroscience* 10:3106-3317.

Sullivan EV et al. (1995) Anterior hippocampal volume deficits in nonamnesic, aging chronic alcoholics. *Alcohol Clin Exp Res* 19:110-122.

Sullivan EV et al. (1996a) Relationship between alcohol withdrawal seizures and temporal lobe white matter volume deficits. *Alcohol Clin Exp Res* 20:348-354.

Sullivan EV et al. (1996b) Hippocampal but not cortical volumes distinguish amnesic and nonamnesic alcoholics. *Soc Neurosci Abstr*.

Sullivan EV et al. (1998) Patterns of regional cortical dysmorphology distinguishing schizophrenia and chronic alcoholism. *Biol Psych* 43:118-131.

Sullivan EV et al. (2000) Cerebellar volume decline in normal aging, alcoholism, and Korsakoff's syndrome: Relation to ataxia. *Neuropsychology* 14:341-352.

Sullivan EV et al. (1999) *In vivo* mammillary body volume deficits in amnesic and nonamnesic alcoholics. *Alcohol Clin Exp Res* 23:1629-1636.

Vaidyanathan M et al. (1997) Normal brain volume measurements using multispectral MRI segmentation. *Mag Res Im* 15:87-97.

Victor M, Adams RD and Collins GH (1971) The Wernicke-Korsakoff syndrome: a clinical and pathological study of 245 patients, 82 with postmortem examinations. *Contemp Neurol Ser* 7:1-206.

Vinitski S et al. (1997) Improved intracranial lesion characterization by tissue segmentation based on a 3D feature map. *Mag Res Med* 37:457-469.

Wilkinson DA (1982) Examination of alcoholics by computed tomographic (CT) scans: a critical review. *Alcohol Clin Exp Res* 6:31-45.

Wong EC et al. Single slab high resolution 3D whole brain imaging using spiral FSE. *Proc Int Soc Mag Res in Med 2000*;8:683.

Zijdenbos AP and Dawant BM (1994) Brain segmentation and white matter lesion detection in MR images. *Crit Rev Biomed Engrg* 22(5/6):401-465.

Zipursky RB, Lim KC and Pfefferbaum A (1989) MRI study of brain changes with short-term abstinence from alcohol. *Alcohol Clin Exp Res* 13:664-666.

14 Biochemical, Functional and Microstructural Magnetic Resonance Imaging (MRI)

Elfar Adalsteinsson, Edith V. Sullivan and Adolf Pfefferbaum

CONTENTS

14.1 INTRODUCTION

Chronic alcoholism is associated with significant widespread shrinkage of brain parenchyma, preferentially affecting the frontal lobes and partially reversible with abstinence. However, even long-abstinent alcoholics exhibit residual tissue volume deficit, leading to the hypothesis that alcohol-related abnormalities comprise both reversible and permanent components (Carlen et al., 1984; Harper, 1998; Lishman, 1990; Ron, 1987). *In vivo* imaging tools can be used to examine the condition of the brain of chronic alcoholics during the natural course of detoxification through abstinence and/or relapse (also see Chapter 13).

Magnetic resonance imaging (MRI) has revolutionized the clinical examination and experimental study of human brain disease. MRI is based on the fundamental discovery that certain atoms, when placed in a strong magnetic field, behave like little magnets themselves and can be detected by the manipulation of the magnetic field. Clinical structural MRI relies on the detection and mapping of protons (^{1}H), primarily from water but also with contributions from fat, and the tissue contrast in the image reflects differences in the local environment of the protons. In this chapter, we explore further extensions of brain MR proton imaging to detect chemical constituents (spectroscopy), activation-induced blood flow (functional MR–fMRI) and diffusion properties (diffusion tensor imaging–DTI) as they can be applied to the study of alcoholism. The descriptions of the underlying mechanisms and some examples of application are meant to provide the reader with an appreciation of the strengths and limitations of the techniques.

14.2 MR SPECTROSCOPY AND SPECTROSCOPIC IMAGING

Although MRI primarily depicts the distribution of water protons, ^{1}H MR can also be used to obtain information about chemical constituents other than water, primarily due to a small frequency shift, or "chemical shift," relative to the water signal. The magnitude of the chemical shift, i.e., the degree to which the resonance frequency of the detected compounds is modulated, is typically on the order of parts per million relative to the water resonance. The acquisition of MR-detectable signals other than those of water and fat is referred to as MR spectroscopy (MRS), MR spectroscopic imaging (MRSI), or chemical shift imaging (CSI), and is an *in vivo* application of traditional laboratory-based NMR spectroscopy.

In addition to proton spectroscopy, other nuclei such as ^{31}P, ^{23}Na, ^{19}F, and ^{13}C produce resonance in a magnetic field. However, the bulk of the spectroscopy work dedicated to the study of alcoholism has been with ^{1}H MRS (Fein et al., 1995; Martin et al., 1995; Mendelson et al., 1990; Schweinsburg et al., 2000; Seitz et al., 1999), applied primarily to the brain, a few studies used ^{31}P MRS (Estilaei et al., 2001; Meyerhoff et al., 1995). Here we focus on ^{1}H MR spectroscopy.

As with conventional MRI, radio-frequency excitation of protons (or "spins") changes their alignment in the magnetic field and two relaxation parameters, T_1 and T_2, determine the relative strength of signals in the acquired data. The T_1 relaxation time determines the rate at which the excited spins recover their resting longitudinal magnetization, while T_2 dictates the rate at which the observable transverse magnetization decays. The choice of echo time, TE, determines the degree to which the data are T_2-weighted, while the repetition period, TR, is the most common source of T_1 weighting. Further T_1 weighting can be obtained by inversion recovery, where an $180°$ pulse is applied prior to excitation. The time from inversion to excitation,

TI, allows recovery of longitudinal magnetization at a rate determined by T_1, and can be optimized to null signals characterized by a particular T_1 value. In addition to T_1 and T_2 relaxation, spin coupling can further modulate the signals from metabolites observed with *in vivo* proton spectroscopy.

14.2.1 Technical Challenges

The primary technical challenges associated with MR spectroscopy are the low signal-to-noise ratio (SNR) of the desired components, contaminating signals from the much stronger water and lipid resonance, and main field inhomogeneities.

The concentration of the spectroscopically visible metabolites is much lower than that of the primary constituent of brain, which is water. Neat water is 55 M, while brain metabolite concentrations are 1–10 mM — a ratio > 5000:1. Thus, spectroscopy faces significant challenges in overcoming low SNR. Because the SNR in MR is proportional to the resolved voxel size and the square root of imaging time (Macovski, 1996), either the imaging time or voxel volume need to be increased to acquire data with sufficient SNR. Thus, spectroscopic images of low-SNR metabolites are of lower spatial resolution, and take longer to acquire than conventional MRI.

The relatively high intensities of water and lipids pose an additional problem because their signals overwhelm that of the metabolites of interest. These dominant resonances need to be suppressed to detect the desired spectroscopic signals. Furthermore, signal components that are co-resonant with either fat or water require special acquisition methods.

A problem of crucial importance is main field inhomogeneities that are inevitable deviations in the main magnetic field, B_0, from an ideally uniform field. Because the peaks in the proton spectrum are identified based on their chemical shift, and resonance frequency is proportional to field strength, undesired variation in the main field causes a shift along the frequency axis that makes reconstruction and identification of metabolite components difficult. In addition, line broadening due to local variation of the field causes loss of SNR and overlap of spectral peaks. Main field inhomogeneities at the field strength used in most commercial scanners are mainly due to susceptibility effects arising from air–tissue interfaces in the body and thus vary from one subject to another. Because of these limiting factors, spectroscopy has been most widely applied deep in the brain because lipid signals are mostly limited to the subcutaneous fat, main field homogeneity is relatively good, and motion is not an issue.

14.2.2 Metabolites Observed *In Vivo*

The most common field strength for commercial *in vivo* human magnets is 1.5 T. At this field, several resonances are readily observed in proton spectra from the brain, the largest of which arise from the N-acetyl moiety, creatine and phosphocreatine (Cr), choline compounds (Cho), and myo-inositiol (mI), and sometimes lactate (lac). With carefully optimized acquisition protocols, several signals with lower SNR can also be detected, e.g., glutamine, glutamate, and gamma-aminobutyric acid (GABA).

The visibility of these metabolites with proton MRS depends on their relaxation and coupling characteristics. For instance, N-acetylaspartate (NAA), Cr, and Cho have relatively long T_2, in the range of 200–400 ms, while mI has much shorter T_2. NAA, Cr and Cho are visible with both short and long echo-time protocols, but mI can be detected only at relatively short TE, typically with TE less than 40 ms. Due to the strong coupling of glutamine and glutamate at 1.5 T, those signals are difficult to detect, even at short echo times.

A brief summary of each of the four most prominent ^{1}H metabolites follows.

- NAA — The predominant *in vivo* proton signal is NAA, with contributions from other N-acetyl compounds, especially N-acetyl aspartyl glutamate (NAAg). First discovered in 1956 (Tallan, 1956), NAA is found almost exclusively in neurons (Petroff et al., 1995; Urenjak et al., 1992, 1993), and thus is considered a measure of neuronal integrity. Postmortem high pressure liquid chromatography (Koller et al., 1984; Nadler and Cooper, 1972) and MRS (Kwo-On-Yuen et al., 1994; Petroff et al., 1995) studies have shown NAA levels to be higher in gray matter than in white matter in healthy subjects, as have *in vivo* studies (Doyle et al., 1995; Lim et al., 1998; Lim and Spielman, 1997; Moyher et al., 1995; Narayana et al., 1989; Pouwels and Frahm, 1998; Schuff et al., 1998; Wang and Li, 1998). Exceptions (Hetherington et al., 1996; Knufman et al., 1992; Tedeschi et al., 1995) to this pattern include a report that NAA/Cr is higher in white than gray matter (Lopez-Villegas et al., 1996), although this difference may be due to higher Cr rather than lower NAA in gray matter (Lopez-Villegas et al., 1996).
- Cr — The Cr signal, generated by creatine and phosphocreatine, is influenced by the state of high-energy phosphate metabolism (Tedeschi et al., 1995). In spectroscopy studies, it is often used as a reference for other peaks on the assumption that its concentration is relatively constant.
- Cho — The *in vivo* MRS-visible Cho peak is generated primarily by water-soluble Cho-containing compounds — free choline, phosphocholine, and glycerophosphocholine (Barker et al., 1994) — and is associated with cell membrane synthesis and turnover. Cho is also an index of cellular density in brain tumors (Gupta et al., 1999) and may be a marker of increases in glial density with age and disease. MRS-measured Cho concentration is higher in white than gray matter (Pfefferbaum et al., 1999b) and increases with normal aging (Chang et al., 1996; Kreis et al., 1993; Moats et al., 1994; Pfefferbaum et al., 1999b; Soher et al., 1996).
- mI — mI is present in glial but not neuronal cell cultures (Brand et al., 1993; Petroff et al., 1995) and is a precursor of the inositol-triphosphate second-messenger pathway, where it functions to maintain cell volume (Ernst et al., 1997; Lien et al., 1990). Thus, it has been described as a glial marker, an intracellular osmolyte, a precursor of myelin phosphatidyl inositol, and a progenitor of the widespread inositol polyphosphate messenger cascade (Shonk et al., 1995) and the breakdown product of phosphatidyl inositol (Parnetti et al., 1997); mI concentration is higher in gray matter than in white matter (Michaelis et al., 1993; Pouwels & Frahm, 1998).

14.2.3 Single-Voxel MR Spectroscopy

There are two basic approaches to *in vivo* proton spectroscopy in clinical investigations: single-voxel MRS and MRSI. Most of *in vivo* spectroscopy of the brain has been performed through single-voxel spectroscopy, either point resolved spectroscopy (PRESS) (Bottomley, 1987) or stimulated echo acquisition mode (STEAM) (Frahm et al., 1989). The relatively small excitation volume reduces problems associated with main field inhomogeneities, and contaminating lipid signals from subcutaneous fat can be avoided by limiting the excitation region to brain tissue only. The voxel size, typically 2–12 cc, is chosen such that the spectra have an adequate SNR in reasonable acquisition times, often in the range of 2–8 min.

Single-voxel studies are well suited for detecting diffuse and widespread phenomena that affect the entire brain relatively uniformly such that observations in one, two, or three locations are representative of the brain in that condition as a whole. They are also well suited for large-scale investigations of many individuals and for repeated studies over time in the same individual.

PRESS and *STEAM* — The principle behind PRESS is the application of three spatially selective radio-frequency (RF) pulses along three orthogonal axes. The first pulse is a slice-selective 90° excitation, while the second and third pulses are slice-selective spin-echo 180° pulses with symmetric gradient lobes that refocus the transverse magnetization first to a column and then to a rectangular solid. The spectroscopic signal is sampled as it decays during the readout period and a spectrum is formed by a fast Fourier transform (FFT) of the time-domain samples. The excitation and readout steps are repeated multiple times and the data averaged together for increased SNR.

As in PRESS, STEAM applies three RF pulses selectively along three orthogonal axes to define a rectangular solid. However, the flip angle of each RF pulse is 90°, thus exciting a stimulated echo as opposed to a spin-echo for PRESS. The advantage of STEAM over PRESS is that shorter echo times are more easily implemented. A drawback of the stimulated echo compared to a spin echo at the same echo time is that only half of the available transverse magnetization is retained, resulting in lower SNR.

Both the PRESS and STEAM acquisition methods are generally preceded by chemically shift selective (CHESS) water suppression (Haase et al., 1985). The CHESS sequence consists of three spectrally selective narrow-band RF saturation pulses, where each pulse excites the water resonance, followed by a gradient lobe that eliminates coherent transverse magnetization. The flip angles of the excitation pulses can be optimized to minimize the residual water signal, and, in commercial implementations on clinical scanners, this optimization is performed automatically as part of a calibration procedure prior to the spectroscopic acquisition.

Shimming — Main field inhomogeneity plays an important role in determining the quality of the acquired spectra. Thus, acquisitions with either PRESS or STEAM are usually preceded by shimming, a procedure to adjust linear, and perhaps also nonlinear, shim currents to optimize the B_0 homogeneity over the volume of interest.

Water Reference — Spectral estimation and quantitation procedures may be confounded by the low SNR of the metabolite data, particularly if the calculations involve nonlinear parameters. To overcome this constraint by gathering better *a priori* information, the high-SNR water signal may be used as a reference signal to aid in the reconstruction and interpretation of the low-SNR metabolite spectrum. The water signal can be used as a chemical shift reference, to estimate constant and time-varying phase terms, and to aid in the quantitation of the metabolites. Thus, a separate short acquisition of unsuppressed water from the excited voxel is often acquired along with the longer metabolite scan, and the resulting reference signal is incorporated into the reconstruction software (Webb et al., 1994).

Quantitation — Absolute metabolite quantitation with *in vivo* MR spectroscopy is desirable and has been demonstrated in several studies (Christiansen et al., 1993; Narayana et al., 1989; Thulborn and Ackerman, 1983). However, absolute quantitation is complicated by a number of factors, and a large body of the literature uses resonance ratios or other relative measures to quantify their results. The area of a resonance peak in a phased MR spectrum is proportional to the absolute concentration of the underlying metabolite, but the proportionality constant contains several factors, some of which may be difficult to estimate. Relaxation times, T_1 and T_2, as well as coupling among spins, influence the signal strength via the choice of TR, TE and possibly TI. Local main field inhomogeneities have an impact on metabolite lineshapes and are influenced by the applied shimming procedures. The fraction of the excited volume that contributes to the metabolite signal, e.g., the ratio of brain tissue vs. cerebral spinal fluid, also influences the measured signal strength. Finally, the spectroscopic data may be affected by the spatial and spectral profile of the excitation pulses, as well as by residual eddy currents from pulsed gradients.

In the absence of complete knowledge of metabolite relaxation times, a quantitative approach that measures signal intensity per unit volume of tissue can be achieved. The power gain required to achieve a 90° flip angle can be used as an estimate of coil loading to standardize the amplitude of the received spectroscopic data (Soher et al., 1996). In combination with information about tissue compartments for each voxel, e.g., from segmented MRI, metabolite intensity for data acquired on a given scanner can be compared across subjects, and over time, although not in units of mM. This approach obviates the need for metabolite ratios and can be extended to the full absolute quantitation, given information about relaxation times.

Acquisition Parameters — PRESS and STEAM are typically used to excite brain volumes on the order of 2–12 cc with echo times ranging from 20 to 288 ms. Because SNR is proportional to voxel size and the square root of acquisition time, the larger voxels require fewer averages. A common set of parameters for an early echo PRESS study at the 1.5 T field strength, are TE = 35 ms, TR = 2 s, 8 cc voxel ($2 \times 2 \times 2$ cm) with 64 averages for a scan time of little over 2 min. Figure 14.1 shows spectra acquired at short- (TE = 35 ms) and long-echo (TE = 144 ms) times from the same subject.

14.2.4 MR Spectroscopic Imaging

As with conventional MRI, the region of interest is often not a small rectangular solid, but rather a larger field of view (FOV), resolved into multiple voxels. The most straightforward extension of single-voxel PRESS, or STEAM to chemical shift

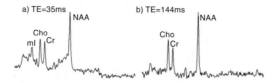

FIGURE 14.1 Examples of *in vivo* spectra from a healthy 58-year-old man. These spectra were acquired with single-voxel PRESS spectroscopy at (a) short- (TE = 35 ms) and (b) long- (TE = 144 ms) echo times from the same subject.

imaging (CSI), or MRSI, is to enlarge the excitation volume to the desired FOV and encode the spatial information with phase encoding gradient lobes on two or three axes prior to the readout period. Thus, for a single slice, 16×16-voxel MRSI, 256 repetition periods are needed, each acquired with a unique pair of phase encoding gradients. A fast Fourier transform of the resulting data in both the temporal and spatial dimensions results in a 16×16-voxel image where each voxel contains a metabolite spectrum.

The advantage of MRSI is that one can ask questions about the regional distribution of metabolites and their differential concentration in gray matter and white matter (there is essentially no metabolite signal from CSF). MRSI is well suited for studies that test hypotheses concerning the regional distribution of metabolites, especially because the concentrations in gray and white matter can be separately determined.

Lipid Suppression — The application of CSI, as opposed to single-voxel spectroscopy, places more stringent demands on the lipid and water suppression, as well as on the shimming procedure. If the excited volume for a brain scan includes the subcutaneous fat, this large contaminating signal needs to be suppressed to reliably observe and quantify, e.g., the NAA resonance. Several lipid suppression schemes have been proposed for *in vivo* CSI of the brain. Outer-volume suppression (OVS) (Duyn et al., 1993; Posse et al., 1994) is very effective at reducing the lipid signals by placing multiple spatial saturation RF pulses around the brain. Often eight such pulses are used to eliminate signals that are outside an octagonal-shaped region

An alternative lipid suppression scheme is to rely on the difference in T_1 between lipids and several of the commonly observed metabolite signals and apply an inversion recovery pulse prior to excitation. As lipids have shorter T_1 than, e.g., NAA, Cr, and Cho, an inversion recovery with TI = 170 ms will reduce the lipid signal by two orders of magnitude while imposing approximately 30% loss for NAA, Cr, and Cho. While the 30% loss of metabolite signals is significant, this tradeoff for a large lipid suppression factor is often valuable. This is particularly the case where cortical gray matter is of interest, since the outer-volume suppression approach may trade off some of the cortical rim for robust lipid suppression. The inversion-recovery lipid suppression in combination with late-echo (TE = 144 ms or TE = 288 ms) has been used successively in a number of volumetric brain CSI studies to map the long-T_2 metabolites NAA, Cr, and Cho (Adalsteinsson et al., 2000; Hetherington et al., 1994; Pfefferbaum et al., 1999a; Spielman et al., 1992).

Shimming — Shimming a large volume is significantly more challenging than a smaller single voxel. Proposed shimming algorithms include a three-plane linear-only shim, "Fastmap" shims (Gruetter, 1993), and linear + nonlinear 3D shims (Kim et al., 2000; Webb and Macovski, 1991).

Time-Varying Readout Gradients — As the matrix size of phase-encoded MRSI increases, so does the imaging time. While the single-slice, 16×16-voxel example cited above requires a minimum imaging time of 8.5 minutes (TR = 2 s), a comparable three-dimensional $16 \times 16 \times 16$-voxel sampling would take over 2 h. Such excessive imaging time is clearly prohibitive for the *in vivo* setting. Multislice CSI is one method to increase the efficiency of the scan; however, data can be acquired only from a limited number of slices (typically three or four), and with finite slice gaps.

Developments in the application of time-varying readout gradients to CSI have provided methods to dramatically reduce the minimum scan time, and thus increase the spatial coverage of *in vivo* MRSI. While early work (Mansfield, 1984) described the basic principles of time-varying readout gradients in MRSI, the more recent combination of high-fidelity gradients and reconstruction techniques for non-uniformly sampled data have enabled the practical use of these methods.

Two examples of the use of readout gradients in MRSI are as follows. In the first, one oscillating gradient, an "echo-planar" gradient, is turned on during the readout period to simultaneously encode the temporal and one spatial dimension. This approach, in combination with conventional phase encoding on the two remaining spatial axes, enables the acquisition of multiple contiguous slices in the same acquisition time as is normally required for a single slice, thereby greatly increasing the time efficiency beyond that of conventional CSI (Adalsteinsson et al., 1995; Posse et al., 1994). For a given voxel size, the imaging time and SNR are the same as for a conventional phase-encoded acquisition (Macovski, 1996).

A natural extension of echo-planar CSI is to oscillate along two (or more) spatial axes during the readout period. Spiral CSI has been introduced as a fast spectroscopic imaging technique based on readout gradients simultaneously oscillating on two axes (Adalsteinsson et al., 1998; Adalsteinsson et al., 1999). Figure 14.2 shows a

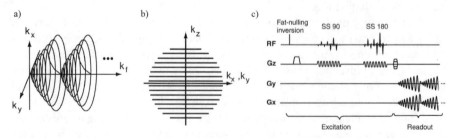

FIGURE 14.2 Spiral CSI. The readout gradient repeatedly covers a spiral trajectory in the k_x-k_y plane (a) while k_z is covered using phase encoding (b). The overall spatial k-space coverage is spherical (or ellipsoidal for anisotropic resolution). (c) An example CSI pulse sequence using a spiral readout in combination with spectral-spatial pulses for water suppression and inversion recovery for lipid suppression.

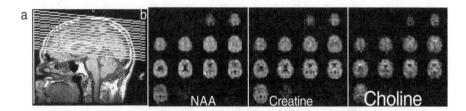

FIGURE 14.3 Volumetric proton CSI data set from a normal volunteer (16 slices, 1.1 cc voxels, 17 min acquisition; TR/TI/TE = 2000/170/144 ms). a) This sagittal high resolution MR image displays the spatial coverage of the 3D CSI. b) The metabolite maps for NAA, creatine, and choline were formed by spectral estimation of peak areas.

pulse sequence and the corresponding k-space trajectory for a spiral-based proton CSI pulse sequence used to collect three spatial and one spectral dimension, and Figure 14.3 shows a typical *in vivo* data set. Using this gradient trajectory, 4-D k-space can be adequately covered in 28 acquisitions using our gradients to generate an $18 \times 18 \times 16 \times 256$ data set (1.1 cc voxels).

14.3 FUNCTIONAL MRI

14.3.1 THE BOLD EFFECT

The use of functional MRI (fMRI) to study brain function in health and disease has increased exponentially in the past decade, especially because, unlike positron emission tomography (PET) and single photon emission tomography (SPECT), there is no exposure to radiation. Only recently has fMRI been applied to the study of alcoholism (George et al., 2001; Pfefferbaum et al., 2001; Tapert et al., 2001). fMRI data can be acquired via several approaches, including bolus injection of contrast agents and arterial spin-tagging schemes, but, currently, the primary method exploits the blood oxygen level dependent (BOLD) contrast mechanism, first described by Ogawa and Lee (1990). Shortly thereafter, the feasibility of applying the BOLD approach to the study of human brain function was demonstrated (Kwong et al., 1992; Ogawa et al., 1992).

The BOLD contrast mechanism does not provide a direct measure of brain activity or metabolism but, instead, is due to the hemodynamic response attendant to an increase or decrease in activity of a given brain region. The relative ratio of oxygenated to deoxygenated hemoglobin in the microvasculature of the brain affects the relaxation rate of nearby protons that in turn affect the intensity of the MR signal.

Oxygenated hemoglobin (oxyhemoglobin, HbO_2) is diamagnetic, that is, it is unaffected by the magnetic field and has little effect on the field itself (Figure 14.4). By contrast, deoxygenated hemoglobin (deoxyhemoglobin, Hb) is paramagnetic (Pauling and Coryell, 1936; Pauling, 1977), that is, it behaves as a tiny magnet when exposed to a magnetic field, and, in turn, affects the local magnetic field. Thus, deoxygenated hemoglobin shortens the time it takes for local dephasing of surrounding protons and causes a decrease in MR signal intensity. This change in the local magnetic field is referred to as the magnetic susceptibility or T_2^* effect and appears

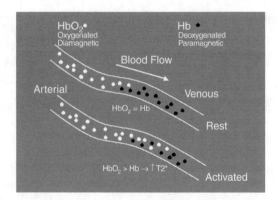

FIGURE 14.4 Schematic diagram of red cells as they pass through the capillary bed and change from the oxygenated to the deoxygenated state. In the activated state, more cells in the oxygenated state reach the venous side resulting in less local susceptibility and therefore activation-induced increased T2* and higher signal strength.

to be the primary mechanism of BOLD contrast. Other effects are due to changes in T_2, as well as flow and perfusion effects on T_1 but the T_2* effect is probably the dominant BOLD mechanism. The T_2* signal modulation, in turn, most likely depends on the oxygenation state of the venous side of the brain's capillary and venule systems, which contain more than 70% of the brain's blood. Although the capillary volume itself is rather small, its surface area, and therefore its potential for affecting the surrounding tissue, is large (Moseley and Glover, 1995).

Contrary to what might have been expected, the BOLD contrast mechanism relies on the fact that brain activation, usually associated with increased blood flow and increased metabolism, is associated with increased, rather than decreased, amounts of oxygenated hemoglobin on the venous side. The activated brain region appears to be supplied with more oxygen than is needed. This in turn leads to more HbO_2 and less Hb in the microvasculature of the activated region, with the increase in HbO_2/Hb producing less T_2* shortening, and therefore larger MR signal intensity.

14.3.2 IMAGE ACQUISITION

Although activation tasks produce only a 2 to 10% change in the signal intensity seen in the human brain, the effects can be detected with repeated cycling of the task and image acquisition. Long echo-time gradient-recalled echo sequences are particularly sensitive to the T_2* effect and are commonly used in BOLD experiments.

Key to BOLD imaging is the ability to image the brain very rapidly. This is usually done in variations of "single-shot" (i.e., collection of all the data necessary to produce the image of a slice after one RF excitation) or "interleaved" (i.e., use of 4 to 8 RF excitations) techniques that can produce an image of a single slice of brain in as short a time as 50–100 milliseconds, depending on the resolution desired.

The two most popular approaches to this rapid acquisition are "echo planar" and "spiral" techniques (Moseley and Glover, 1995), both of which traverse *k*-space (the spatial frequency domain) very rapidly. Both approaches have strengths and

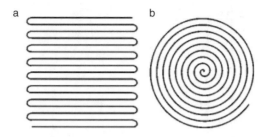

FIGURE 14.5 A schematic illustration of (a) echo-planar and (b) spiral k-space sampling.

weaknesses and suffer from artifacts due to main field inhomogeneities that manifest in different ways for the two techniques. In general, the echo planar images suffer from spatial distortions, especially at tissue–air interfaces where the main field is highly inhomogeneous, whereas spiral techniques do not have this distortion but the tissue–air interfaces produce blurring (Figure 14.5).

A typical fMRI experiment that attempts to cover most of the brain with moderate slice thicknesses (e.g., 5 or 6 mm) produces several thousand images. Their computer storage size depends on the image resolution collected, but a typical native dataset can be on the order of 100 megabytes. In all fMRI experiments there is a trade-off among spatial resolution, extent of coverage, signal-to-noise ratio, and acquisition time, and the choices depend upon the specific questions being addressed.

14.3.3 IMAGE ANALYSIS

It is crucial to appreciate the fact that the BOLD effect is a differential effect, in that it relies on the comparison of images acquired in two different conditions.

In its simplest form, the BOLD effect can be demonstrated by subtracting the image(s) acquired in one condition (e.g., rest) from the image(s) acquired under an activated condition (e.g., photic stimulation). With multiple acquisitions from two different states or experimental conditions, t-tests can also be employed and the results presented as a t-statistic map. This type of analysis is particularly applicable to data collected in a few large blocks of stimuli (i.e., many repetitions of condition A followed by many repetitions of condition B).

A more elaborate approach involves multiple alternations of blocks of two or more conditions and the correlation of the input function with the resultant images. This, of course, must take into account the delay in the hemodynamic response, which lags the input function. In this approach, referred to as a "block design," multiple images are collected during each condition with image acquisition linked to the onset of each activation condition block (Figure 14.6). Still more elaborate is the "event-related" design adapted from a long tradition in human-event-related potential electrophysiology experiments. In this design, each image acquisition is time-locked to a state or condition change (e.g., stimulus presentation), and the resultant data can be examined on a trial-by-trial basis.

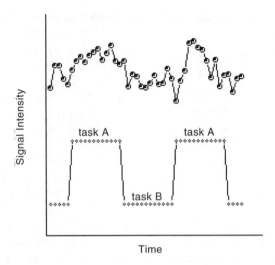

FIGURE 14.6 Response of a pixel in the temporal lobe (top) to two different conditions of a memory task as they alternate over time (each point represents a sample, taken every 3 sec).

A number of factors affect image quality and can produce artifacts in the BOLD signal. Among these factors are physiological (cardiorespiratory) motion and voluntary and involuntary patient movement. Some of these artifacts can be minimized by techniques such as the use of a bite-bar to hold the subject still (Menon et al., 1997) and postacquisition realignment of the images (Woods et al., 1998).

Statistical analyses are usually done on a pixel-by-pixel basis so that the resultant data have the same spatial resolution as the acquired image data. The pixel-by-pixel approach also permits the results to be projected onto a high resolution structural image for visualization of the anatomic specificity of the activation. While many laboratories have developed their own analysis software, a common platform used is Statistical Parametric Mapping or SPM, developed by Karl Friston of the Wellcome Department of Cognitive Neurology, which is freely available on the Internet at http://www.fil.ion.ucl.ac.uk/spm/. Another statistical analysis package, AFNI, has been developed by Robert Cox at the NIMH and is made available through the Internet at http://afni.nimh.nih.gov/afni/. These packages contain options for correction or control of sources of artifact and provide a standardized output format readily appreciated by most investigators in the field. Principal features of these analysis packages include methods to correct for image misregistration arising from movement throughout an imaging session and image smoothing after motion correction for reducing spatial distortion, especially prominent at structure interfaces. These spatially based maneuvers set the stage for co-registration of the time series and the hemodynamically filtered waveform. In light of the thousands of possible comparisons within these large image data sets, SPM and AFNI also provide methods to correct for multiple voxel contrasts, including application of a cluster size method to exclude potentially significant voxels that are tiny spatial islands.

Random effects analysis is a common approach used to compare groups of subjects (e.g., alcoholics vs. controls). Through the statistical analysis packages, activated images for each experimental condition comparison (e.g., selective attention task vs. working memory task) are computed for each subject. The images are spatially normalized to a template using an affine normalization (x, y, z translation, rotations, and scaling) and thus transformed into a common coordinate system with each subject warped to a common brain size. This allows pixel-by-pixel statistical comparisons performed across subjects or groups.

For additional information about fMRI acquisition and processing, we refer the reader to specialized books on the topic (Papanicolaou, 1998; Toga and Mazziotta, 2000).

14.3.4 EXPERIMENTAL DESIGN

A rigorous and elegant design uses the exact same stimulus conditions in the experimental and control conditions but simply changes the task instructions. This approach provides complete control over the stimulus load in the two conditions and permits the experimenter to assume that any difference in behavior or activation is attributable to the task and not to the stimuli. For example, in a cognitive BOLD experiment using the "n-back task" to examine spatial working memory as demonstrated in Figure 14.7 (Pfefferbaum et al., 2001), a "0," which changes across presentations, is presented in one of nine locations. In the working memory task, subjects must press a response key to a newly presented "0" when it is in the same location as one seen two items back. In the comparison attention task condition, subjects see the same series of 0s presented in nine different locations, but this time, they must press a response key when a 0 appears in the middle location; in the latter task, subjects must be vigilant but they do not have working memory demands.

Significant activation clusters are typically projected onto a high resolution structural image that is representative of the study subjects. An example of such activation data and group differences between the attention task described above and a rest condition is presented in Figure 14.8.

Theoretically, the BOLD effect generated by the comparison of two conditions *is* the experimental response. However, it is advisable to collect concurrent behavioral data in the fMRI environment to provide assurance that subjects were performing the tasks as assumed and, in the case of two group designs, such behavioral data that may itself account for observed group differences in brain activation, can address potential differences in behavioral responses.

14.4 DIFFUSION TENSOR IMAGING (DTI)

Conventional structural MR image intensity and contrast is determined by the local environment of free water protons and is adjusted by manipulating a variety of image acquisition parameters. At a macrostructural level, physical boundaries are readily visible because of differences in the local environment experienced by water protons in different tissue types. Among the local environmental factors that influence the MR signal is the degree to which water protons freely diffuse, and the intrinsic properties of water diffusion in human tissue can also provide information

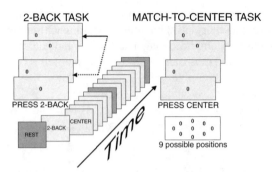

FIGURE 14.7 A pictorial representation of a block design, spatial working memory experiment and its three conditions: rest (green), 2-back task (yellow), and match-to-center attentional task (aqua). In the 2-back task, subjects were instructed to press a response key when a 0 appeared in the same position as one presented two presentations back. In the match-to-center task, subjects pressed a response key when a 0 appeared in the middle. (Figures 14.7 and 14.8 also shown in color insert following page 202.)

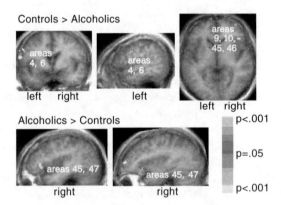

FIGURE 14.8 An example of SPM-derived significant activation clusters of differences between the attention task and a rest condition. Loci where the activation of the control group was significantly greater than that of the alcoholic group are depicted in the red to yellow tones. Loci where the activation of the alcoholic group was significantly greater than that of the control group are depicted in the blue to purple tones. Activation of control group was greater than that of alcoholic group in anterior medial and superior prefrontal cortex, where a large region of activation included Brodmann areas 9, 10, 45, and 46 bilaterally, although more widespread in the right than left hemisphere; the left motor cortex (areas 4 and 6) was also significantly more activated. By contrast, activation of alcoholic group was greater than that of control group in the right prefrontal cortex (areas 45 and 47), positioned more inferiorly and posteriorly than prefrontal regions significant in the controls. The differences in activations between the groups were due to greater activation in one group rather than deactivation in the other group (Figure 4 in Pfefferbaum et al., *Neuroimage* 14: 7-20, 2001).

about the microstructural characteristics of tissue as is accomplished with diffusion tensor imaging.

The diffusion of water molecules in tissue with unconstrained microstructure, such as cerebrospinal fluid, is characterized by Brownian motion, and the molecular displacement has a Gaussian distribution. The diffusion is, therefore, "isotropic" — the molecules move equally in all directions. In tissues with a regular and orderly microstructure, such as brain white matter, the water molecules behave in a more constrained fashion with a preponderant motion in a given orientation; the diffusion is "anisotropic" (Figure 14.9).

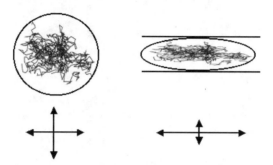

FIGURE 14.9 A schematic illustration of isotropic diffusion (left) and anisotropic diffusion (right) that is preferential to the left/right orientation (Lim et al., *Arch Gen Psych*, 56:367-374, 1999, Figure 1).

Diffusion in tissue can be quantified by three major indices: average magnitude or apparent diffusion coefficient (ADC), the predominant orientation of the diffusion, and its degree of anisotropy. The more unrestricted the water molecules are in a given tissue, the higher will be the average ADC and the lower the anisotropy, with the orientation approaching a random direction; these are the characteristics of CSF. By contrast, white matter is highly organized in fiber bundles that restrict the diffusion of water; ADC is lower than in CSF, anisotropy is higher, and the orientation depends on the orientation of the specific fiber tracts observed. Gray matter falls between CSF and white matter in terms of anisotropy and the randomness of orientation, but has a similar ADC to white matter. These properties can be quantified by acquiring diffusion weighted MR images, the signal intensities of which are sensitive to the diffusion of water molecules. A number of recent advances have made it practical to perform diffusion imaging in a clinical research setting, and alterations in one or more of these diffusion indices have been demonstrated in multiple sclerosis (Arnold et al., 1998; Miller et al., 1998), stroke (Spielman et al., 1996), schizophrenia (Buchsbaum et al., 1998; Lim et al., 1999), and chronic alcoholism (Pfefferbaum et al., 2000a).

14.4.1 ANISOTROPY

Among the approaches for quantifying diffusion is the computation of the degree of anisotropy on a voxel-by-voxel basis. The fractional anisotropy (FA) (Pierpaoli and Basser, 1996) is among the most numerically stable of these measures. The FA is computed from a mathematical manipulation of a matrix of numbers, referred to as a "tensor," hence the currently popular term "Diffusion Tensor Imaging." The FA of CSF is near 0 and the FA of the corpus callosum approaches 1.

FA is independent of the orientation of the fibers in the voxel, but reflects their deviation from isotropic diffusion. Because the FA of white matter is high, CSF is low, and gray matter is in between, an FA image of a slice of the brain looks like a picture of its white matter skeleton (Figure 14.10). Some regions of white matter

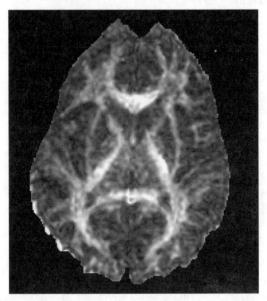

FIGURE 14.10 An example of a map of intravoxel coherence (FA). Note the high FA in the regions of white matter (Pfefferbaum et al., *Mag Res Med*, 44:259-268, 2000, Figure 3a).

normally have considerably lower FA than others even though they are fully volumed, and this probably represents architectural differences in fiber tract organization at the intravoxel level, i.e., intact fibers crossing within a voxel. Thus, the corpus callosum has a very high FA, but the crossing white matter tracts leading to the frontal lobes normally have considerably lower FA than the corpus callosum. Nonetheless, processes that cause changes at the microstructural level, such as demyelination or the deep WMHIs seen in normal aging and hypertension, can cause a significant measurable decrease in FA.

14.4.2 IMAGE ACQUISITION

Diffusion tensor imaging involves collecting several proton MR images with magnetic gradients applied in different directions during acquisition. When these diffusion gradients are applied in a minimum of six non-collinear orientations, the information about water diffusion can be determined in all orientations for a single ellipsoid model. Essentially, seven images must be collected — one with no diffusion gradient and one each with a gradient applied in each of the six directions — to compute fractional anisotropy and vector direction (Figure 14.11). On conventional 1.0 to 1.5 T clinical scanners, diffusion gradients of b = 800–1000 sec/mm^2 are used. Echo-planar imaging (EPI) techniques allow acquisition of multislice data in minutes, making it practical to apply this technique to clinical investigation. To compensate for the low SNR of EPI, image acquisition can be repeated for each gradient orientation and then averaged.

The plane of image acquisition depends on the fibers of primary interest for the posed research question. For example, callosal fibers that traverse the hemispheres may be best visualized in the axial plane; longitudinal fasciculi, which are oriented anteriorly to posteriorly may be visualized in the sagittal plane and quantified in the coronal plane. Acquisition of isotropic voxels affords flexibility in postacquisition processing and quantification.

In addition to the DTI acquisition, it is advisable to collect high-resolution MR images robust to main field inhomogeneity that are aligned identically with the diffusion images. The high-resolution protocols are used as the template to unwarp the distortion of EPI images arising from main field inhomogeneity and eddy currents, to identify the boundaries of circumscribed structures on images not containing the dependent measure, and to correct for age-related or disease-related selective tissue or brain structural volume loss that could influence the DTI outcome measure.

14.4.3 QUANTIFICATION

A general scheme for DTI quantification is as follows. The diffusion gradients produce different patterns of eddy current distortion that can be corrected based on inversion recovery (IR) images collected without diffusion gradients. The IR images are used as basis images to unwarp eddy current-induced image distortions that vary from one diffusion direction to the next in the diffusion weighted images. This procedure can use an affine correction on a slice-by-slice basis (Woods et al., 1998). The IR image is used in the unwarping procedure to avoid inclusion of CSF in the reference image, which would result in excessive expansion of the diffusion weighted images (de Crespigny and Moseley, 1998).

Fractional Anisotropy (FA) — Using the averaged images with b = 0 and b = 800–1000 sec/mm^2, six maps of the ADC can be calculated, each being a sum of two diagonal elements and one off-diagonal element of the diffusion tensor. For example, the ADC map corresponding to (Gx,Gy,Gz) = (1,1,0) is given by ADC_{110} = ADC_{xx} + ADC_{yy} + $2ADC_{xy}$. Solving the 6 independent equations with respect to ADC_{xx}, ADC_{xy} etc. yields the elements of the diffusion tensor $\underline{\underline{ADC}}$.

$$\underline{\underline{ADC}} = \begin{vmatrix} ADC_{xx} & ADC_{xy} & ADC_{xy} \\ ADC_{xy} & ADC_{yy} & ADC_{yz} \\ ADC_{xz} & ADC_{yz} & ADC_{zz} \end{vmatrix}$$

The general diffusion tensor $\underline{\underline{ADC}}$ can then be diagonalized, yielding eigenvalues λ_1, λ_2, λ_3, αs well as three corresponding eigenvectors that define the predominant diffusion direction. Based on the eigenvalues, the degree of anisotropy can be calculated on a voxel-by-voxel basis. A number of scalar measures

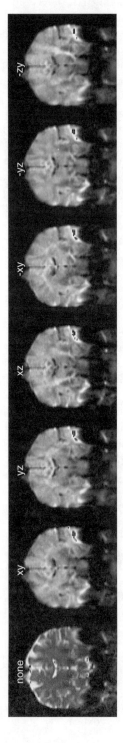

FIGURE 14.11 Native image components of a six-term diffusion tensor acquisition. The decay of the signal is high if the orientation of fiber tracts is along an applied gradient direction. Note, for instance, the difference in signal intensity in the corpus callosum when the magnetic gradient is in the xy direction, compared with the yz direction.

of anisotropy are derived from the tensor (Basser, 1995); FA is a robust intravoxel measure that yields values between 0 (perfectly isotropic diffusion) and 1 (the hypothetical case of a cylinder infinitely long and infinitely thin) and represents the magnitude of the anisotropic component of the tensor as a percentage of the magnitude of the total diffusion tensor:

$$FA = \sqrt{\frac{3}{2}} \frac{\sqrt{(\lambda_1 - \lambda)^2 + (\lambda_2 - \lambda)^2 + (\lambda_3 - \lambda)^2}}{\sqrt{\lambda_1^2 + \lambda_2^2 + \lambda_3^2}}, \lambda = \frac{(\lambda_1 + \lambda_2 + \lambda_3)}{3}$$

Thus, each diffusion weighted study can be reduced to a set of three images for each slice (FA; ADC or trace = the sum of the diagonal elements $[\lambda_1 + \lambda_2 + \lambda_3]$; b = 0) to be used for subsequent analysis in conjunction with the anatomical images.

DTI Image Distortion — Echo planar imaging inherently suffers from geometric distortions due to magnetic field susceptibility differences, especially at the borders between tissue and air, making this particularly bothersome in the sinuses and in the frontal lobes. Approaches to improve distortion include:

- Applying ramp sampling and full echo as opposed to fractional echo sampling, which decreases geometric distortions but at the expense of increased echo time and thus decreased SNR and increased scan time
- Acquisition of a magnetic field inhomogeneity map for postacquisition correction (Jezzard and Balaban, 1995)
- Application of 2-D and 3-D high-order polynomial correction method (Woods et al., 1992), in which a set of EPI images obtained under similar conditions as the FSE images (e.g., T_2-weighted) are geometrically corrected to match the FSE images (Figure 14.12). The FSE structural images are stripped of non-brain elements and then segmented into gray matter,

before warping   after warping

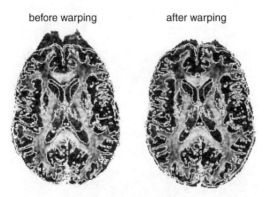

FIGURE 14.12 Echo-planar-derived FA maps overlaid with spin-echo-derived segmentation contours before and after spatial warping (Pfefferbaum et al., *Mag Res Med*, 44:259-268, 2000, Figure 5 b, c).

white matter and CSF compartments. For DTI analysis, the regions of interest are determined on the native FSE images or the segmentation maps, and the DTI data are classified as white matter, gray matter or CSF based on the segmentation map, not the FA maps, thus avoiding the problem of using the dependent variable (e.g., FA) to define the anatomic region of interest.

Although there are non-echo-planar acquisition approaches to DTI (e.g., line scan methods (Makris et al., 1997)), they are time consuming and are most appropriate for scanners that do not have the high-speed gradient hardware required for single-shot EPI, or for acquisitions in regions of very high main field inhomogeneity. Currently, much research effort is being directed at overcoming the spatial distortion problem inherent in short-acquisition-time, long-echo-time acquisitions. Approaches involve interleaved acquisition with navigators, unique patterns of coverage of k-space (e.g., propeller, [Pipe, 1999] spirals [Meyer et al., 1992]) and phased array coils (Bammer et al., 1999).

Intervoxel Coherence and Tractometry — The diffusion tensor contains information about spatial orientation of fiber tracts, and the eigenvectors can be used to define the direction of such tracts. The connectivity and coherence between different regions of the brain is easily appreciated by visual inspection of such vector maps. However, these maps are difficult to quantify, as no obvious method of averaging exists for vectors. Various "connectivity maps" for analysis of structural connectivity of white matter have been proposed including the lattice index (Pierpaoli and Basser, 1996), "dot product" maps (Tang et al., 1997), fiber-tract trajectories (Basser, 1998) and maps of the degree of "alignment" between two neighboring vectors (Jones et al., 1999). These in turn can be used to generate image of white matter tracts as they course through the brain (Figure 14.13).

Intervoxel Coherence (C) — We have proposed the creation of white matter coherence images (Pfefferbaum et al., 2000b), applying an approach similar to that described by Jones et al. (1999) and Pajevic and Pierpaoli (1999), using the average of the angle (α) between the eigenvector of the largest eigenvalue of a given voxel and its neighbors, which represents the extent to which the vectors have the same

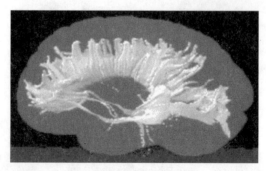

FIGURE 14.13 White matter tractography from diffusion tensor analysis, courtesy of Peter J. Basser (http://eclipse.nichd.nih.gov/nichd/DTMRI/mri/home.html).

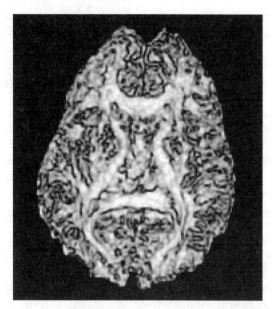

FIGURE 14.14 An example of a map of intervoxel coherence (Pfefferbaum et al., *Mag Res Med*, 44:259-268, 2000, Figure 3b).

orientation and are, therefore, coherent. The maximum angular distance between any two unit length vectors is π. Because diffusion along a given fiber is equal in both directions along the long axis of the fiber, as the angle between voxels progresses from 0 to $\pi/2$ it is less and less coherent; as it exceeds $\pi/2$ it becomes more coherent again. Thus, for a 3-by-3, symmetrical kernel the angle α can be expressed as the arc cosine of the magnitude of the dot product of the vector of the reference voxel and neighboring voxels [$\alpha = \text{acos}(|(x_1x_2 + y_1y_2 + z_1z_2)|)$]. A perfectly coherent kernel of a reference voxel and its eight nearest neighbors would have an average $\alpha = 0$. These values are expressed as a coherence metric, $C = [(\pi/2-\alpha)/(\pi/2)]$, with a range of 0 to 1.0, with 1.0 being perfect coherence. Random noise produces a mean and mode α of .318 π, $C = .363$, and an axial slice at the level of the ventricles produced a mean α of .106 π, $C = .789$. A final coherence image can be computed for each voxel in each image with a 3-by-3 kernel (Figure 14.14). Low C in normal appearing white matter has recently been shown to be related to MS symptomatology (Cercignani et al., 2001), attentional deficits in alcoholism (Pfefferbaum et al., 2000a), and to postural instability in normal aging (Sullivan et al., 2001).

14.5 APPLICATION TO HUMAN STUDIES OF ALCOHOLISM

The imaging modalities described in this chapter have been used to study human alcoholism at different stages of its course. To date, most studies of alcoholism have used proton MRS because it has been available for the longest time and several have reported longitudinal data. Studies based on single voxel MRS generally report lower

NAA/Cr values in alcoholic relative to control groups. Other studies have identified abnormally higher Cho/Cr ratios in alcoholic groups. Regionally, these metabolism differences were present in frontal regions (Fein et al., 1994; Jagannathan et al., 1996) and cerebellar vermis (Jagannathan et al., 1996; Mann et al., 1998). Cross-sectional study using short-echo single voxel MRS reported abnormally high mI in the thalamus and anterior cingulate cortex of recently detoxified, but not necessarily in long-term-abstinent alcoholic men, suggesting a transient gliotic response or hyperosmolarity during early recovery from alcoholism (Schweinsburg et al., 2000). Longitudinal study has largely confirmed the implications of the cross-sectional studies and indicate improvement in NAA/Cho ratios with abstinence and enhancement of the abnormality with relapse (Martin et al., 1995). Future studies employing MRS imaging will be able to address potential regional patterns of sparing and abnormality of selective metabolites present in each tissue type in alcoholics who are at different points in their recovery.

Application of fMRI to the study of alcoholism is relatively new. Two of the three published controlled studies used fMRI to address the nonverbal spatial working memory deficit characteristic of uncomplicated alcoholism. Tapert and colleagues (Tapert et al., 2001) examined young alcohol-dependent women, who showed less activation than controls on the spatial working memory task vs. vigilance task comparison in selective right hemisphere cortical regions hypothesized to subserve spatial working memory, including superior and inferior parietal, middle frontal and postcentral cortex. Our group observed a similar pattern of activation abnormality in older alcoholic men compared with age- and sex-matched controls in a visuospatial working memory paradigm (Pfefferbaum et al., 2001). The group differences in patterns of brain activation, despite equivalent performance on the behavioral task itself, could reflect functional reorganization of the brain systems invoked by alcoholics to perform the task at hand. The third study examined regional brain activation to alcohol-specific cues in active alcoholics compared with social drinkers (George et al., 2001). In addition to reporting higher craving ratings, the alcoholic group showed increased activation in left dorsolateral prefrontal cortex and anterior thalamus in alcohol cue vs. nonalcoholic beverage cue conditions, whereas controls showed significant brain activation in the nonalcohol vs. control conditions but not when alcohol cues were present.

Currently, only one DTI study has been published on alcoholism (Pfefferbaum et al., 2000a). In that study, alcoholic men had lower FA (intravoxel coherence) than controls in the corpus callosum and centrum semiovale. Although C (intervoxel coherence) was lower in the alcoholics than controls, the difference was not statistically significant. These indices of regional white matter microstructure may be functionally meaningful because working memory performance correlated selectively with FA in the splenium, whereas scores on attention correlated selectively with C in the genu. Considering the rich tradition of neuropathological reports of selective white matter degradation in alcoholism (e.g., Harper and Kril, 1990), DTI offers an especially relevant and safe imaging method to track the condition of white matter microstructure over the course of alcoholism. When combined with fMRI, DTI may contribute to understanding why functions may employ different neural processors and processing routes in alcoholism relative to controls.

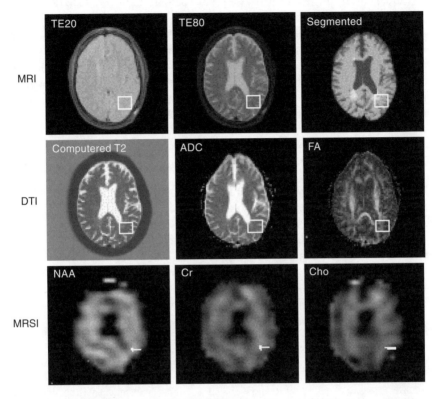

FIGURE 14.15 An example of anatomical, diffusion and metabolite images acquired in a 57-year-old alcoholic man. The top row displays high-resolution MR images, highlighting a white matter hyperintensity (WHMI) enclosed in a box. The WHMI is accompanied by increased computed T2 and (middle row of images) ADC, and a "hole" in the FA image. The bottom row presents the MRSI data from the same slice.

14.6 CONCLUSIONS

Each of the three MRI modalities addressed in this chapter quantitates different aspects of brain integrity: tissue metabolites, induced activation, and microstructure. Combining these modalities, while technically challenging, offers a potentially powerful approach to characterizing the dynamic course of human alcoholism at different levels of analysis and affords the opportunity of posing hypotheses about the progression of improvement with prolonged abstinence from alcohol and deterioration with resumption of drinking. For example, a pattern of brain tissue deterioration would include decreases in FA and NAA along with increases in ADC, Cho, and mI; in white matter these changes may signify reversible demyelination and axonal regression or permanent axonal degeneration. A pattern of brain recovery, or improvement, would include increases in FA and NAA along with decreases in ADC, Cho and mI; in white matter, these changes may signify remyelination and regrowth of neuronal processes. Concurrent studies with fMRI could target functions of cortical regions connected by the white matter pathways found to be disrupted by alcoholism and then improved with abstinence. In the example displayed in Figure 14.15, a 57-year-old chronic alcoholic man, sober for 21 d and with large bilateral posterior white matter signal hyperintensities (WMHI), was imaged with FSE, DTI and MRSI, all acquired at the same slice thickness and anatomical location. Patterns of recovery and deterioration derived from such *in vivo* neuroimaging studies may provide clues to cellular mechanisms underlying reversible and permanent brain structural and functional changes occurring during the course of alcoholism.

ACKNOWLEDGMENTS

The authors thank Dr. Peter J. Basser for providing the example of DTI white matter tractography. This work was supported by grants from the National Institute on Alcohol Abuse and Alcoholism (AA05965, AA12388, and AA10723), the National Institute of Aging (AG18942, AG17919), the National Center for Research Resources (RR09784), and the National Cancer Institute (CA48269).

REFERENCES

Adalsteinsson, E. et al. (1995) Three-dimensional spectroscopic imaging with time-varying gradients. *Mag Reson Med* 33: 461-466.

Adalsteinsson, E. et al. (1998) Volumetric spectroscopic imaging with spiral-based k-space trajectories. *Mag Reson Med* 39: 889-898.

Adalsteinsson, E. et al. (1999) Reduced spatial side lobes in chemical-shift imaging. *Mag Reson Med* 42: 314-323.

Adalsteinsson, E. et al. (2000) Longitudinal decline of the neuronal marker N-acetyl aspartate in Alzheimer's Disease. *Lancet* 355: 1696-1697.

Arnold, D.L. et al. (1998) The use of magnetic resonance spectroscopy in the evaluation of the natural history of multiple sclerosis. *J Neurol Neurosurg Psychiatry* 64: S94-S101.

Bammer, P. et al. (1999) Diffusion-weighted imaging with navigated interleaved echo-planar imaging and a conventional gradient system. *Radiology* 211: 799-806.

Barker, P.B., Breiter, S.N. and Soher, B.J. (1994) Quantitative proton spectroscopy of canine brain: *in vivo* and *in vitro* correlations. *Mag Reson Med* 32: 157-163.

Basser, P.J. (1995) Inferring microstructural features and the physiological state of tissues from diffusion-weighted images. *NMR in Biomedicine* 8: 333-344.

Basser, P.J. (1998) Fiber-Tractography via Diffusion Tensor MRI (DT-MRI). *Proc Int Soc Mag Res Med,* 6th Meeting 1226.

Bottomley, P. (1987) Spatial localization in NM spectroscopy *in vivo. Ann New York Acad Sci* 508: 333.

Brand, A., Richter-Landsberg, C. and Leibfritz, D. (1993) NMR studies on the energy metabolism of glial and neuronal cells. *Devel Neurosci* 15: 289-298.

Buchsbaum, M.S. et al. (1998) MRI white matter diffusion anisotropy and PET metabolic rate in schizophrenia. *Neuroreport* 9: 425-430.

Carlen, P.L. et al. (1984) Partially reversible cerebral atrophy and functional improvement in recently abstinent alcoholics. *Can J Neurol Sci* 11: 441-446.

Cercignani, M. et al. (2001) Intra-voxel and inter-voxel coherence in patients with multiple sclerosis assessed by diffusion tensor MRI (abs). *Proc Int Soc Mag Res Med* 9: 149.

Chang, L. et al. (1996) *In vivo* proton magnetic resonance spectroscopy of the normal aging human brain. *Life Sci* 58: 2049-2056.

Christiansen, P. et al. (1993) *In vivo* quantification of brain metabolites by 1H-MRS using water as an internal standard. *Mag Res Im* 11: 107-118.

de Crespigny, A. and Moseley, M. (1998) Eddy current induced image warping in diffusion weighted EPI (abs). International Society of Magnetic Resonance in Medicine 661.

Doyle, T.J., Beddell, B.J. and Narayana, P.A. (1995) Relative concentration of proton MR visible neurochemicals in gray and white matter in human brain. *Mag Reson Med* 33: 755-759.

Duyn, J.H. et al. (1993) Multisection proton MR spectroscopic imaging of the brain. *Radiology* 188: 277-282.

Ernst, T. et al. (1997) Frontotemporal dementia and early Alzheimer disease: Differentiation with frontal lobe H-1 MR spectroscopy. *Radiology* 203: 829-836.

Estilaei, M.R. et al. (2001) Effects of chronic alcohol consumption on the broad phospholipid signal in human brain: An *in vivo* 31P MRS study. *Alcohol Clin Exp Res* 25: 89-97.

Fein, G. et al. (1994) 1H magnetic resonance spectroscopic imaging separates neuronal from glial changes in alcohol-related brain atrophy. In *Alcohol and Glial Cells,* NIAAA Research Monograph # 27. Lancaster, F., Ed., Bethesda, MD: U.S. Government Printing Office, pp 227-241.

Fein, G., Meyerhoff, D.J. and Weiner, M.W. (1995) Magnetic resonance spectroscopy of the brain in alcohol abuse. *Alc Hlth Res Wrld* 19: 3056-3314.

Frahm, J. et al. (1989) Localized high-resolution proton NMR spectroscopy using stimulated echos: Initial application to human brain *in vivo. Mag Reson Med* 9: 79-93.

George, M.S. et al. (2001) Activation of prefrontal cortex and anterior thalamus in alcoholic subjects on exposure to alcohol-specific cues. *Arch Gen Psych* 58: 345-352.

Gruetter, R. (1993) Automatic, localized *in vivo* adjustment of all first- and second-order shim coils. *Mag Res Med* 29: 804-811.

Gupta, R.K. et al. (1999) Inverse correlation between choline magnetic resonance spectroscopy signal intensity and the apparent diffusion coefficient in human glioma. *Mag Res Med* 41: 2-7.

Haase, A. et al. (1985) 1H NMR chemical shift selective (CHESS) imaging. *Phys Med Biol* 30: 341-344.

Harper, C. (1998) The neuropathology of alcohol-specific brain damage, or does alcohol damage the brain? *Neuropath Exp Neurol* 57: 101-110.

Harper, C G. and Kril, J.J. (1990) Neuropathology of alcoholism. *Alcohol Alcohol* 25: 207-216.

Hetherington, H.P. et al. (1996) Quantitative 1H spectroscopic imaging of human brain at 4.1 T using image segmentation. *Mag Reson Med* 36: 21-29.

Hetherington, H.P. et al. (1994) 1H spectroscopic imaging of the human brain at 4.1T. *Mag Res Med* 32: 530-534.

Jagannathan, N.R., Desai, N.G. and Raghunathan, P. (1996) Brain metabolite changes in alcoholism: An *in vivo* proton magnetic resonance spectroscopy (MRS) study. *Mag Res Im* 14: 553-557.

Jezzard, P. and Balaban, D. (1995) Correction for geometric distortion in echo planar images from B0 field variations, MRM 1995, 34, 65-73. *Mag Res Med* 34: 65-73.

Jones, D. et al. (1999) Non-invasive assessment of axonal fiber connectivity in the human brain via diffusion tensor MRI. *Mag Reson Med* 42: 37-41.

Kim, D.H. et al. (2000): SVD regularization algorithm for improved high-order shimming (abs) International Society for Magnetic Resonance in Medicine 3:1685.

Knufman, N.M. et al. (1992) N-Acetyl-aspartate differences between gray and white matter as observed by proton spectroscopic imaging in normal subjects (abs). *Proc Ann Mtg Soc Mag Res Med* 1905.

Koller, K.J., Zaczek, R. and Coyle, J.T. (1984) N-Acetyl-aspartyl-glutamate: Regional levels in rat brain and the effects of brain lesions as determined by a new HPLC method. *J Neurochem* 43: 1136-1142.

Kreis, R., Ernst, T. and Ross, B.D. (1993) Absolute quantitation of water and metabolites in the human brain II. Metabolite concentrations. *J Mag Res* 102: 9-19.

Kwo-On-Yuen, P.F et al. (1994) Brain N-acetyl-L-aspartic acid in Alzheimer's disease: A proton magnetic resonance spectroscopy study. *Brain Res* 667: 167-174.

Kwong, K.K. et al. (1992) Dynamic magnetic resonance imaging of human brain activity during primary sensory stimulation. *Proc Natl Acad Sci USA* 89: 5675-5679.

Lien, Y., Shapiro, J. and Chan, L. (1990) Effects of hypernatremia on organic brain osmoles. *J Clin Invest* 85: 1427-1435.

Lim, K.O. et al. (1998) Proton magnetic resonance spectroscopic imaging of cortical gray and white matter in schizophrenia. *Arch Gen Psych* 55: 346-352.

Lim, K.O. et al. (1999) Compromised white matter tract integrity in schizophrenia inferred from diffusion tensor imaging. *Arch Gen Psych* 56: 367-374.

Lim, K.O. and Spielman, D.M. (1997) Estimating NAA in cortical gray matter with applications for measuring changes due to aging. *Mag Res Med* 37: 372-377.

Lishman, W.A. (1990) Alcohol and the brain. *Brit J Psych* 156: 635-644.

Lopez-Villegas, D. et al. (1996) High spatial resolution MRI and proton MRS of human frontal cortex. *NMR in Biomedicine* 9: 297-304.

Macovski, A. (1996) Noise in MRI. *Mag Res Med* 36: 494-497.

Makris, N., et al. (1997) Morphometry of *in vivo* human white matter association pathways with diffusion-weighted magnetic resonance imaging. *Ann Neurol* 42: 951-962.

Mann, K. et al. (1998) Proton MR spectroscopy of the cerebellum in detoxified alcoholics and healthy controls (abs). American College of Neuropsychopharmacology.

Mansfield, P. (1984) Spatial mapping of the chemical shift in NMR. *Mag Res Med* 1: 370-386.

Martin, P.R. et al. (1995) Brain proton magnetic resonance spectroscopy studies in recently abstinent alcoholics. *Alcohol Clin Exp Res* 19: 1078-1082.

Mendelson, J.H. et al. (1990) *In vivo* proton magnetic resonance spectroscopy of alcohol in human brain. *Alcohol* 7: 443-448.

Menon, V. et al. (1997) Design and efficacy of a head coil bite bar for reducing movement-related artifacts during functional MRI scanning. *Behav Res Meth, Instr, & Comp* 29: 589-594.

Meyer, C. et al. (1992) Fast spiral coronary artery imaging. *Mag Res Med* 28: 202-213.

Meyerhoff, D.J. et al. (1995) Effects of chronic alcohol abuse and HIV infection on brain phosphorus metabolites. *Alcohol Clin Exp Res* 19: 685-692.

Michaelis, T. et al. (1993) Absolute concentrations of metabolites in the adult human brain *in vivo* - quantification of localized proton MR spectra. *Radiology* 187: 219-227.

Miller, D.H. et al. (1998) The role of magnetic resonance techniques in understanding and managing multiple sclerosis. *Brain* 121: 3-24.

Moats, R.A. et al. (1994) Abnormal cerebral metabolite concentrations in patients with probable Alzheimer's disease. *Mag Res Med* 32: 110-115.

Moseley, M.E. and Glover, G.H. (1995) Functional MRI. *Neuroim Clin N Am* 5: 161-191.

Moyher, S.E. et al. (1995) High spatial resolution MRS and segmented MRI to study NAA in cortical gray matter and white matter of the human brain (abs). *Proc Soc Mag Res* 1: 332.

Nadler, J.V. and Cooper, J.R. (1972) N-acetyl-aspartic acid content of human neural tumours and bovine peripheral nervous tissue. *J Neurochem* 19: 313-319.

Narayana, P.A. et al. (1989) Regional *in vivo* proton magnetic resonance spectroscopy of brain. *J Mag Res* 83: 44-52.

Ogawa, S. and Lee, T. (1990) Magnetic resonance imaging of blood vessels at high fields: *in vivo* and *in vitro* measurements and image simulation. *Mag Res Med* 16: 9-18.

Ogawa, S. et al. (1992) Intrinsic signal changes accompanying sensory stimulation: functional brain mapping with magnetic resonance imaging. *Proc Nat Acad Sci USA* 89: 5951-5955.

Pajevic, S. and Pierpaoli, C. (1999) Color schemes to represent the orientation of anisotropic tissues from diffusion tensor data: Application to white matter fiber tract mapping in the human brain. *Mag Res Med* 42: 526-540.

Papanicolaou, A. (1998) *Fundamentals of Functional Brain Imaging: A Guide to the Methods and their Applications to Psychology and Behavioral Neuroscience*,. Lisse, Netherlands: Swets & Zeitlinger, B.V.

Parnetti, L. et al. (1997) Proton magnetic resonance spectroscopy can differentiate Alzheimer's disease from normal aging. *Mech Aging Devel* 97: 9-14.

Pauling, L. (1977) Magnetic properties and structure of oxyhemoglobin. *Proc Nat Acad Sci USA* 74: 2612-2613.

Pauling, L. and Coryell, C. (1936) The magnetic properties and structure of hemoglobin, oxyhemoglobin and carbonmonoxyhemoglobin. *Proc Nat Acad Sci USA* 22: 210-216.

Petroff, O.A.C., Pleban, L.A. and Spencer, D.D. (1995) Symbiosis between *in vivo* and *in vitro* NMR spectroscopy: The creatine, N-acetylaspartate, glutamate and GABA content of the epileptic human brain. *Mag Res Im* 13: 1197-1211.

Pfefferbaum, A. et al. (1999a) *In vivo* brain concentrations of N-acetyl compounds, creatine and choline in Alzheimer's disease. *Arch Gen Psych* 56: 185-192.

Pfefferbaum, A. et al. (1999b) *In vivo* spectroscopic quantification of the N-acetyl moiety, creatine and choline from large volumes of gray and white matter: Effects of normal aging. *Mag Res Med* 41: 276-284.

Pfefferbaum, A. et al. (2001) Reorganization of frontal systems used by alcoholics for spatial working memory: an fMRI study. *NeuroImage* 14: 7-20.

Pfefferbaum, A. et al. (2000a) *In vivo* detection and functional correlates of white matter microstructural disruption in chronic alcoholism. *Alcohol Clin Exp Res* 24: 1214-1221.

Pfefferbaum, A. et al. (2000b) Age-related decline in brain white matter anisotropy measured with spatially corrected echo-planar diffusion tensor imaging. *Mag Res Im Med* 44: 259-268.

Pierpaoli, C. and Basser, P.J. (1996) Toward a quantitative assessment of diffusion anisotropy. *Mag Res Med* 36: 893-906.

Pipe, J.G. (1999) Motion correction with PROPELLER MRI: application to head motion and free-breathing cardiac imaging. *Mag Res Med* 42: 963-969.

Posse, S., DeCarli, C. and Le Bihan, D. (1994) Three-dimensional echo-planar MR spectroscopic imaging at short echo times in the human brain. *Radiology* 192: 733-738.

Pouwels, P.J.W. and Frahm, J. (1998) Regional metabolite concentrations in human brain as determined by quantitative localized proton MRS. *Mag Res Med* 39: 53-60.

Ron, M.A. (1987) *The Brain of Alcoholics: An Overview*, New York: Guilford Press.

Schuff, N. et al. (1998) Alzheimer disease: Quantitative H-1 MR spectroscopic imaging of frontoparietal brain. *Radiology* 91-102.

Schweinsburg, B. et al. (2000) Elevated myo-inositol in gray matter of recently detoxified but not long-term abstinent alcoholics: A preliminary MR spectroscopy study. *Alcohol Clin Exp Res* 24: 699-705.

Seitz, D et al. (1999) Localized proton magnetic resonance spectroscopy of the cerebellum in detoxifying alcoholics. *Alcohol Clin Exp Res* 23: 158-163.

Shonk, T.K. et al. (1995) Probable Alzheimer disease: Diagnosis with proton MR spectroscopy. *Radiology* 195: 65-72.

Soher, B.J. et al. (1996) Quantitative proton MR spectroscopic imaging of the human brain. *Mag Res Med* 35: 356-363.

Spielman, D. et al. (1996) Diffusion-weighted imaging of clinical stroke. *Int J Neurorad* 1: 44-55.

Spielman, D. et al. (1992) Lipid-suppressed single- and multisection proton spectroscopic imaging of the human brain. *J Mag Res Im* 2: 253-262.

Sullivan, E.V. et al. (2001) Equivalent disruption of regional white matter microstructure in aging healthy men and women. *Neuroreport* 12: 99-104.

Tallan, H.H. (1956) Studies on the distribution of N-acetyl-L-aspartate acid in brain. *J Biolog Chem* 224: 41-45.

Tang, C.Y. et al. (1997) Image processing techniques for the eigenvectors of the diffusion tensor (abs). International Society for Magnetic Resonance in Medicine 5th Meeting 2054.

Tapert, S.F. et al. (2001) fMRI measurement of brain dysfunction in alcohol-dependent young women. *Alcohol Clin Exp Res* 25: 236-245.

Tedeschi, G. et al. (1995) Brain regional distribution pattern of metabolite signal intensities in young adults by proton magnetic resonance spectroscopic imaging. *Neurology* 45: 1384-1391.

Thulborn, K.R. and Ackerman, J.J.H. (1983) Absolute molar concentrations by NMR in inhomogenous B1. A scheme for analysis of *in vivo* metabolites. *J Mag Res* 53: 357-371.

Toga, A. and Mazziotta, J., Eds. (2000) *Brain Mapping: The Systems*, San Diego: Academic Press.

Urenjak, J. et al. (1992) Specific expression of N-acetylaspartate in neurons, oligodendrocyte-type-2 astrocyte progenitors and immature oligodendrocytes *in vitro*. *J Neurochem* 59: 55 - 61.

Urenjak, J. et al. (1993) Proton nuclear magnetic resonance spectroscopy unambiguously identifies different neural cell types. *J Neurosci* 13: 981-089.

Wang, Y. and Li, S.-J. (1998) Differentiation of metabolic concentrations between gray matter and white matter of human brain by *in vivo* 1H magnetic resonance spectroscopy. *Mag Res Med* 39: 28-33.

Webb, P. and Macovski, A. (1991) Rapid, fully automatic, arbitrary volume, *in-vivo* shimming. *Mag Res Med* 20: 113-122.

Webb, P. et al. (1994) Automated Single-Voxel Proton MRS: Technical Development and Multisite Verification. *Mag Res Med* 31: 365-373.

Woods, R. et al. (1998) Automated image registration: I. General methods and intrasubject, intramodality validation. *J Comput Assist Tomogr* 22: 139-152.

Woods, R.P., Cherry, S.R. and Mazziotta, J.C. (1992) Rapid automated algorithm for aligning and reslicing PET images. *J Comput Assist Tomogr* 16: 620-633.

Index

X